Francis Gilbert (Ed.)

# Insect Life Cycles

## Genetics, Evolution and Co-ordination

With 91 Figures

Springer-Verlag
London Berlin Heidelberg New York
Paris Tokyo Hong Kong

Francis Gilbert, PhD
Department of Zoology,
Nottingham University,
Nottingham NG7 2RD, UK

ISBN 978-1-4471-3466-4 ISBN 978-1-4471-3464-0 (eBook)
DOI 10.1007/978-1-4471-3464-0

British Library Cataloguing in Publication data
Insect life cycles.
1. Insects. Life cycle
I. Gilbert, Francis S. (Francis Sylvest), 1956–
595. 703

Library of Congress Cataloging-in-Publication Data
Insect life cycles: genetics, evolution and coordination/Francis Gilbert (ed.).
p. cm.
Selection of papers from the International Congress of Entomology held in Vancouver, B.C., Canada, in July 1988.
Includes indexes.

1. Insects–Life cycles–Congresses. 2. Insects–Genetics–Congresses. 3. Insects–Evolution–Congresses.
I. Gilbert, Francis S. (Francis Sylvest), 1956– . II. International Congress of Entomology (1988: Vancouver, B.C.)
QL461.I52 1990
595.7′03–dc20 90-9947
CIP


Softcover reprint of the hardcover 1st edition 1990



Typeset by Wilmaset, Birkenhead, Wirral
Printed by Page Bros (Norwich) Ltd., Mile Cross Lane, Norwich, Norfolk
2128/3916–543210 Printed on acid-free paper

*I dedicate my efforts in this work to my parents, in thanks*

# Preface

No International Congress of Entomology would now be complete without a symposium on insect life-cycles. The latest Congress, held at Vancouver, BC (Canada), in July 1988, was no exception, with a symposium on the genetics, evolution, and coordination of insect life-cycles organized by Bill Bradshaw and Valerie Brown. The present volume arose from papers contributed by most of the speakers at the symposium, together with papers from other invited authors.

In editing the book, I have been assisted greatly by the other authors, particularly Bill Bradshaw, Val Brown and Fritz Taylor. All contributors agreed to referee two other chapters, a system that worked efficiently and effectively: I thank all authors for performing this task in the face of other demands on their time. I would also like to thank Philip Corbet, John Greenslade, Bryan Clarke, and Gillian Thompson of Springer for their help.

Nottingham
January 1990

Francis Gilbert

# Contents

# Contributors

An asterisk indicates a speaker at the Symposium

John F. Addicott
Department of Zoology, University of Alberta, Edmonton, Alberta T6G 2E9, Canada

David B. Anderson
Department of Plant and Forest Protection, Swedish University of Agricultural Sciences, Box 7044, S–750 07 Uppsala, Sweden

May R. Berenbaum*
Department of Entomology, University of Illinois, Urbana, IL 61801, USA

William E. Bradshaw*
Department of Biology, College of Arts and Sciences, University of Oregon, Eugene, OR 97403, USA

Judith Bronstein*
Department of Ecology and Evolutionary Biology, University of Arizona, Tucson, AZ 85721, USA

Valerie K. Brown*
Department of Zoology, Imperial College, Silwood Park, Ascot, Berks SL5 7PY, UK

Hugh Dingle*
Department of Entomology, University of California, Davis, CA 95616, USA

Jonas Förare
Department of Plant and Forest Protection, Swedish University of Agricultural Sciences, Box 7044, S–750 07 Uppsala, Sweden

Francis Gilbert*
Department of Zoology, Nottingham University, Nottingham NG7 2RD, UK

Joseph L. Graves
Department of Ecology and Evolutionary Biology, School of Biological Sciences, University of California, Irvine, CA 92717, USA

Christina M. Holzapfel
Department of Biology, College of Arts and Sciences, University of Oregon, Eugene, OR 97403, USA

Edward W. Hutchinson
Institute for Behavioral Genetics, Box 447, University of Colorado, Boulder, CO 80309, USA

Finn Kjellberg
CEPE, Centre National de la Recherche Scientifique, Montpellier 34033, France

Sinzo Masaki*
Department of Entomology, Faculty of Agriculture, Hirosaki University, Hirosaka 036, Japan

Derek A. Roff*
Department of Biology, McGill University, 1205 Dr Penfield Avenue, Montreal, Quebec, Canada H3A 1B1

Michael R. Rose*
Department of Ecology and Evolutionary Biology, School of Biological Sciences, University of California, Irvine, CA 92717, USA

Susan M. Scott
Department of Entomology, University of California, Davis, CA 95616, USA

Bryan Shorrocks
Department of Pure and Applied Biology, University of Leeds, Leeds LS2 9JT, UK

Christer Solbreck*
Department of Plant and Forest Protection, Swedish University of Agricultural Sciences, Box 7044, S–750 07 Uppsala, Sweden

Fritz Taylor*
Department of Biology, University of New Mexico, Albuquerque, NM 87131, USA

Wolfgang Wipking
Zoologisches Institut, Universität zu Köln, Weyertal 119, D–5000 Köln 41, FRG

*Section I*

# Genetics of Life-Cycle Traits

# Introduction

William E. Bradshaw

Significant and sometimes substantial additive genetic variation underlies life-history as well as morphological, behavioural and physiological traits. Heritabilities of fitness traits have not universally been driven to extremely low or zero values by ongoing selection. Rather, a variety of processes, both internal and environmental, may serve to maintain substantial heritability of high fitness traits. These processes are considered in general in Chap. 1, by Derek Roff. Roff contrasts the sorts of information we are able to gain from optimality with that from genetic models. He then continues by examining the relative contribution of mutation, frequency-dependent selection, environmental heterogeneity and antagonistic pleiotropy to the maintenance of genetic variation in "ecologically relevant characters".

Individual traits do not exist in isolation from the remainder of the phenotype. Rather, correlated life-history traits exist and the bases for correlated traits are analysed experimentally by Michael Rose, Joseph Graves and Edward Hutchinson in Chap. 2. Rose, Graves and Hutchinson consider correlation among characters that may have arisen as a consequence of correlated patterns of selection or as a consequence of pleiotropy. Since genetic correlations are "critically dependent on the environment in which the characters are measured", these authors employ long-standing laboratory populations of *Drosophila melanogaster*. Their populations may be presumed to have approached genetic equilibrium. Consequently, genetic correlations should reflect pleiotropy rather than linkage disequilibrium and be free from the effects of novel environments. While admittedly the authors employ laboratory populations, their results demonstrate convincingly and conclusively that antagonistic pleiotropy exists but is not the exclusive basis for genetic variability or for trade-offs among conflicting contributions to fitness of life-history traits.

*Chapter 1*

# Understanding the Evolution of Insect Life-Cycles: The Role of Genetic Analysis

Derek A. Roff

## 1.1 Introduction

Phenotypic variation is ubiquitous in natural populations, both within and between populations of the same species. Variation in body size is an obvious example of phenotypic variation and in some insects can reach extreme limits, intrapopulational differences in weight being as great as six-fold in the coleopteran *Asemum striatum* (Andersen and Nilssen 1983), and five-fold in female *Aedes sierrensis* (Hawley 1985). Variation in phenotype within a population has caused considerable taxonomic confusion. In many insect species, for example, populations may comprise two distinct morphs, winged and wingless, the two morphs also frequently showing other substantial morphological differences. Such variation has led many taxonomists to classify these morphs as separate species.

While morphological variation is frequently dramatic there may also be extreme variation in life-cycle events. For example, off the coast of Great Britain, the flatfish *Hippoglossoides platessoides* reaches maturity in its third year at a length of 20 cm and has a lifespan of no more than 6 years, but on the Grand Banks off the Newfoundland coast it does not mature until it is 15 years old at a length of 40 cm and may live to 32 years (Roff 1981). Intrapopulation variation may also be found in life-cycle events. For example age-specific fecundity varies between wing morphs in pterygomorphic insects, the winged form showing a delay in the start of reproduction and a reduction (in females) of overall fecundity (Chap. 11; Roff 1986a; Roff and Fairbairn 1990).

The presence of phenotypic variation in both morphology and life-cycle requires explanation. In this chapter I shall examine how two different methods of analysis are able to account for, and predict, phenotypic variation within populations. The first approach assumes that natural selection maximizes individual fitness and, on the basis of some measure of fitness such as the Malthusian parameter $r$, it seeks that combination of traits maximizing the fitness criterion. I shall refer to such models as optimality models, since they search for the "optimal" set of traits. The second approach makes no assumption about what is being maximized (although in some cases some measure of fitness, such as $r$, may be maximized, and in all cases each phenotype has an associated selection coefficient), but considers the evolution of a trait or set of traits on the basis of the genetic control of those traits. This approach can be divided according to the type of genetic model adopted: models based upon the enumerated effects of individual loci (population genetics) or the statistical interaction of many loci (quantitative genetics).

## 1.2 Are the Predictions of the Two Approaches the Same?

If both methods of analysis lead to the same conclusion, the only problem is to select that approach which is most convenient for the problem under study. However, if different conclusions may be reached then careful consideration must be given to the method of analysis. Any demonstration of the two approaches to different predictions must be based upon a realistic scenario; the following example is drawn from a consideration of the temporal change in migration rate following colonization of a new habitat, a common circumstance for insects and organisms in general.

Consider an insect that inhabits early successional habitats such as meadows. The optimal strategy following initial colonization is to produce a large frequency of non-dispersing offspring, and, as succession proceeds, to produce an increasing frequency of migratory offspring. If the insect cannot assay the state of succession, the optimal strategy is to produce a constant fraction of dispersing offspring.

What predictions are made by models that assume dispersal to be genetically determined? The simplest of such models predict exactly the opposite result to that of the optimality model, namely that the frequency of non-dispersing offspring increases with time and that the highest level of dispersal away from the habitat actually occurs immediately following colonization (Roff 1986b). The reason for this is that at each generation the dispersal morph leaves the population, and so over time the frequency of the genetically determined non-dispersing morph increases. Obviously more complicated genetic models could be devised in which the propensity to disperse is density dependent, but these assume that densities increase and are indicative of the stage of succession. The important point is that the optimality model assumes that genetic variation is not limiting, while the genetic model explicitly sets limits on the strategies available.

Tests of the above two predictions can be made in two ways: firstly, and best, one can follow the temporal changes in frequency at a given site (vertical analysis) and, secondly, one can compare the frequency of dispersers in habitats at different stages of succession (horizontal analysis).

Stein (1977) monitored changes in the proportion of the winged morph of four species of weevil in recently seeded meadows. In two species, *Sitona hispidulus* and *Apion virens*, the control of wing morph is known to be due to a single locus with two alleles, with brachyptery dominant (Roff 1986b). Therefore, providing dispersal rate is correlated with the frequency of winged individuals, we would expect, on the basis of the previously cited genetic model, the frequency of the brachypterous morph to increase with time. Because insufficient data are provided it is not possible to predict the trajectory quantitatively. The frequency of brachypterous individuals increases over time (Fig. 1.1) and because of the genetic control of wing form the final population will be entirely brachypterous and unable to produce dispersing offspring when required; clearly a prediction contrary to the optimality model. Data suitable for a horizontal analysis were provided by Wallace (1953) for the hemipteran *Coranus subapterus*; as before the increase in brachyptery is as predicted by the genetic model (Fig. 1.2). Because the suitability of the sites for the growth and reproduction of the species shown in Figs. 1.1 and 1.2 is not given, support for the predictions of the genetic models over the optimality approach have to be viewed cautiously, though it seems reasonable to suppose that in both cases the eventual demise of the population will be by extinction rather than by dispersal. The important point is that a failure to take into account the genetic control of the trait can lead to spurious predictions.

The above example is not sufficient grounds for rejecting optimality models; rather it suggests caution in the interpretation of such models. Could we in fact scrap optimality models and rely entirely upon genetic models? There are two reasons why I think not. Firstly, our knowledge of the genetic control of morphogenesis, behaviour and physiology is woefully inadequate. What is urgently required is empirical analysis of the genetical structure of ecologically important characters. The second reason for not rejecting optimality models is that they have been very successful in predicting variation in life-history or morphological traits: this success suggests that in the absence of a knowledge of the genetic architecture of a suite of traits such an approach is worth examining.

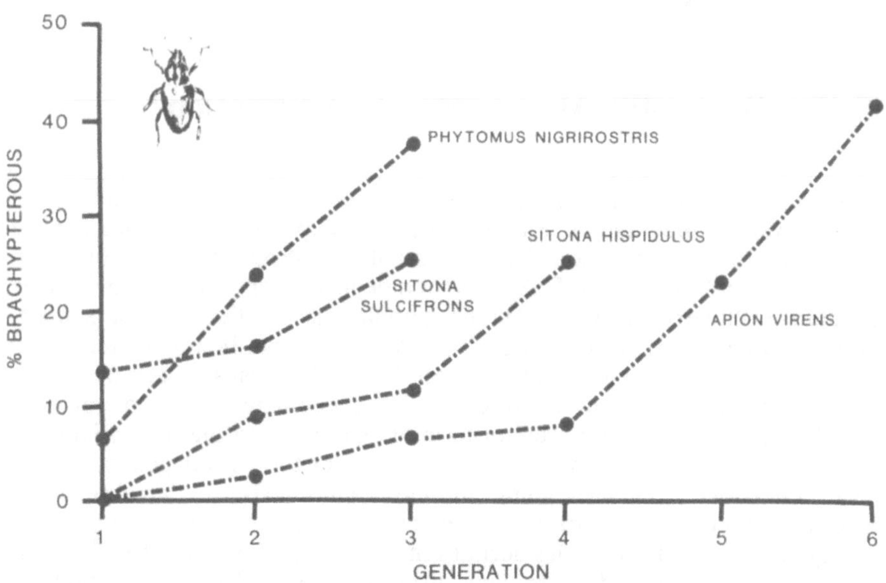


**Fig. 1.1.** Change in percentage of brachypterous morphs of four species of weevils in newly seeded meadows. (Data from Stein 1977.)

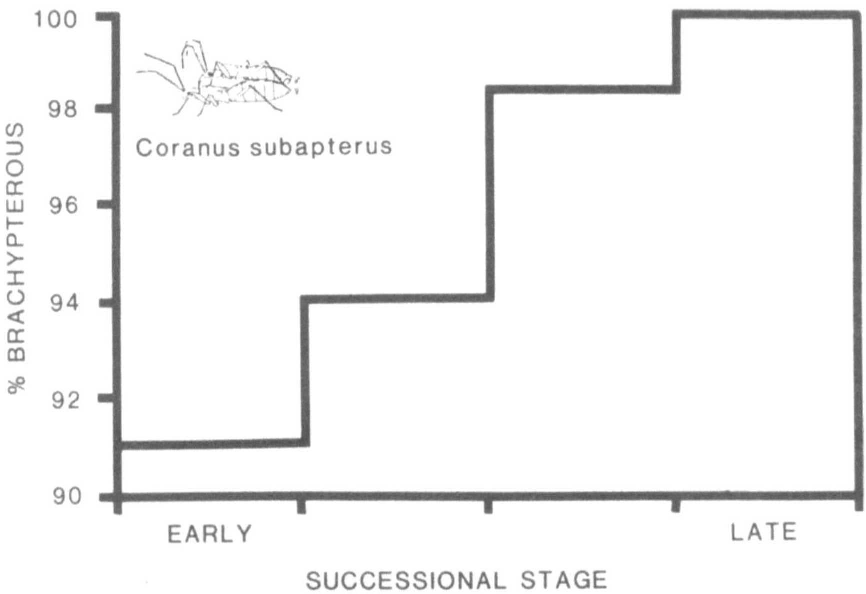


**Fig. 1.2.** Change in the percentage of the brachypterous morph of the hemipteran *Coranus subapterus*, in sites differing in successional stage. (Data from Wallace 1953.)

I suspect that we shall find quantitative genetic analysis most useful in the analysis of variation within populations, but that interspecific variation, and perhaps interpopulational variation, will be best understood by an analysis of physiological and ecological constraints, a hypothesis that I shall discuss in more detail later.

## 1.3 Can Optimality Models Maintain Variation?

There are two circumstances in which optimality models predict that phenotypic variation will be maintained. Firstly, frequency-dependent strategies can be maintained within a population. The analysis of such strategies, generally called evolutionary stable strategies (ESS), has become increasingly popular and can be illustrated with two examples. Within populations of the bluegill sunfish, *Lepomis machrochirus*, three types of breeding male can be found: territorial males, sneakers and satellite males (Gross 1982). The last two can be conveniently grouped together as cuckolders. The cuckolding males do not hold territories or make nests but attempt to intervene between the territorial male and the receptive female and spread their milt over the eggs deposited by the female. Of course, the cuckolding strategy cannot exist in the absence of territorial males (without territorial males there would be no nests) and hence as the frequency of territorial males declines the advantages of cuckoldry diminish. An analysis of these breeding strategies using an ESS model indicated that the observed proportion of territorial males and cuckolders in Lake Opinicon was concordant with theory (Gross 1982). The theory does not make any prediction concerning the basis of the trait: it could be entirely genetic in the sense that the two forms are genetically distinct; or it might be entirely ontogenetic, some or all males going through the cuckolding stage before reaching a size that enables them to keep territories; or it might represent a norm of reaction in that males that for whatever reason grow slowly become cuckolders and fast-growing males become territorial.

The second example of a frequency-dependent strategy concerns the cricket species *Gryllus integer*. Cade (1979) suggested that there are two types of male in the population, callers and satellites. As with the sunfish example, the calling males attract the females, while the satellite males attempt to intercept the female as she approaches: the evidence for these two types is not, however, as convincing as for the sunfish, and Cade presents no data demonstrating that a true bimodality exists. The two strategies are hypothesized to be maintained in the population because, although calling males are more successful in obtaining females, they are also more susceptible to parasitism by the fly *Euphasiopteryx ochracea*, which locates the males by phonotaxis (Cade 1979). Again the mechanism by which the two forms are created does not have to be specified, although Cade (1981) has shown that it is possible to alter the calling rate by selection.

The second mechanism by which optimality theory predicts the maintenance of phenotypic variation is environmental heterogeneity. For many, if not most, insects the year can be divided into a season for growth and reproduction and a season of quiescence; in the temperate region the latter is usually the winter period and in the tropics it is frequently the dry summer season. The period of inactivity is passed in a state of diapause and in most cases only one stage is capable of entering this state. Thus there are strict temporal limits on the life-cycle of an insect. However, the amount of time available for growth and reproduction varies from year to year and because of a positive phenotypic correlation between fecundity, body size and development time it would be advantageous for an insect to adjust its development time in accordance with the predicted time available for growth and reproduction. Although a variable adjustment would be best, insects are likely to be constrained to a fixed increment by virtue of being able to increase development time only by the addition of one or more instars. Even within this constraint it can be shown that the fitness of an insect can be increased in some circumstances by the simple rule "add an instar if the number of degree-days exceeds *Y* on Julian date *X*". This rule is not beyond the physical capabilities of an insect, since degree-days can be measured indirectly by size achieved and Julian date by photoperiod. Such a rule appears to operate in the red-legged grasshopper, *Melanoplus femurrubrum*. In this species the number of instars can be estimated from the number of antennal segments in the adult (Bellinger and Pienkowski 1987). The number of instars increases with rearing temperature. Field surveys indicate that the number of nymphal instars varies both geographically and temporally. Furthermore, the number of instars in a particular year is positively correlated with the number of degree-days achieved in May (Bellinger and Pienkowski 1987). It is not possible to determine from the available data whether this norm of reaction is optimal or if individuals vary in their propensity to add instars.

The important message from the above examples is that genetic variation is not a necessary condition for the generation of phenotypic variation, although it is assumed in the optimality approach that sufficient genetic variation exists to allow organisms to evolve to the optimal suite of traits. Obviously conditions are continually changing, in the short term about some average value, but in the long term this average value will itself change. Therefore, optimality theory must necessarily assume that genetic variation permitting long-term shifts in strategy must be found in populations. Thus, both approaches ultimately invoke genetic variation: is such variation observed in natural populations?

## 1.4 Does Phenotypic Variation Reflect Genetic Variation?

Although phenotypic variation does not imply genetic variation, one would certainly expect a portion of such variation to be ascribable to genetic differences. Life-cycle traits, such as development time, propensity to enter diapause, fecundity, etc., show continuous variation and hence the methods of quantitative genetics are appropriate. A simple measure of genetic variation is the heritability (in the narrow sense) of the trait. Classical theory predicts that traits closely connected to fitness, such as the traits presently under consideration, will have a relatively low heritability (Fisher 1958; Falconer 1981).

Heritability estimates for a variety of life-cycle traits in insects are summarized in Table 1.1. For the purposes of comparing traits closely connected with fitness (life-cycle traits) with those further removed I also compiled heritability estimates for morphological traits. I combined estimates under general categories such as "fecundity" and also combined estimates for the same trait from different studies on the same species, except for fecundity in *Drosophila melanogaster* in which different studies gave dramatically different results. The average standard error (SE) was computed as the square root of the mean variance, and where more than one estimate was given the standard deviation between estimates also was calculated. This latter measure of variation includes variation ascribable to error and to differences due to stock origin. Although as expected the standard deviation of estimates is larger on average than the mean SE ($P<0.025$, Wilcoxon matched-pairs test, one-tailed) the scatter is very large and the overall difference slight (Fig. 1.3). Therefore, much of the variation in heritability estimates is ascribable to error and not the stock source: most importantly, where SE values are not available the standard deviation of a series of estimates can be substituted.

The SE of heritability estimates is frequently very large and estimates less than about 0.20 are generally not significantly different from zero (Fig. 1.3), though the inclusion of zero in the 95% confidence region of the estimate should not be taken to imply that the heritabilities are necessarily negligible; it simply points out the difficulty of obtaining accurate estimates. Among insects, the mean heritability is 0.30 (SE=0.05) for life-cycle traits and 0.44 (0.04) for morphological traits. These results are very similar for animals in general in which heritabilities of life-history traits average 0.27 (0.03) and heritabilities of morphological traits average 0.51 (0.02) (data from Mousseau and Roff 1987). Analysis of a single genus, *Drosophila*, and animals in general indicates that the heritabilities of life-history traits have significantly lower values than the heritabilities of morphological traits (Roff and Mousseau 1987; Mousseau and Roff 1987). Although the heritabilities of life-cycle traits may be relatively low there is still sufficient variation to enable quite rapid response to selection. Thus, the data suggest that

**Table 1.1.** Heritability estimates for life-cycle and morphological traits among insects

| Species | $h^2$ | SE | SD | $n$ | Ref. |
|---|---|---|---|---|---|
| **Fecundity** | | | | | |
| *Apis mellifera* | 0.29 | 0.01 | — | 1 | 1 |
| *Dysdercus bimaculatus* | 0.27 | 0.09 | 0.07 | 3 | 2 |
| *Oncopeltus fasciatus* | 0.24 | 0.11 | 0.14 | 6 | 3, 4 |
| *Tribolium casteneum* | 0.33 | 0.04 | 0.04 | 6 | 5 |
| *Drosophila melanogaster* | 0.18 | — | — | 1 | 6 |
| *Drosophila melanogaster* | 0.04 | 0.04 | 0.02 | 4 | 7, 8 |
| *Drosophila melanogaster* | 0.35 | — | 0.18 | 7 | 9 |
| *Drosophila simulans* | 0.11 | 0.07 | — | 1 | 8 |
| **Time to first clutch** | | | | | |
| *Apis mellifera* | 0.57 | 0.02 | — | 1 | 1 |
| *Dysdercus bimaculatus* | 0.32 | 0.08 | 0.11 | 3 | 2 |
| *Oncopeltus fasciatus* | 0.42 | 0.18 | 0.27 | 7 | 3, 4, 10, 11 |
| **Development time** | | | | | |
| *Oncopeltus fasciatus* | 0.61 | 0.24 | 0.31 | 3 | 3, 4 |
| *Tribolium casteneum* | 0.32 | 0.04 | 0.08 | 9 | 12, 13 |
| *Drosophila melanogaster* | 0.16 | 0.04 | 0.09 | 10 | 7, 14, 15, 16 |
| *Drosophila subobscura* | 0.12 | 0.03 | 0.09 | 2 | 17 |
| *Gryllus firmus* | 0.46 | 0.17 | 0.18 | 4 | 18 |
| **Miscellaneous** | | | | | |
| *Apis mellifera* | 0.32 | 0.27 | — | 1 | 19 |
| *Dysdercus bimaculatus* | 0.00 | 0.14 | — | 1 | 2 |
| *Dysdercus bimaculatus* | 0.11 | 0.06 | — | 1 | 2 |
| *Allonemobius fasciatus* | 0.72 | 0.16 | 0.27 | 9 | 20 |
| *Drosophila melanogaster* | 0.06 | 0.06 | 0.05 | 3 | 7, 8, 9 |
| *Drosophila simulans* | 0.15 | 0.09 | — | 1 | 8 |
| *Hyphantria cunea* | 0.62 | 0.22 | 0.14 | 6 | 21 |
| *Oncopeltus fasciatus* | 0.70 | — | — | 1 | 3 |
| *Oncopeltus fasciatus* | 0.04 | 0.04 | — | 1 | 3 |
| **Morphological traits** | | | | | |
| *Dysdercus bimaculatus* | 0.51 | 0.21 | — | 1 | 2 |
| *Musca domestica* | 0.48 | 0.15 | 0.18 | 11 | 22, 23 |
| *Oncopeltus fasciatus* | 0.50 | 0.17 | 0.22 | 3 | 3, 4 |
| *Pieris rapae* | 0.43 | 0.24 | 0.22 | 17 | 24, 25 |
| *Tribolium casteneum* | 0.59 | 0.09 | 0.28 | 7 | 26 |
| *Gryllus firmus* | 0.27 | 0.07 | 0.06 | 4 | 18 |
| *Gryllus firmus* | 0.65 | 0.08 | 0.04 | 2 | 27 |
| *Ephestia kuhniella* | 0.60 | — | 0.21 | 4 | 28 |
| *Drosophila melanogaster* | 0.32 | 0.05 | 0.12 | 107 | 29 |
| *Drosophila melanogaster* | 0.33 | 0.08 | 0.18 | 218 | 29 |
| *Drosophila simulans* | 0.29 | 0.06 | 0.05 | 25 | 29 |
| *Drosophila pseudoobscura* | 0.34 | — | 0.14 | 12 | 29 |
| *Chorthippus brunneus* | 0.41 | — | 0.21 | 11 | 30 |
| *Chorthippus brunneus* | 0.20 | — | 0.31 | 11 | 30 |
| *Allonemobius fasciatus* | 0.66 | 0.28 | 0.25 | 18 | 20 |
| *Allonemobius fasciatus* | 0.52 | 0.14 | 0.15 | 18 | 20 |

$h^2$, heritability estimate; SE, standard error of the mean; SD, standard deviation; $n$, sample size.

References: 1, Hillesheim 1984; 2, Derr 1980; 3, Hegmann and Dingle 1982; 4, Palmer and Dingle 1986; 5, Orozco 1976; 6, Robertson 1957; 7, Tantawy and El-Helw 1966; 8, Tantawy and Rakha 1964; 9, Rose and Charlesworth 1981a; 10, Dingle et al. 1977; 11, Groeters and Dingle 1987; 12, Englert and Bell 1970; 13, Dawson 1975; 14, Sang and Clayton 1957; 15, Sang 1962; 16, Prout 1962; 17, Clarke et al. 1961; 18, D. A. Roff unpublished data; 19, Rinder et al. 1983; 20, Mousseau and Roff 1987; 21, Morris and Fulton 1970; 22, Bryant 1977; 23, Bryant and Turner 1978; 24, Gilbert 1984; 25, Tanaka 1987a; 26, Halliburton and Gall 1981; 27, Roff 1986c; 28, Imura 1980; 29, various authors, see Roff and Mousseau 1987; 30, Butlin and Hewitt 1986.

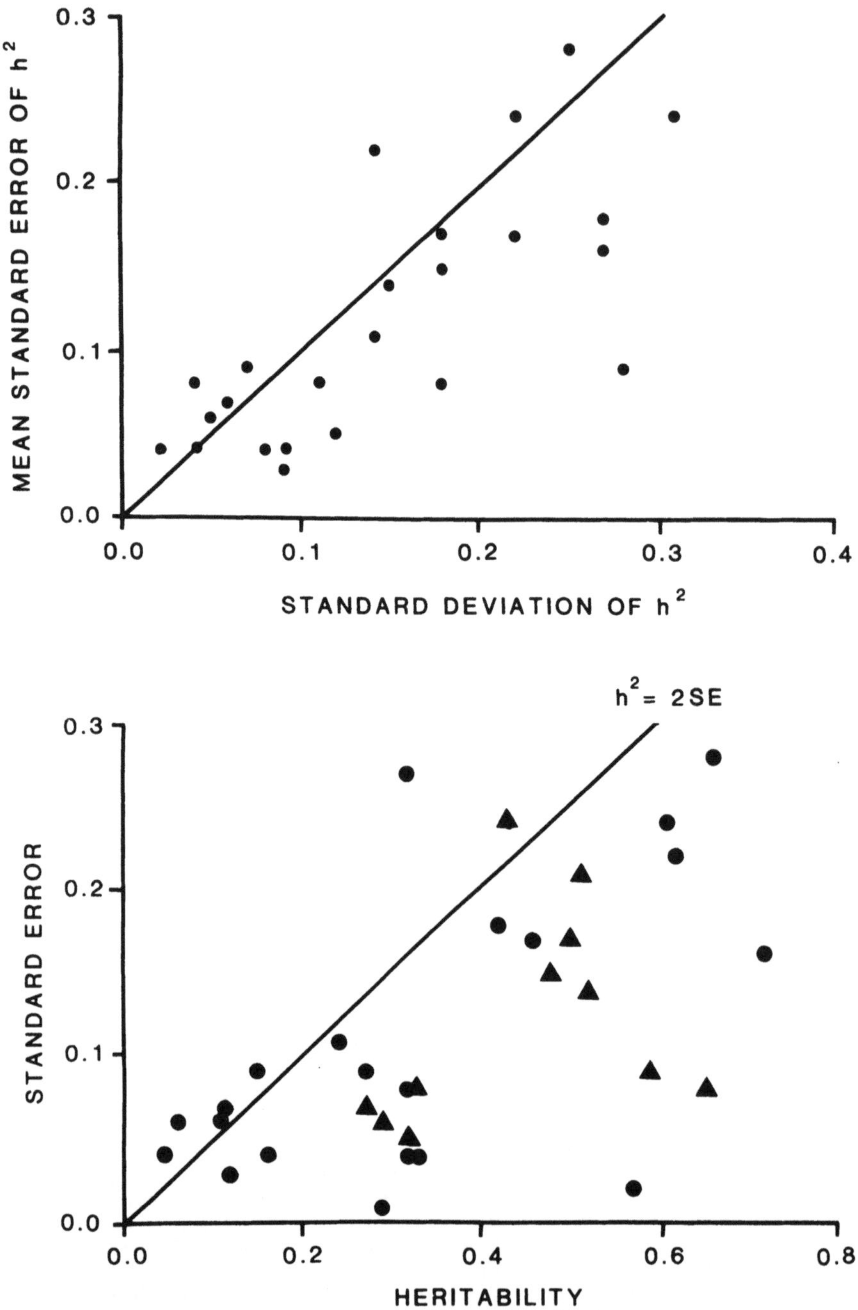


**Fig. 1.3.** *Upper panel*: mean standard error of $h^2$ for insect life-cycle traits plotted against the standard deviation of $h^2$. Data presented in Table 1.1. For methods of estimation, see the text.

*Lower panel*: distribution of heritability estimates for insect life-cycle traits (●) and morphological traits (▲). Data presented in Table 1.1. Points lying above the line $h^2 = 2SE$ include zero within their 95% confidence intervals.

there is a significant genetic component to phenotypic variation and, therefore, models which ignore this component of variation could be seriously misleading.

## 1.5 What Maintains Genetic Variation?

As described above, some optimality models predict that phenotypic variation can be maintained in a population but do not specify the mechanism. The presence of significant genetic variation underlying phenotypic variation suggests that we should examine more closely the genetic models that could maintain variation:

1. Mutation
2. Environmental heterogeneity (including genotype × environment interaction)
3. Antagonistic pleiotropy

In this chapter I shall discuss only the last two factors (for a discussion of the first, see Charlesworth 1988).

### 1.5.1 Environmental Heterogeneity

Consider the example, previously discussed, of an insect adjusting its development time in accordance with cues that predict the relative length of the season favourable for growth and reproduction: if the available cues have no predictive power then no adaptive norm of reaction is possible. An optimality model would predict that the best strategy is to adopt the development rate that maximized the geometric rate of increase (Cohen 1970), and that genetic variation is maladaptive. On the other hand, a model that incorporates a genetic basis for the trait can, under certain conditions, maintain genetic variation (Gillespie and Turelli 1989). Environmental heterogeneity, even if not capable in the very long term (i.e. at equilibrium) of preventing the erosion of genetic variation, may, nevertheless, slow the rate of erosion down considerably, and could be an important factor in determining the level of genetic variation observed. Variation in climatic conditions over space and time in conjunction with migration could be critical in the maintenance of variation for diapause termination in the fall webworm, *Hyphantria cunea*.

*Hyphantria cunea* is a forest lepidopteran of widespread distribution in North America. In Canada it is univoltine, the moths emerging in June and July, the larvae maturing in late summer and fall and the pupae diapausing until the following spring. Diapause is broken after the pupae have been chilled and subjected to temperatures above 51 °F (10.6 °C): the total number of degree-days accumulated from the time of chilling to adult eclosion is defined as $K_p$ (Morris and Fulton 1970). There is considerable variation in $K_p$, both across years and among different geographical sites, and offspring on parent regression indicates that this variation is largely genetic in origin (Fig. 1.4), the heritability of the trait being 0.62 (SE=0.22) (Morris 1971). In cold years, the progeny from late-

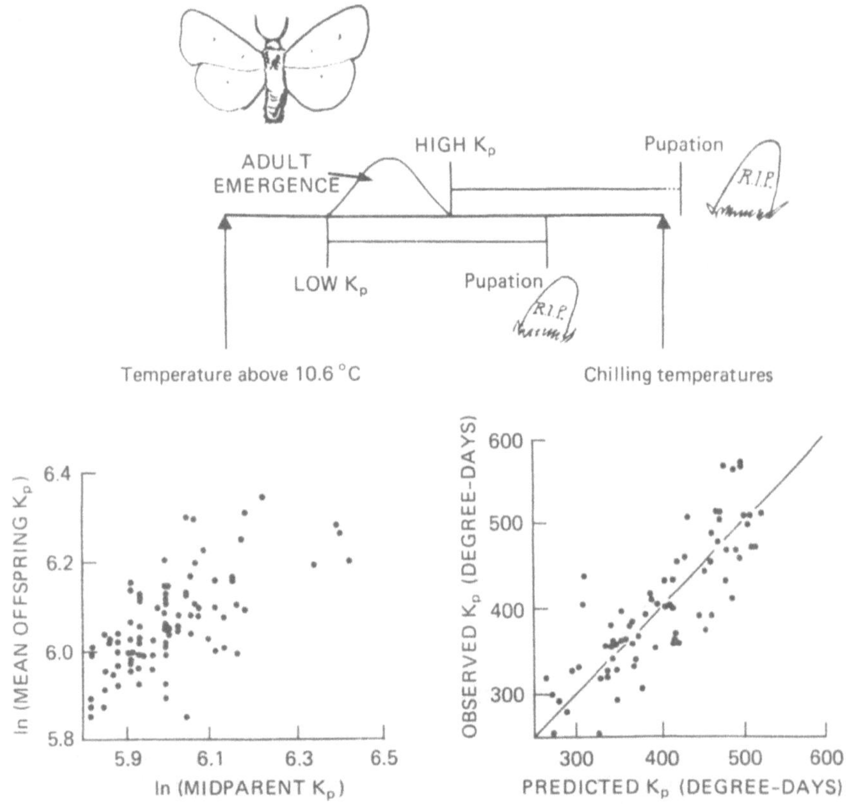


**Fig. 1.4.** *Upper panel*: schematic illustration of selection on *Hyphantria cunea*. Moths that eclose too early (low $K_p$) deplete their fat reserves before chilling temperatures induce diapause. Moths that emerge too late (high $K_p$) fail to reach the pupal stage before chilling temperatures. Both types are eliminated from the population (stabilizing selection). Changes in the average annual temperature favour shifts in the mean $K_p$ (fluctuating directional selection), thereby, with migration, maintaining genetic variation in $K_p$.

*Lower panel, left*: regression of mean offspring $K_p$ on midparent $K_p$.

*Lower panel, right*: observed $K_p$ on value predicted from a simulation model incorporating the genetic relationship illustrated on the left. (Model and analysis from Morris 1971.)

emerging (high $K_p$) moths fail to attain the pupal stage by fall and hence die, while in warm years the progeny of early-emerging (low $K_p$) moths suffer high mortality because they pupate early and use up their fat reserves before the onset of chilling temperatures (Morris 1971). From extensive laboratory studies, Morris (1971) constructed a simulation model predicting the changes in $K_p$ based upon climatic factors and a model of the inheritance of $K_p$. Predictions were made for eight sites from 1957 to 1968, the prediction for each year being based on the predicted value from the previous year (omnibus prediction) not the observed value (stepwise prediction). This method of analysis is the most rigorous. After adjustment for a hypothesized effect due to temperature range, the match between prediction and observation was excellent (Fig. 1.4). While I disagree with several components of the simulation model (in particular with the direct use of the offspring–parent regression), this work has to be highly commended as being one of the few, if not only, studies that has taken quantitative genetics into the field.

At any given site the effect of temporal variation may be to decrease genetic variation. However, selection varies between sites and hence migration could restore genetic variation and across the range of the species little or no loss might occur. Other, more complex interactions such as that modelled by Gillespie and Turelli (1989) might also operate, but it is not clear how these can be studied experimentally.

The above example is based on the hypothesis that temporal variation is not predictable by cues available to the organism and hence no fully adaptive norm of reaction is possible. Variation among habitats may require different responses that, due to morphological or physiological restrictions, also cannot be met by a single norm of reaction. Jago (1973), in a review of savanna grasshoppers, notes that "in 99 percent of the cases, there is a different subspecies or related species" on either side of the equator and suggests that this is a consequence of changing photoperiods and seasons: "Species keyed to decreasing day-length north of the equator experience difficulty in adjusting maturation and timing of hatching emergence to fit rains that are 6 months out of phase across the equator" (Jago 1973). This problem may also be experienced by populations that straddle the transition zone between different phenologies (e.g. univoltine and bivoltine life-cycle); the required adjustments may be so great that no optimal solution exists and genetic variation either for distinctly different phenologies may persist and/or genotypes that vary with respect to their norms of reaction. Because climatic variables are relatively easy to measure and because shifts between phenologies are very common, this particular problem is ideal for the analysis of the relative roles of genetic variation and phenotypic plasticity in natural populations.

## 1.5.2 Antagonistic Pleiotropy

The concept of a trade-off between traits is the basis of most optimality models. While in some cases there may be obvious physical constraints underlying such trade-offs (a hummingbird cannot lay an ostrich-type egg, for example), in others the underlying causal relationship may be due to unknown physiological processes. It is reasonable to suppose that traits sharing some part of a common developmental pathway or drawing upon a common nutrient store will not be able to vary independently if changes occur within the shared element (unless there is a major realignment of the physiological processes themselves). If the shared component is genetically determined, the two or more derived traits will demonstrate pleiotropy, perhaps measurable as a genetic correlation. It is theoretically possible for high additive genetic variation to be maintained in one trait because of the antagonistic pleiotropic effects of another trait (Dickerson 1955; Rose 1982, 1983, 1985). For example, the high heritability of fecundity in *Drosophila melanogaster* observed by Rose and Charlesworth (1981a) is postulated to be maintained by a negative genetic correlation with longevity, the potential contribution of high fecundity to fitness being offset by a short lifespan. Selection experiments verified this negative correlation (Rose and Charlesworth 1981b).

Genetic correlations between morphological traits in insects are generally positive and high (Table 1.2) and are similar in magnitude to the phenotypic correlation. The genetic correlation between morphological and life-cycle traits may be either positive or negative (Table 1.3). If body size is correlated with development time, a positive correlation between fecundity and body size implies antagonistic pleiotropy, since the larger size is achieved at the expense of a longer development time and hence a smaller intrinsic rate of increase. The data for *Drosophila melanogaster* and *Oncopeltus fasciatus* (Tables 1.2 and 1.3) are ambiguous, and more accurate data are required, since in most cases the estimates include zero within their confidence region. Genetic correlations between life-cycle traits are equally difficult to interpret because of large SE values (Table 1.4).

**Table 1.2.** Some representative phenotypic and genetic correlations between morphological traits within the Insecta

| Species | Phenotypic | | Genetic | | Ref. |
|---|---|---|---|---|---|
| | Range | Mean | Range | Mean | |
| *Apis mellifica* | 0.22–0.61 | 0.39 | 0.32–0.97 | 0.54 | 1 |
| *Pieris rapae* | 0.56–0.95 | 0.79 | 0.00–1.00 | 0.54 | 2 |
| *Oncopeltus fasciatus* | 0.39–0.76 | 0.53 | 0.16–0.92 | 0.46 | 3 |
| *Musca domestica* | Not given | | 0.79–0.89 | 0.82 | 4 |
| *Allonemobius fasciatus* | 0.66–0.82 | 0.73 | 0.57–0.95 | 0.77 | 5 |
| *Drosophila melanogaster* | One value | 0.58 | One value | 0.79 | 6 |
| *Drosophila melanogaster* | 0.03–0.14 | 0.09 | 0.07–0.62 | 0.35 | 7 |
| *Drosophila simulans* | Not given | | 0.75–0.80 | 0.77 | 8 |

References: 1, Zawilski 1972; 2, Tanaka 1987b; 3, Palmer and Dingle 1986; 4, Bryant 1977; 5, Mousseau and Roff 1987; 6, Tantawy and El-Helw 1966; 7, Sheridan et al. 1968. All measurements are dimensions of body parts except for those in ref. 7, which are bristle numbers.

As described above, genetic correlations may arise because of some common element in growth and morphogenesis. In my analysis of the quantitative genetics of growth and development in the cricket *Gryllus firmus* I have taken the approach of making predictions concerning correlated responses due to pleiotropic interactions on the basis of the hypothesized hormonal control of the various traits. The main experiment undertaken was selection for macroptery and microptery. A consequence of the protocol used was to select also for a reduction in development time and hence the control lines of this experiment can be used to examine correlated responses and the possible existence of antagonistic pleiotropy.

The sand cricket, *G. firmus*, is a moderate-sized cricket (approximately 0.7 g adult weight) found in sandy areas from Florida to North Carolina. It inhabits sites in the early successional stage of plant colonization and hence dispersal is an integral part of its life-history strategy. Two distinct morphs occur, a long-winged (macropterous) form capable of flight and a short-winged (micropterous) form that cannot fly. Although the macropterous morph has the advantage of dispersal, there is a cost in that macropterous females delay reproduction and suffer reduced (40%) fecundity (Roff 1984). These results suggest that even in circumstances where high dispersal rates are optimal, the trade-off between dispersal and reproduction could maintain significant genetic variation in both these and other correlated traits. This hypothesis depends also on the details of

**Table 1.3.** Genetic correlations between life-cycle and morphological traits in *Drosophila melanogaster* and *Oncopeltus fasciatus*

| Character 1 | Character 2 | Genetic correlation | Antagonistic pleiotropy[a] |
|---|---|---|---|
| *Drosophila melanogaster* | | | |
| Fecundity (days 5–10) | Wing | 0.03 | Yes |
| | Thorax | 0.04 | Yes |
| Fecundity (days 4–8) | Thorax | 0.00 | No |
| Development time | Thorax | 0.32[b] | Yes |
| Pupal duration | Thorax | 0.21[b] | Yes |
| *Oncopeltus fasciatus* | | | |
| | Iowa population (Hegmann and Dingle 1982) | | |
| Clutch size i | Wing length | 0.73[b] | Yes |
| | Body length | 0.87[b] | Yes |
| Development time | Wing length | −0.55[b] | No |
| | Body length | −0.56[b] | No |
| | Iowa population (Palmer and Dingle 1986) | | |
| Development time | Wing length | 0.09 | Yes |
| | Body length | 0.57 | Yes |
| | Head capsule width | −0.01 | No |
| | Pronotal width | 0.32 | Yes |
| Fecundity | Wing length | −0.57[b] | No |
| | Body length | 0.59[b] | Yes |
| | Head capsule width | 0.62[b] | Yes |
| | Pronotal width | 0.48[b] | Yes |
| | Puerto Rico population (Dingle et al. 1988) | | |
| Fecundity (1st 5 days) | Wing length | −0.25 | No |
| | Head capsule width | −0.45 | No |
| Fecundity (2nd 5 days) | Wing length | 0.35 | Yes |
| | Head capsule width | −0.03 | No |
| Clutch size (1st 5 days) | Wing length | 0.31 | Yes |
| | Head capsule width | 0.03 | Yes |
| Clutch size (2nd 5 days) | Wing length | −0.03 | No |
| | Head capsule width | −0.20 | No |

*Drosophila* data from Tantawy and Rakha (1964) and Tantawy and El-Helw (1966, 1970).
[a] Presence of antagonistic pleiotropy depends on relationship between morphological trait and development time.
[b] At least one comparison $h^2-2$ SE $>0$.

the genetic mechanism underlying the antagonistic pleiotropy, though the precise limitations are as yet unclear (Rose 1982, 1983, 1985).

Southwood (1961) and Wigglesworth (1961) hypothesized that wing dimorphism is controlled by the titre of some hormone, possibly juvenile hormone (JH) during development. They suggested that high titres of JH block the development of wings, thereby producing the micropterous morph, while low titres do not inhibit development and the "normal" macropterous morph is produced. Recent biochemical work on a related cricket, *Gryllus rubens*, supports this hypothesis (Zera and Tiebel 1989). Juvenile hormone is also involved in the regulation of nutrient metabolism (Slama 1985; Riddiford 1985) and thus may play a role in the regulation of growth. Other hormones, such as ecdysone are undoubtedly important in this regard and may be involved in the control of wing morphology (A. J. Zera, personal communication). The general hypothesis I derive from these observations is that life-cycle traits such as development time, and morphological traits such as wing form, probably share common physiological

**Table 1.4.** Genetic correlations between life-cycle traits in *Drosophila melanogaster* and *Oncopeltus fasciatus*

| Character 1 | Character 2 | Genetic correlation | Antagonistic pleiotropy |
|---|---|---|---|
| *Drosophila melanogaster* | | | |
| Longevity | Fecundity (days 5–10) | 0.07 | No |
| | (*simulans*) | 0.03 | No |
| | (days 1–5) | −1.43?[a] | Yes |
| | (days 6–10) | 0.30? | No |
| | (days 11–15) | −0.71? | Yes |
| Fecundity (1 day) | Starvation resistance | −0.91[b] | Yes |
| Development time | Pupal duration | 0.40[b] | No |
| *Oncopeltus fasciatus* | | | |
| | Iowa population | | |
| Age at reproduction | Clutch size (1, 2 or 3) | −0.49[b] | No |
| | % hatching (1, 2 or 3) | 0.12 | Yes |
| | Interclutch interval | 0.07 | No |
| | Development time | −0.13 | No |
| Clutch size i | Clutch size i | 0.93[b] | No |
| | % hatching (1, 2 or 3) | 0.72[b] | No |
| | Interclutch interval | −0.57[b] | No |
| | Development time | −0.64[b] | No |
| % hatching in i | % hatching in i | 1.00? | No |
| | Interclutch interval | −0.57[b] | No |
| | Development time | −0.43[b] | No |
| Interclutch interval | Development time | 0.51 | No |
| Development time | Fecundity | 0.45 | Yes |
| | Puerto Rico population | | |
| Fecundity (1st 5 days) | Age at reproduction | −0.52 | No |
| Fedundity (2nd 5 days) | Age at reproduction | −0.36 | No |
| Clutch size (1st 5 days) | Age at reproduction | 0.10 | Yes |
| Clutch size (2nd 5 days) | Age at reproduction | 0.14 | Yes |

*Drosophila* data from Tantawy and Rakha (1964) and Tantawy and El-Helw (1966, 1970), Rose and Charlesworth (1981b) and Service and Rose (1985). *Oncopeltus* data are from Hegmann and Dingle (1982), Palmer and Dingle (1986) and Dingle et al. (1988).
[a] A question mark indicates no SE.
[b] At least one comparison $h^2$−2 SE >0.

pathways and hence are genetically correlated. Further, since the micropterous form arises via the action of a hormone or hormones that maintain the juvenile morphology we might also expect that these hormones will maintain the juvenile period longer than in the case of the macropterous form, and hence the development time of the micropterous morph should be longer than that of the macropterous morph.

In *G. firmus*, development time and growth rate are highly correlated, while body size is only weakly correlated with development time (D. A. Roff unpublished data). I propose the following model for the physiological and genetic basis of growth and development in *G. firmus*.

1. Eclosion to the adult occurs when a critical size is achieved. Such a mechanism has been proposed for other insects, including the house cricket, *Acheta domesticus* (Wigglesworth 1934; Nijhout 1975, 1979; Blakley and Goodner 1978; Lounibos 1979; Woodring 1983). The critical size for metamorphosis in

*G. firmus* is inherited, with a heritability for the final body size of approximately 0.27±0.07 (D. A. Roff, unpublished data).
2. The rate or timing of production of the hormone(s) controlling development and wing form is inherited, a high titre producing slow growth and a micropterous morph. Since the critical size for metamorphosis is independent of the rate of growth, this morph will have a relatively long development time.

In principle, selection for early development could result from a decrease in the critical size at metamorphosis or from a change in the titre of the hormone(s) controlling development. Since there is only a weak correlation between development time and body size in *G. firmus*, but a high correlation between growth rate and development time, selection is predicted to reduce development time by the latter mechanism. This should result in:

1. Increased growth rate
2. No change in body size
3. Increased percentage macroptery

As previously described, the main aim of the selection experiment was to select for an increase in the frequency of each type of wing morph. Males and females of a desired morph, or both morphs in the case of the control lines, were selected upon eclosion into adults until approximately 100 of each sex were obtained. The relatively large number of parents was used to prevent inbreeding depression. This protocol necessarily tended to select also for early development. However, it was unavoidable both because of book-keeping difficulties and because females cannot be left unmated for too long, since they become bloated with unfertilized eggs and die. To reduce the selection for early development as much as possible, the six cages containing the 360 nymphs used in each replicated line were not all set up on the same day and nymphs were collected from the parents only 1 week after the inclusion of the last selected parent, the earlier-laid eggs being discarded. In one control line this procedure did not eliminate selection for early development, but it was successful in the second control line (Fig. 1.5). The first control line will hereafter be designated a selected line (line 1), and the second control line will be designated a control line (line 2) that was not selected for changes in development time. The lines selected for increased or decreased incidence of macroptery will not be considered.

Development time in the selected line (line 1) decreased substantially from 55.6 days (midpoint of mean development times of macropterous and micropterous morphs) to 50.3 days. As predicted, in both lines the micropterous morph always took longer to develop than the macropterous morph. The realized heritabilities of development time are similar to those obtained from offspring–parent regression and full-sibling analysis (Table 1.5).

Growth rate in the line selected for decreased development time increased as predicted whereas in the control line no response was observed (Fig. 1.6). Growth rate and development rate are highly correlated (Fig. 1.6) and within morphs there is no difference between lines in the relationship between them ($P>0.05$ covariance analysis). As predicted, despite the substantial decline in development time there was no significant change in body size (as measured by femur length) for example in line 1 (see Fig. 1.7).

In the "control line" (line 2) there was no change in the percentage of

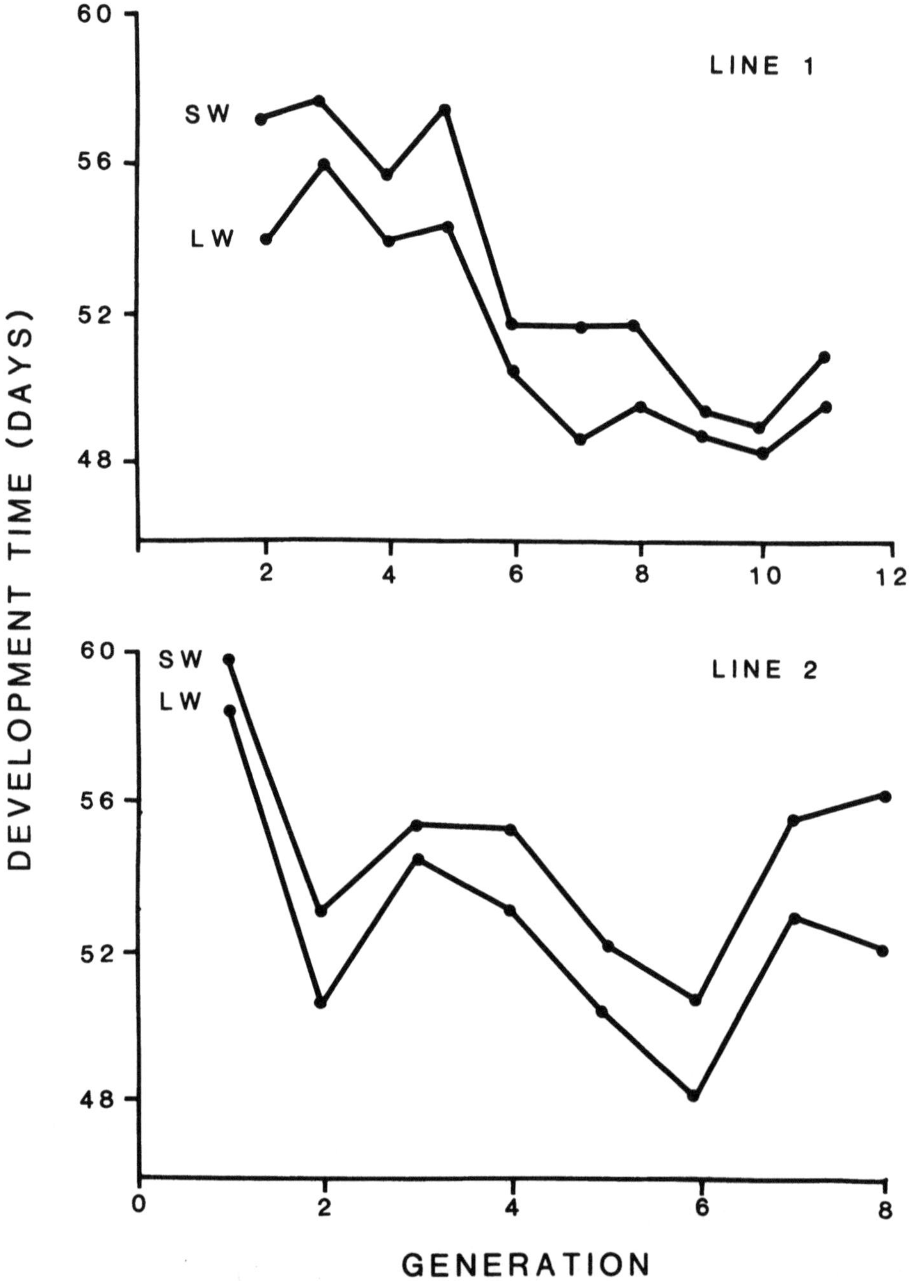


**Fig. 1.5.** Change in mean development time of macropterous (LW) and micropterous (SW) *Gryllus firmus* in a selected (line 1) and a "control" (line 2) line.

**Table 1.5.** Heritability estimates for development time in *Gryllus firmus*

| Morph | Method | $h^2$ | SE |
|---|---|---|---|
| L | Offspring–parent regression[a] | 0.19 | 0.19 |
| S | Offspring–parent regression | 0.49 | 0.24* |
| L+S | Covariance | 0.40 | — |
| L | Nested ANOVA[b] | 0.39 | 0.09* |
| S | Nested ANOVA | 0.37 | 0.09* |
| L+S | Line 1[c] | 0.30 | 0.11* |
| L+S | Line 2[d] | 0.22 | 0.21 |

* $P<0.05$ at least. L, long-winged; S, short-winged; ANOVA, analysis of variance.
[a] Based on 17 families.
[b] Based on 34 families (including above 17 families).
[c] Line 1 selected for decreased development time.
[d] Line 2 not selected, hence the large SE in the estimate.

macropterous crickets, but in the selected line (line 1) the proportion increased significantly (Fig. 1.8, covariance analysis corrected for sex; $F_{1,19} = 8.1$, $P<0.05$). Fecundity in the first week after the final moult was measured in generation 5. As found in previous experiments (see Roff 1984, 1989) the macropterous females produced statistically fewer eggs in the first week than did the micropterous females. Therefore, an increase in the incidence of macroptery correlated to the reduction in development time will lead to a corresponding reduction in the mean fecundity of the faster-developing crickets.

In the field, development time, percentage macroptery and fecundity will be traits that contribute significantly to fitness; the observed pleiotropy between them could play a role in maintaining genetic variation in both characters. It is important to note that this conclusion does in no way mean that antagonistic pleiotropy is important in the wild, only that it is a plausible mechanism. I hope that further analysis of the lines selected for differences in wing morph may shed more light on the genetic and physiological interaction between these and other life-cycle traits. Finally, it is important to note that any experiment on antagonistic pleiotropy must not only demonstrate the possibility of antagonistic interaction in experiments in which a single trait is selected but also that selection on both traits in the direction of increased fitness (e.g. increased fecundity and increased longevity) is unsuccessful or at least very slow. Furthermore, the reversed dominance necessary in present theoretical models of antagonistic pleiotropy must be demonstrated. These questions remain to be addressed in the case of *G. firmus*.

## 1.6 Concluding Comments

Both optimality and genetic models predict that phenotypic variation will be present in natural populations. Neither approach is entirely satisfactory for the analysis of life-cycle evolution: the former because it ignores genetic constraints and the latter because our ignorance of empirical relationships does not permit us at present to construct models incorporating more than one or two characters. It is, however, evident that genetic variation is present in populations, and

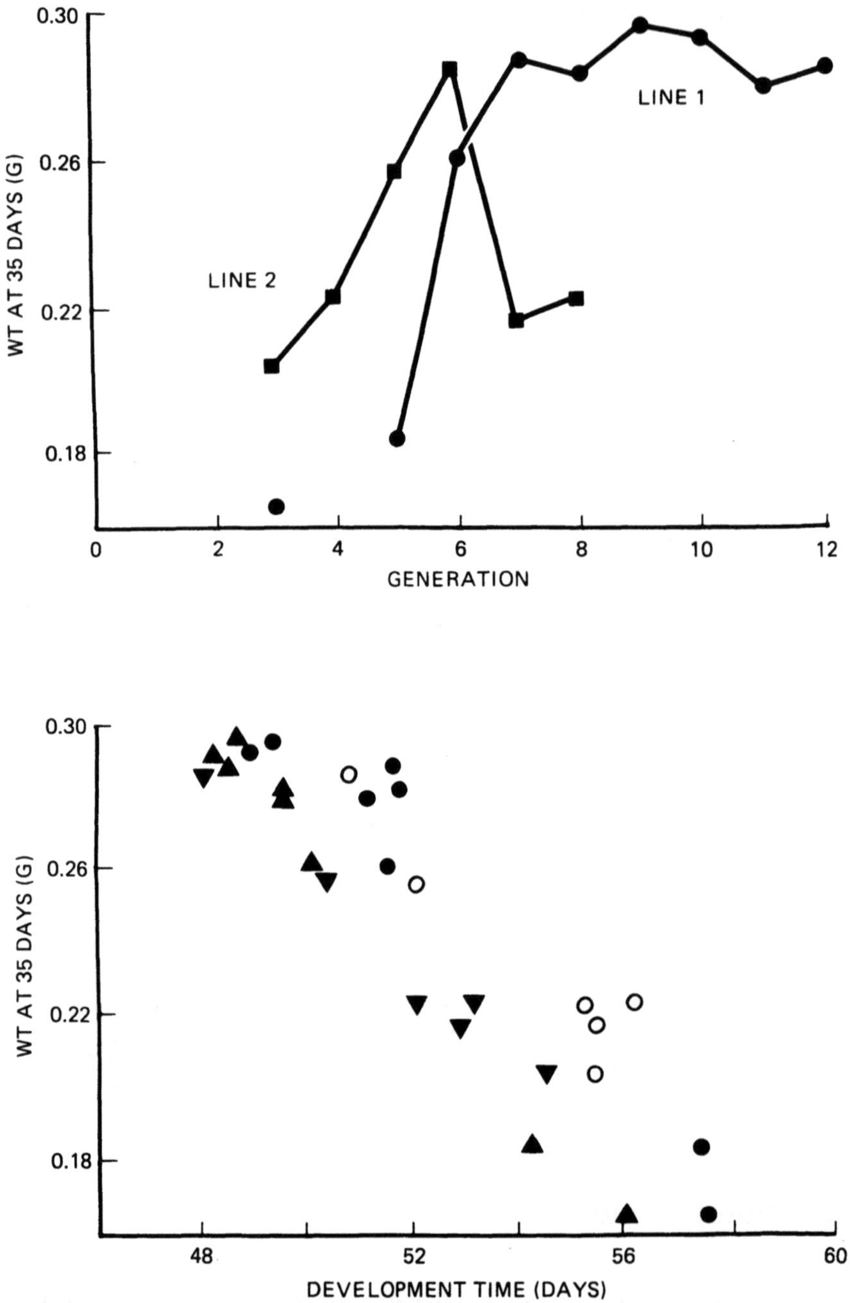


**Fig. 1.6.** *Upper panel*: correlated response of development rate (Weight (*WT*) at 35 days) to selection for reduced development time. Line 1, selected; line 2, unselected.

*Lower panel*: correlation between mean development rate and mean development time. Individuals cannot be sexed at 35 days and hence the mean weight at 35 days is the same for macropterous (LW) and micropterous (SW) crickets within lines. Data from all generations. (●) SW line 1, (○) SW line 2, (▲) LW line 1, (▼) LW line 2.

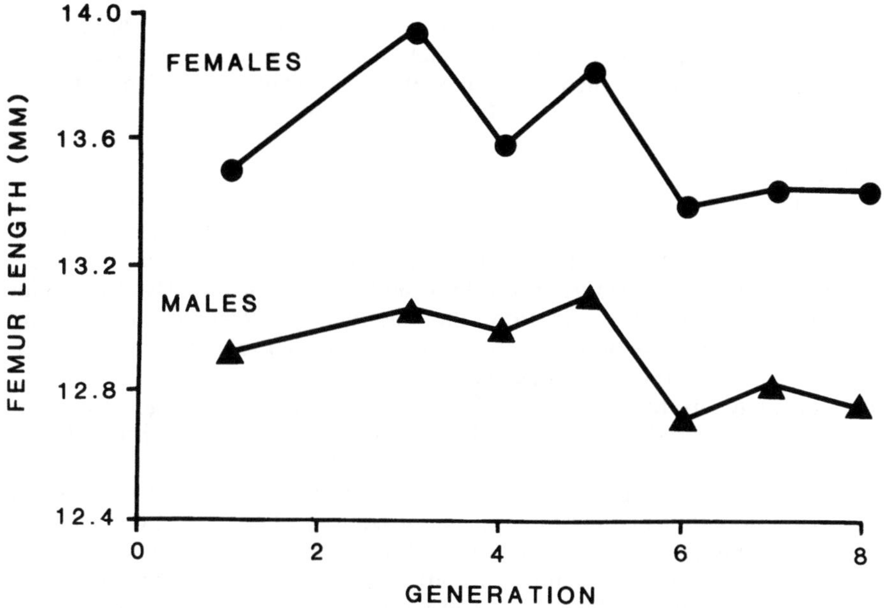


Fig. 1.7. Correlated response in femur length to selection for reduced development time (line 1).

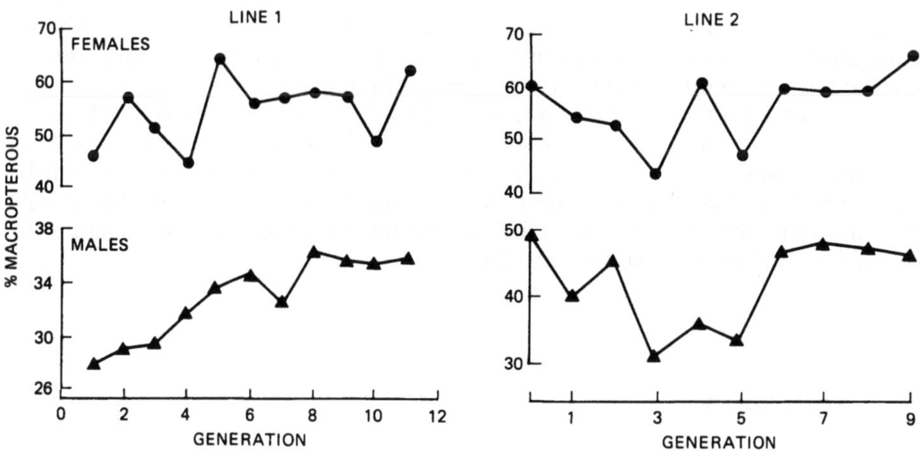


| | r | slope |
|---|---|---|
| females | 0.41 | 0.47 |
| males | 0.92 | 0.53 |

| | r | slope |
|---|---|---|
| females | 0.47 | 0.65 |
| males | 0.25 | 0.33 |

Fig. 1.8. Correlated response in percentage macroptery in lines selected (line 1) and unselected (line 2) for reduced development time.

optimality modelling alone is unlikely to direct research into the appropriate areas. We are in need of more empirical analyses of the genetic structure of life-cycle traits. Unfortunately, such research is labour-intensive, expensive and time-consuming. In the absence of a satisfactory theory of quantitative genetics for long-term changes in character states, such data may also turn out to be unintelligible.

I have examined what I consider to be the two most important factors likely to maintain genetic variation in natural populations, environmental heterogeneity and antagonistic pleiotropy. How do these factors interact and is one more important? In the case of fecundity and wing morphology I suspect that there is an unavoidable trade-off resulting from the high-energy demands of the wing musculature and reproduction; it is simply not possible in the wild to support both concurrently. In this regard it is perhaps significant that wing-muscle histolysis accompanies egg production (Roff 1989) as it does in a wide range of insects. Furthermore, the variability between the degree of persistence of different sites even within the same local area favours different magnitudes of dispersal rate. Genetic variation for wing form and the consequent variation in age schedule of reproduction may thus be maintained. The physiological control of wing form is due, in part at least, to the titre of JH during development: as a result, variation in percentage macroptery is reflected in variation in hormone titre, which may also control both the age schedule of reproduction and the rate of development. Selection on the latter character produces the expected correlated response in wing form, but a direct effect on reproduction is yet to be demonstrated. If energy were abundant, selection might produce a macropterous cricket that did not have to trade-off wing muscles for reproduction. This will depend upon whether processing and assimilation are themselves limiting. If these processes are not limiting then the antagonistic pleiotropy arises because of the constraints of physiology and ecology. Thus, within a species, environmental heterogeneity and antagonistic pleiotropy may be linked. However, between species, significant changes in ecology are likely to alter the constraints due to physiology and hence antagonistic pleiotropy determined in one species may not be relevant for predicting the combination of traits expected in another.

To understand the evolution of insect life-cycles no single approach is likely to succeed, at least at the present state of our knowledge; there is a place for both optimality modelling and genetic analysis. The latter, however, should not be simply a statistical analysis of heritabilities and genetic correlations but should attempt to understand the physiological and genetic processes that generate the variability and covariance between traits.

## 1.7 Summary

A central question of evolutionary biology is, "What maintains phenotypic variation?" In this chapter I examine the solutions suggested by models based on the assumption that fitness is maximized (optimality models), and models based upon genetic architecture (genetic models). It is shown that the solutions may be very different and hence predictions of optimality models must be viewed cautiously and, if possible, supplemented by a genetic analysis. Optimality models predict that phenotypic variation will be maintained if there is frequency-dependent selection or environmental heterogeneity. To attain the optimal combination of traits, optimality models assume that genetic variation for these traits exists. The heritability of life-cycle traits among insects and animals in general is significant and, in part at least, underlies the phenotypic variation. Factors maintaining additive genetic variation are mutation, environmental heterogeneity and antagonistic pleiotropy. The potential importance of the last two phenomena is explored.

*Acknowledgements.* I am most grateful to the participants of the symposium for discussions of the ideas presented in this paper. The comments of Drs W. Bradshaw, B. Riska and M. Bradford on earlier drafts were exceedingly helpful. This work was supported by an operating grant from NSERC.

## References

Andersen J, Nilssen AC (1983) Intrapopulation size variation of free-living and tree-boring coleoptera. Can Entomol 115:1453–1464

Bellinger RG, Pienkowski RL (1987) Developmental polymorphism in the red-legged grasshopper, *Melanoplus femurrubrum* (DeGeer) (Orthoptera: Acrididae). Environ Entomol 16: 120–125

Blakley N, Goodner SR (1978). Size-dependent timing of metamorphosis in milkweed bugs (*Oncopeltus*) and its life history implications. Biol Bull 155: 499–510

Bryant E (1977) Morphometric adaptation of the house fly, *Musca domestica*, in the United States. Evolution 31: 580–596

Bryant E, Turner C (1978) Comparative morphometric adaptation of the housefly and the face fly in the United States. Evolution 32: 759–770

Butlin RK, Hewitt GM (1986) Heritability estimates for characters under sexual selection in the grasshopper, *Chorthippus brunneus*. Anim Behav 34: 1256–1261

Cade W (1979) The evolution of alternative male reproduction strategies in field crickets. In: Blum MS, Blum NA (eds) Sexual selection and reproductive competition in insects. Academic Press, New York, pp 343–379

Cade W (1981) Alternative male strategies: genetic differences in crickets. Science 212: 563–564

Charlesworth B (1988) The heritability of fitness. In: Bradbury JW, Andersson HB (eds) Sexual selection: testing the alternatives. Wiley, New York, pp 21–40 (Dahlem Workshop)

Clarke JM, Maynard Smith J, Sondhi KC (1961). Asymmetrical response to selection for rate of development in *Drosophila subobscura*. Genet Res 2: 70–81

Cohen D (1970) A theoretical model for the optimal timing of diapause. Am Nat 104: 389–400

Dawson PS (1975) Directional versus stabilizing selection for development time in natural and laboratory populations of flour beetles. Genetics 80: 773–783

Derr J (1980) The nature of variation in life history characters of *Dysdercus bimaculatus*, a colonizing species. Evolution 34: 548–557

Dickerson GE (1955) Genetic slippage in response to selection for multiple objectives. Cold Spring Harb Symp Quant Biol 20: 213–224

Dingle H, Brown C, Hegmann JP (1977) The nature of genetic variance influencing photoperiodic diapause in a migrant insect, *Oncopeltus fasciatus*. Am Nat 111: 1047–1059

Dingle H, Evans KE, Palmer JO (1988) Responses to selection among life history traits in a non-migratory population of milkweed bugs (*Oncopeltus fasciatus*). Evolution 42: 79–92

Englert DC, Bell AE (1970) Selection for time of pupation in *Tribolium casteneum*. Genetics 64: 541

Falconer DS (1981) Introduction to quantitative genetics, 2nd edn. Longman, London New York

Fisher RA (1958) The genetical theory of natural selection. Dover Publications, New York

Gilbert N (1984) Control of fecundity in *Pieris rapae*. I. The problem. J Anim Ecol 53: 581–588

Gillespie JH, Turelli M (1989) Genotype–environment interactions and the maintenance of polygenic variation. Genetics 121: 124–138

Groeters FR, Dingle H (1987) Genetic and maternal influences on life history plasticity in response to photoperiod by milkweed bugs (*Oncopeltus fasciatus*). Am Nat 129: 332–346

Gross M (1982) Sneakers, satellites and parentals: polymorphic mating strategies in North American sunfishes. Z Tierpsychol 60: 1–26.
Halliburton R, Gall G (1981). Disruptive selection and assertive mating in *Tribolium casteneum*. Evolution 35: 829–843
Hawley WA (1985) The effect of larval density on adult longevity of a mosquito, *Aedes sierrensis*: epidemiological consequences. J Anim Ecol 54: 955–964
Hegmann J, Dingle H (1982) Phenotypic and genotypic covariance structure in milkweed bug life history traits. In: Dingle H, Hegmann J (eds) Evolution and genetics of life histories. Springer, Berlin Heidelberg New York, pp 177–188
Hillesheim E (1984) Heritability of physiological characters in the Cape honeybee, *Apis mellifera capensis*. Apidologie 15: 271–273
Imura O (1980) The color variation in larvae of *Ephestria kuehniella* (Lepidoptera, Phycitdae). I. On the inheritance of color variation. Konchu 48: 248–258
Jago ND (1973) The genesis and nature of tropical forest and savanna grasshopper faunas, with special reference to Africa. In: Meggers BJ, Ayensu ES, Duckworth D (eds.) Tropical forest ecosystems in Africa and South America: a comparative review. Smithsonian Institution Press, Washington, DC, pp 187–196
Lounibos LP (1979) Temporal and spatial distribution, growth and predatory behaviour of *Toxorhynchites brevipalpis* (Diptera: Culicidae) on the Kenya coast. J Anim Ecol 48: 213–236
Morris RF (1971) Observed and simulated changes in genetic quality in natural populations of *Hyphantria cunea*. Can Entomol 103: 893–906
Morris R, Fulton W (1970) Heritability of diapause intensity in *Hyphantria cunea* and correlated fitness responses. Can Entomol 102: 927–938
Mousseau TA, Roff DA (1987) Natural selection and the heritability of fitness components. Heredity 59: 181–197
Nijhout HF (1975) A threshold size for metamorphosis in the tobacco hornworm, *Manduca sexta* (L.). Biol Bull 149: 214–225
Nijhout HF (1979) Stretch-induced moulting in *Oncopeltus fasciatus*. J Insect Physiol 25: 277–281
Orozco F (1976) A dynamic study of genotype environment interaction with egg laying of *Tribolium casteneum*. Heredity 37: 157–171
Palmer JO, Dingle H (1986) Direct and correlated responses to selection among life history traits in milkweed bugs (*Oncopeltus fasciatus*). Evolution 40: 767–777
Prout T (1962) The effects of stabilizing selection on the time of development in *Drosophila melanogaster*. Genet Res 3: 364–382
Riddiford LM (1985) Hormone action at the cellular level. In: Kerkut GA, Gilbert LI (eds) Comprehensive insect physiology, biochemistry and pharmacology. Pergamon Press, New York, pp 37–83
Rinder TE, Collins AM, Brown MA (1983) Heritabilities and correlations of honeybee response to *Nosema apis* longevity and alarm response to isopentylacetate. Apidologie 14: 79–86
Robertson FW (1957) Studies in quantitative inheritance. XI. Genetic and environmental correlation between body size and egg production in *Drosophila melanogaster*. J Genet 55: 428–443
Roff DA (1981) Reproductive uncertainty and the evolution of iteroparity: why don't flatfish put all their eggs in one basket? Can J Fish Aquat Sci 38: 968–977
Roff DA (1984) On the cost of being able to fly: a study of wing polymorphism in two species of crickets. Oecologia 63: 30–37
Roff DA (1986a) The evolution of wing dimorphism in insects. Evolution 40: 1009–1020
Roff DA (1986b) The evolution of wing polymorphism and its impact on life cycle adaptation in insects. In: Taylor F, Karban R (eds) The evolution of insect life cycles. Springer, Berlin Heidelberg New York, pp 209–221
Roff DA (1986c) The genetic basis of wing dimorphism in the sand cricket, *Gryllus firmus*, and its relevance to the evolution of wing dimorphisms in insects. Heredity 57: 221–231
Roff DA (1989) Exaptation and the evolution of dealation in insects. J Evol Biol 2: 109–123
Roff DA, Fairbairn DJ (1990). Wing dimorphisms and the evolution of migratory polymorphisms among the insecta. Am Zool (in press)
Roff DA, Mousseau TA (1987) Quantitative genetics and fitness: lessons from *Drosophila*. Heredity 58: 103–118
Rose MR (1982) Antagonistic pleiotropy, dominance and genetic variation. Heredity 48: 63–78
Rose MR (1983) Theories of life-history evolution. Am Zool 23: 15–23.
Rose MR (1985) Life history evolution with antagonistic pleiotropy and overlapping generations. Theor Popul Biol 28: 342–358

Rose MR, Charlesworth B (1981a) Genetics of life history in *Drosophila melanogaster*. I. Sib analysis of adult females. Genetics 97: 173–186

Rose MR, Charlesworth B (1981b) Genetics of life history in *Drosophila melanogaster*. II. Exploratory selection experiments. Genetics 97: 187–196

Sang JH (1962) Selection for rate of larval development using *Drosophila melanogaster* cultured axenically on deficient diets. Genet Res 3: 70–99

Sang JH, Clayton GA (1957) Selection for larval development time in *Drosophila*. J Hered 48: 265–270

Service PM, Rose MR (1985) Genetic covariation among life history components: the effect of novel environments. Evolution 39: 943–945

Sheridan AK, Frankham R, Jones LP, Rathie KA, Barker JSF (1968) Partitioning of variance and estimation of genetic parameters for various bristle number characters of *Drosophila melanogaster*. Theor Appl Genet 38: 179–187

Slama K (1985) Pharmacology of insect juvenile hormones. In: Kerkut GA, Gilbert LI (eds) Comprehensive insect physiology, biochemistry and pharmacology. Pergamon Press, New York, pp 357–394

Southwood TRE (1961) A hormonal theory of the mechanism of wing polymorphism in heteroptera. Proc R Entomol Soc London (A) 36: 63–66

Stein W (1977) Die Beziehung zwischen Biotop-alter und Auftreten der Kurzflügeligkeit bei Populationen dimorpher Rüsselkäfer-arten (Col., Curculionidae). Z Angew Entomol 83: 37–39

Tanaka Y (1987a) Polygenic analyses of morphological characters in *Pieris rapae crucivora* Boisduval (Lepidoptera: Pieridae). I. Heritability estimates. Appl Entomol Zool 22: 125–132

Tanaka Y (1987b) Polygenic analyses of morphological characters in *Pieris rapae crucivora* (Pieridae: Lepidoptera). II. Phenotypic, genetic and environmental correlations. Jpn J Genet 62: 59–67

Tantawy AO, El-Helw MR (1966) Studies on natural populations of *Drosophila*. V. Correlated response to selection in *Drosophila melanogaster*. Genetics 53: 97–110

Tantawy AO, El-Helw MR (1970) Studies on natural populations of *Drosophila*. IX. Some fitness components and their heritabilities in natural and mutant populations of *Drosophila melanogaster*. Genetics 64: 79–91

Tantawy AO, Rakha FA (1964) Studies on natural populations of *Drosophila*. IV. Genetic variances of and correlations between four characters in *D. melanogaster* and *D. simulans*. Genetics 50: 1349–1355

Wallace HR (1953) Notes on the biology of *Coranus subapterus* de Geer (Hemiptera: Reduviidae). Proc R Entomol Soc London (A) 28: 100–110

Wigglesworth VB (1934) The physiology of ecdysis in *Rhodnius prolixus* (Hemiptera). II. Factors controlling moulting and "metamorphosis". Q J Microsc Sci 77: 191–222.

Wigglesworth VB (1961) Insect polymorphism in a tentative synthesis. Symp R Ent Soc London 1: 103–113

Woodring JP (1983) Control of moulting in the house cricket, *Acheta domesticus*. J Insect Physiol 29: 461–464

Zawilski A (1972) Heritability of four morphological characters in honeybees (*Apis mellifica* L.). Genet Pol 13: 41–49

Zera AJ, Tiebel KC (1989) Differences in juvenile hormone esterase activity between presumptive macropterous and brachypterous *Gryllus rubens*: implications for the hormonal control of wing polymorphism. J Insect Physiol 35: 7–18

*Chapter 2*

# The Use of Selection to Probe Patterns of Pleiotropy in Fitness Characters

Michael R. Rose, Joseph L. Graves and Edward W. Hutchinson

## 2.1 Introduction

One of the more interesting questions in evolutionary biology is to what extent different characters are evolutionarily interdependent because of alleles having effects on multiple characters, or pleiotropy. A number of characters may evolve in a correlated manner, as witness the allometric patterns to be found among many vertebrate attributes (see e.g. Sacher 1959), but this does not necessarily implicate pleiotropy. The correlation could instead be due to correlated patterns of selection, such as an ecological association between two different types of selection. It is conceivable that one predator could select on both sprint speed and endurance in a prey species, producing a correlated response that has nothing to do with any common effects of alleles. Such correlated responses are not considered here. The problem of interest is the ways in which fitness characters can be correlated with each other because of pleiotropy.

The reason why pleiotropy should be of interest is that it gives rise to relatively fundamental constraints on evolution. Here "fundamental constraints" refers not to ineluctable constraints, for patterns of allelic pleiotropy are demonstrably modifiable (see Lenski 1988), but to the idea of correlations that will not be readily abrogated. Correlated evolutionary patterns may be due to correlated selective factors, as discussed above. As the correlations between selective factors vary, then so should such patterns of correlation among fitness characters. Similarly, quantitative genetic correlations may be due to linkage disequilibrium, particularly in small populations. If effective population size increases and recombination moves the loci toward linkage equilibrium, then such genetic correlations will decrease in magnitude. In addition, patterns of genetic correlation are critically dependent on the environment in which the characters are measured (see e.g. Service and Rose 1985). However, in studying the potential for multiple effects of allelic variation, we can address basic questions concerning the ways in which loci can affect multiple characters. While an exhaustive survey of all pleiotropic interdependencies may be impractical, particularly from the standpoint of accurate quantitative estimation of the interaction system involved (Wright 1968, 1977), ascertaining which qualitative patterns of pleiotropy arise may reveal those patterns of evolutionary correlation that will be somewhat durable.

This general goal still leaves unspecified the means for its attainment. One way to unravel patterns of pleiotropy is to use mutationally generated extreme alleles, especially null-alleles. The problem with this methodology for fitness characters is that the isolation of such alleles requires inbreeding at least at some points in the genetic analysis, and such inbreeding then biases genetic correlation estimates for fitness characters toward positive values (Rose 1984a). In organisms in which inbreeding depression does not arise, such as *Caenorhabditis elegans* (Johnson and Wood 1982), this procedure can be used to detect allelic effects of all kinds. Friedman and Johnson (1988) have found a *C. elegans* allele that exhibits pronounced "antagonistic pleiotropy" (Rose 1982, 1985) between reproduction and longevity, enhancing longevity while depressing reproductive output. But for most organisms, this method is not practical.

Another potential method is the use of correlations between relatives to estimate additive genetic correlations between fitness characters. We have used this method ourselves (Rose and Charlesworth 1981a), in a study of *Drosophila melanogaster* life-history characters. The correlations estimated by these procedures are dependent on linkage equilibrium and other requirements (Falconer 1981). In addition, it is difficult to achieve much statistical power in the estimation of these parameters with most organisms, particularly insects. The daily estimates of the additive genetic variance of 24 h fecundity in *D. melanogaster* provided by Rose and Charlesworth (1981a) illustrate the pronounced variability between estimates of quantitative genetic parameters obtained in sibling analyses of insect species.

Perhaps the simplest approach to the problem of probing patterns of pleiotropy is to select on the characters of interest and observe correlated responses. Information about pleiotropy allows prediction of the indirect responses of populations to natural selection. Selection on fitness characters under laboratory conditions is an obvious way to test for the kinds of correlated response which might occur in nature. This is not to say that laboratory selection experiments can be regarded as reasonable approximations to natural selective conditions. The idea is that they mimic some of the components of selection which could arise under natural conditions. How the experimental organism responds to these particular types of laboratory selection then becomes the information from which a picture of the potential evolutionary interconnections between fitness characters is drawn.

However, laboratory selection experiments on fitness characters are by no means unproblematic or obvious in their implementation or interpretation. In particular, there are severe trade-offs in the design of these experiments, especially those relating to population size, intensity of selection, inbreeding within lines, replication of lines, sufficient controls, and sufficient duration of selection, to give only a partial listing. One way out of these trade-off dilemmas is to perform selection experiments with different designs. Results that are similar over different experimental designs have some prospect of general validity. We have pursued this strategy in a somewhat haphazard fashion, and accordingly organize our discussion around the three basic types of experimental design that we have used: (a) laboratory natural selection; (b) artificial truncation selection; and (c) filtering, or screening, selection.

All of the experiments to be discussed were performed using a *D. melanogaster* stock which has been maintained in the laboratory since 1975, after sampling by P. T. Ives from the South Amherst, MA, population. Our main concern has been with adult fitness characters, but we will be studying some larval characters in the future. In any case, the present discussion is intended more to illustrate methodological ideas, so that the limitations to the scope of the research being considered should not be too important.

## 2.2 Laboratory Natural Selection

### 2.2.1 Selection Procedures

In our laboratory natural selection experiments, we have been concerned with alternative regimes of demographic selection. That is, we have been interested in the response to natural selection arising from different regimes of culture reproduction in the laboratory. The classic culture regime in *D. melanogaster* is 2-week discrete generations at 25 °C. Since it takes 8–10 days to complete the life-cycle at 25 °C in this species, this amounts to selection for early adult reproduction, since the preservation of discrete generations requires that the adults be discarded after they have been used for the day 14 egg-laying. Mean longevity at 25 °C in this species is normally from 30 to 60 days as adults, or 40–70 days total, providing mating is allowed. Therefore, an obvious culture alternative is to reproduce cultures using older adults, and thereby select for later adult reproduction, and so the requisite survival to later ages as well. In these experiments, it should be noted, the experimenter does not directly measure any phenotypes, nor is mortality directly applied.

Experiments of this kind have been done since the 1960s (Wattiaux 1968a, b; Lints and Hoste 1974, 1977; Rose and Charlesworth 1980, 1981b; Rose 1984b; Luckinbill et al. 1984; Luckinbill and Clare 1985). Here we will discuss the experiment reported by Rose (1984b) in particular. (Inconsistencies between the present results and those of other experiments of this kind have been shown to be due to density artifacts (Luckinbill and Clare 1985; Service et al. 1988).) In this experiment, ten distinct cultures were derived from the base population described above.

*B Lines*. These were reproduced using 2-week discrete generations and were designated $B_1$ to $B_5$.

*O Lines*. The other five were reproduced using progressively older adults, with maintenance of discrete generations. These cultures were designated $O_1$ to $O_5$. After 2 years, O cultures were reproduced using flies that had achieved 70 days of age, from the egg stage. The basic idea of this experiment was to investigate the effects of selection for the capacity to survive to, and reproduce at, later ages.

*R Lines*. In a second experiment (Service et al. 1988), a third set of five cultures was obtained by subculturing a new set of lines, $R_1$ to $R_5$, from the O stocks, each $R_i$ being derived from the corresponding $O_i$. The R's were split off in January 1985, and from then on handled like the B stocks. The intent of this experiment was to observe the effects of switching selection back toward favouring early reproduction (Service et al. 1988).

### 2.2.2 Response to Laboratory Natural Selection for Later Reproduction

A wide range of characters responds to selection for later reproduction. Predictably, late fecundity and longevity increased (Rose 1984b). There were many indirect responses: early fecundity decreased (Rose 1984b); early ovary weight decreased (Rose et al. 1984); resistance to starvation, desiccation and ambient ethanol vapour increased (Service et al. 1985); lipid content and later locomotor activity increased, while early metabolic rate and early locomotor activity decreased (Service 1987); flight duration increased (Graves and Rose, 1990). Evidently, numerous aspects of the phenotype are bound up with the response to selection for increased later reproduction.

The pattern of indirect responses suggests the existence of two distinct types of responding character. Firstly, there are those characters which appear to constitute cases of antagonistic pleiotropy. These characters are probably associated with early reproduction: early fecundity, early ovary weight, early metabolic rate, and early locomotor activity. The latter pair of characters are included in this group because they can reasonably be expected to increase with increased early reproductive activity, although this is only a tentative association at present.

Secondly, there are those indirect responses which seem likely to enhance survival, albeit in ways that are somewhat obscure at present. Resistance to environmental stress, such as ambient ethanol vapour, would seem of obvious benefit, although there is little direct evidence that flies in laboratory culture are subject to acute environmental stress. The pattern of lipid differentiation correlates closely with the enhancement of starvation resistance (Service 1987), so if starvation is a source of mortality, then selection for more lipid is readily explicable. Two locomotor enhancements have been observed: walking at later ages (Service 1987) and flight duration at a wide range of adult ages (Graves et al. 1988; Graves and Rose, 1990). Both improvements may reflect an increased ability to escape from sticky medium or fouled cage surfaces. This second group of characters may thereby underlie the direct response of longevity, at least in part.

### 2.2.3 Response to Reversal of Laboratory Natural Selection

The results obtained from the R cultures help to sort out some of the pleiotropic interconnections among the characters that differentiate the B and O populations. When selection was reversed, with a return to selection for early fertility, the Rs exhibited an increase in early fecundity and a fall in starvation resistance

(Service et al. 1988). No changes in desiccation resistance or ethanol vapour resistance were detected. The indirect response of starvation resistance fits with a negative estimate for the additive genetic correlation between early fecundity and starvation resistance, obtained earlier (Service and Rose 1985). These responses to reversed selection suggest that some of the fitness characters of this system, especially early fecundity and starvation resistance, were bound together by antagonistic pleiotropy. The status of the other characters remained unclear, except that resistance to desiccation and ethanol vapour did not seem to be tightly coupled to the characters of *this* antagonistic pleiotropy complex, nor did they apparently exhibit antagonistic pleiotropy with any other component of early reproduction.

## 2.3 Artificial Truncation Selection

### 2.3.1 Selection Procedures

One of the problems with laboratory natural selection is that the targets of selection can be unclear. The experimenter attempts to contrive selection by indirect means. An alternative is for the experimenter to attempt to control selection as directly as possible. This can be done to some extent using the methods of "artificial selection", as defined by quantitative genetics (Falconer 1981). Reduced to its core, artificial selection occurs when contributions to the next generation are determined according to some direct phenotypic measure(s) of the parent(s), while other forms of selection are minimized. In effect, the experimenter directly determines the overt pattern of selection. This is a considerable contrast in procedure with that of laboratory natural selection.

This approach can be illustrated using a selection experiment on one of the characters known to be critical to the laboratory evolution of the *D. melanogaster* stock discussed thus far: early fecundity. The experimental design, here designated F-1, involved eight lines, four undergoing truncation selection for increased early fecundity and four controls (Rose 1984c), all derived from a sample of the Ives population mentioned at the outset. Each line was maintained in its own population cage, within which mating was not controlled. Females were sampled at 14 days of age and allowed to lay in separate wells of a culture dish for 6 h. In the selection lines, the 14 females that laid most eggs were used to initiate 14 vials, each containing 20 eggs, sampled from a single selected mother. In the control lines, the procedure was identical, except that the females that contributed eggs to the next generation were chosen at random.

This experimental procedure was repeated again with subcultures of three B stocks, $B_1$, $B_2$ and $B_3$, from the laboratory natural selection experiment. The design differed in that egg-laying took place over 24 h, the top 50 females were chosen as mothers, and the rearing density was 30 eggs per vial (E. W. Hutchinson et al., unpublished results). Three control lines were maintained in this experiment, which is designated F-2, as well. In parallel with this selection experiment, artificial truncation selection was applied to early starvation resistance in subcultures of the $O_1$, $O_2$ and $O_3$ populations, again from the laboratory natural selection experiment, together with paired controls from these same populations. In this experiment, the starvation resistance scores of both parents were combined to determine a joint score for each couple, the top 50 couples then being chosen to contribute to the next generation (E. W. Hutchinson et al., unpublished results). This will be referred to as the S selection experiment.

### 2.3.2 Direct Response to Artificial Selection

Statistically significant, reproducible, direct responses to selection on early fecundity and starvation resistance were obtained, at least in the initial generations (Rose 1984c; E. W. Hutchinson et al., unpublished results). However, an interesting feature of the results from the F-1 experiment was a high degree of sensitivity to inbreeding depression. In experiment F-1, after only eight generations of selection with an effective population size of about 30, inbreeding began to undermine the direct response to selection (Rose 1984c). For this reason, it may be doubted that artificial selection will often be the preferred type of selection in experiments on patterns of pleiotropy between fitness components, particularly if selection is to be sustained and a useful stock obtained. On the other hand, in the F-2 and S selection experiments of E. W. Hutchinson et al. (unpublished results), in which effective population sizes were of the order of 100 or more, no evidence of inbreeding depression was found, in spite of considerable tests for it. However, these experiments were extremely labour-intensive.

### 2.3.3 Indirect Response to Artificial Selection

An obvious potential function for artificial selection is that it can be used to test for the presence of pleiotropy that has been inferred using correlations between relatives or the results of laboratory natural selection experiments. Service et al. (1985, 1988) inferred the presence of antagonistic pleiotropy between starvation resistance and early fecundity, using laboratory natural selection experiments, while Service and Rose (1985) used sibling analysis to estimate a significantly negative additive genetic correlation between early fecundity and starvation resistance. Accordingly, it was expected that artificial selection on early fecundity (experiment F-2) and starvation resistance (experiment S) would give antagonistic indirect responses in the other character. The regression of starvation resistance on early fecundity when the latter was under selection was $-0.008\pm0.003$, while the converse regression of fecundity on starvation resistance was $-0.022\pm0.025$. Only the former was statistically significant, though both results are in the correct direction. The point of mentioning this result is that the correlated responses to artificial selection are in fact quite weak, compared with the results of sibling analysis and laboratory natural selection. Yet every other experiment that has been done in our laboratory indicates that there is antagonistic pleiotropy between these characters. Why the artificial selection results should be so much weaker is unclear. Relative to laboratory natural selection experiments, population sizes are smaller and the duration of the experiments is shorter with artificial selection. As both of these problems arise inherently from the labour-intensive nature of artificial selection, together with the vulnerability to inbreeding depression of fitness characters and resultant biases to positive genetic correlations (Rose 1984a), it is uncertain whether artificial selection experiments can normally provide results that will help to dissect patterns of pleiotropy between fitness characters. However, it must be understood that this is due to practical design problems, such as population sizes, not fundamental deficiencies.

The F-2 and S artificial selection experiments with the B and O stocks qualitatively fit the previous inferences of antagonistic pleiotropy between early fecundity and starvation resistance. The stocks that these experiments have generated also provide some evidence bearing on other pleiotropic interconnections. We designate as S the O stocks selected for enhanced starvation resistance, corresponding to the labelling of the experiment. Tests were made of the desiccation and ethanol vapour resistance levels of these S stocks, relative to those of the Os (M. R. Rose et al., unpublished results). The stock means are given in Table 2.1. The Ss have substantial increases in starvation resistance and slight increases in resistance to the other two stresses. Of the two indirect responses, only that of desiccation resistance is statistically significant at the $P<0.05$ level. These results suggest that the three resistance characters are not closely coupled, as also found by Service et al. (1988), although there may be some degree of pleiotropy between them. We also studied the flight duration of S stocks compared with that of O stocks. As shown in Table 2.2, the S and O stocks do not differ significantly with respect to flight duration. This suggests that flight duration is not pleiotropically connected to starvation resistance.

**Table 2.1.** Mean values of the O and S stocks for the indicated stress resistance characters. The values given are the mean hours to death of flies from a given stock subjected to a particular lethal stress. The treatment difference between O and S stocks is statistically significant ($P<0.05$) according to analysis of variance only for starvation and desiccation

| Stock | Population | | |
|---|---|---|---|
| | 1 | 2 | 3 |
| *Starvation resistance* | | | |
| O | 43.8 | 31.9 | 26.4 |
| S | 50.5 | 50.1 | 52.9 |
| *Desiccation resistance* | | | |
| O | 16.4 | 15.7 | 14.3 |
| S | 16.8 | 19.1 | 15.2 |
| *Ethanol vapour resistance* | | | |
| O | 23.0 | 24.6 | 23.6 |
| S | 24.8 | 26.3 | 24.2 |

**Table 2.2.** Mean values for flight duration of the different lines of the O and S experimental treatments. The difference in line means between treatments is not significant at the $P=0.05$ level in a paired *t*-test

| Stock | Population | | |
|---|---|---|---|
| | 1 | 2 | 3 |
| O | 104.6 | 84.5 | 97.9 |
| S | 98.4 | 109.6 | 83.7 |

# 2.4 Laboratory Culling

## 2.4.1 Selection Procedures

The problem with laboratory natural selection experiments is that it is unclear what the actual pattern of selection is. The problem with artificial selection experiments is that they will usually be too small in scale to be sustained with fitness characters. In an attempt to escape from this dilemma, we have tried to develop a third type of selection procedure that we call "laboratory culling". The basic idea of this experiment is much like that of microbial selection procedures: a large number of organisms are subjected to an extreme environmental challenge which only a few can survive. Those few are then used to reproduce the culture. With *Drosophila*, it is straightforward to culture thousands of flies, place them under some potentially lethal stress, and harvest the surviving adults to lay eggs for the next generation. In this way, the source of the selective differential is operationally well defined, while the numbers that undergo selection can be quite large.

We have been subjecting subcultures from the aforementioned O stocks to laboratory culling using desiccation as the culling agent. There are five selected

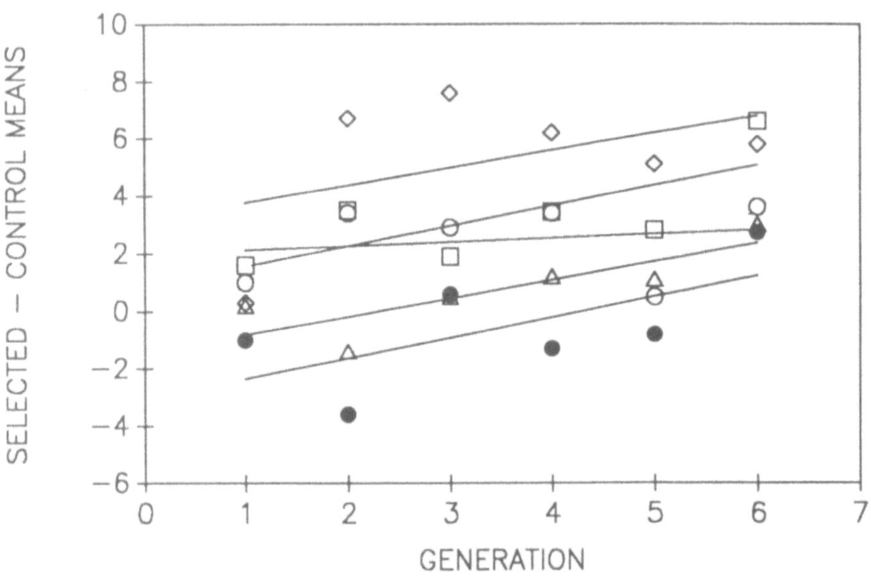


**Fig. 2.1.** The figure shows the difference between mean desiccation resistances for each D line and its paired C line in each generation. The different symbols designate specific lines, so that the circles, for example, give the difference between D and C lines for the same pair of lines throughout. The paired differences of each pair of lines were analysed by least-squares linear regression, and the resultant regression lines are plotted. The regression lines for desiccation resistance have slopes significantly different from zero at the $P$=0.05 level, when tested as a group using a $t$-test.

lines, $D_1$ to $D_5$, and five paired control lines, $C_1$ to $C_5$, derived from $O_1$ to $O_5$, respectively. Two-week discrete generations are used, as with B culture. The culling involves harvesting all the adults of a D stock, placing them in a cage with desiccant only and sealing that cage atmospherically using a plastic bag. Once most of the flies have died or been incapacitated by desiccation, the plastic bag is removed and food is supplied. The eggs laid on the food are used to start the next generation. The Cs undergo the same procedure, except that desiccant is not added to their cages when they are sealed off. Both Ds and Cs are kept without food during the period of sealing.

## 2.4.2 Response to Laboratory Culling

Figs. 2.1, 2.2 and 2.3 show the responses of desiccation resistance, starvation resistance, and ethanol vapour resistance, respectively, to laboratory culling using desiccation resistance. There is a statistically significant direct response of desiccation resistance, as ascertained by tests for the significance over the slopes of least-squares linear regressions on each line's mean resistance. There is at most a marginal response of starvation resistance, while ethanol vapour resistance responds dramatically and significantly. Together with the results of Service et al. (1985, 1988), in which desiccation resistance and ethanol vapour resistance evolved in a similar manner, this seems to indicate that desiccation and ethanol vapour resistance could be closely associated, genetically.

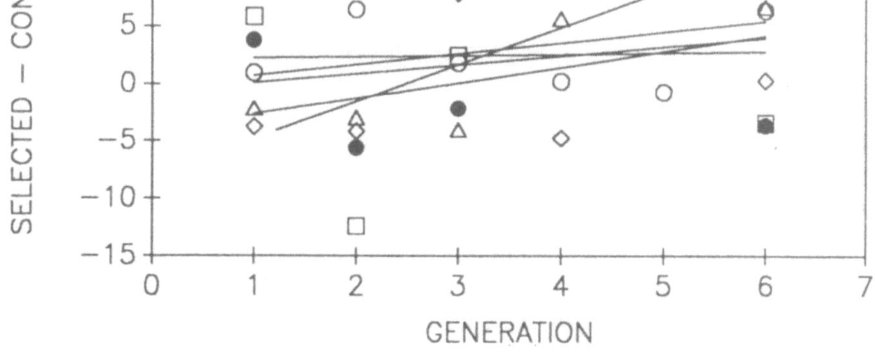


**Fig. 2.2.** The figure shows the difference between mean starvation resistances for each D line and its paired C line in each generation. The different symbols designate specific lines, so that the circles, for example, give the difference between D and C lines for the same pair of lines throughout. The paired differences of each pair of lines were analysed by least-squares linear regression, and the resultant regression lines are plotted. The regression lines for starvation resistance do not have slopes significantly different from zero at the $P=0.05$ level, when tested as a group using a $t$-test.

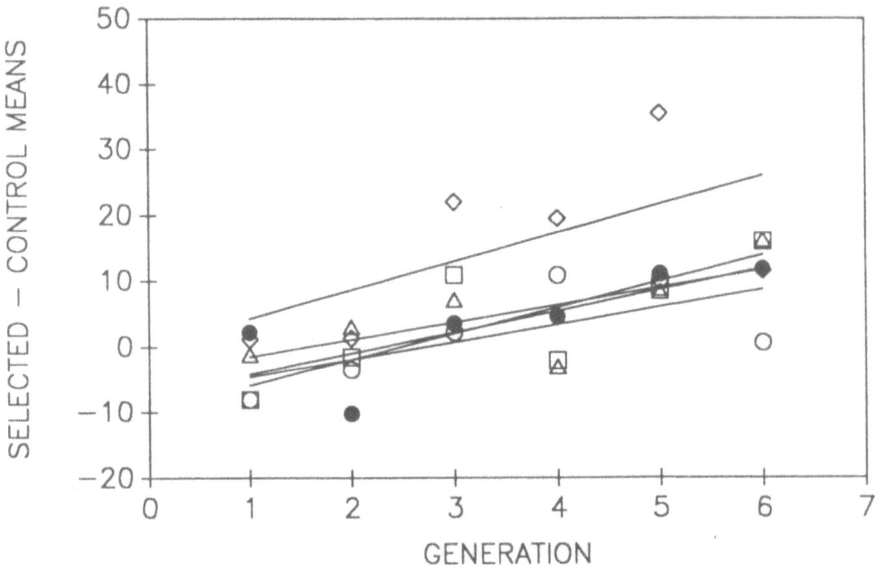


**Fig. 2.3.** The figure shows the difference between mean ethanol vapour resistances for each D line and its paired C line in each generation. The different symbols designate specific lines, so that the circles, for example, give the difference between D and C lines for the same pair of lines throughout. The paired differences of each pair of lines were analysed by least-squares linear regression, and the resultant regression lines are plotted. The regression lines for ethanol vapour resistance have slopes significantly different from zero at the $P$=0.05 level, when tested as a group using a $t$-test.

We have also studied the flight duration response to culling for desiccation resistance. As shown in Table 2.3, there is a statistically significant increase in flight duration. This indicates some pleiotropic connection between desiccation resistance and flight endurance.

## 2.5 Comparison of the Selection Procedures

As Figs. 2.1–2.3 and Table 2.3 show, it is possible to detect rapid direct and indirect responses to laboratory culling. If we compare these results with those of Luckinbill and Clare (1985), who plotted the results of laboratory natural selection with sampling every few generations, it seems clear that laboratory culling can give rise to more readily detectable responses, at least initially. In terms of avoiding inbreeding depression, laboratory culling can allow much larger effective population sizes than does artificial selection. It is easy to sustain selection on thousands of fruit flies per generation per line, compared to the hundreds that artificial selection can be practised upon. Therefore, it seems clear that laboratory culling has indisputable value in laboratory studies of fitness characters. This is not to argue against performing the other two types of selection. Laboratory natural selection will always be interesting in terms of ascertaining just what natural selection might happen to do, given a particular set

**Table 2.3.** Mean values for flight duration of the different lines of the D and C experimental treatments. The difference in line means between treatments is significant at the $P$=0.05 level in a paired $t$-test. Note that the lines were handled as pairs consisting of one line of each stock type, paired according to the line number, which reflects the O stock of origin

| Stock | Line | | | |
|---|---|---|---|---|
| | 1 | 2 | 3 | 4 |
| C | 89.2 | 119.7 | 116.7 | 102.5 |
| D | 107.8 | 137.0 | 120.6 | 112.0 |

of circumstances. Artificial truncation selection has the advantage of being extremely well defined; it is the best for monitoring the responses of the population during selection.

For many problems, some combination of all three selection procedures may be advisable, since any selection design will be susceptible to its own characteristic set of artifacts. When different basic designs are used in a series of experiments, it is less likely that results which are consistent over experiments will be due to artifacts.

## 2.6 Patterns of Pleiotropy in *Drosophila melanogaster* Fitness Characters

### 2.6.1 What We Think We Know

From more than a decade's work on fitness characters in our populations of *D. melanogaster*, we can see fairly clearly that there is some pattern of antagonistic pleiotropy relating early fecundity to adult survival. The fact that other laboratories have obtained similar antagonistic patterns between early fecundity and adult survival in *D. melanogaster* (e.g. Hiraizumi 1985; Luckinbill and Clare 1985), even though they were using different stocks and different methods, suggests that our results are not idiosyncratic. This antagonistic pleiotropy appears to be mediated, at least in part, by allocation of resources between lipid and eggs, such that females that (a) have ovaries distended with eggs early in life, (b) are more active when young, and (c) have higher early metabolic rates, concomitantly have less lipid and die sooner. Presumably this effect arises because of diversion of calories from soma to germ-line, although it might occur because of a change in enzymic pathways which renders germ-line tissue relatively more efficient than somatic tissue, without any re-allocation.

It is also apparent that there are other characters that are bound up with fitness, at late or early stages of adult life, but none the less are not pleiotropically interconnected with the lipid/fecundity suite of characters. From the experiments that we have done to this point, both ethanol vapour resistance and flight duration apparently can respond independently of these other characters (see Service et al. 1988; Tables 2.1 and 2.2). There may be some, as yet undetected, degree of pleiotropy between these two characters and the lipid/fecundity suite, but it would appear to be small. Therefore, we conclude that there can be some degree of evolutionary independence between characters. Pleiotropy is not universal.

### 2.6.2 What We Would Like to Know

There are three obvious lines of work for further investigation of this system. Firstly, what are the details of the lipid/fecundity suite of characters? E. W. Hutchinson and M. R. Rose (unpublished results) and E. W. Hutchinson et al. (unpublished results) have shown that these characters are additive in their inheritance on average, without maternal effects. There is no reason to reject the quantitative-genetic null hypothesis that they are affected by numerous loci, according to tests of segregation indices. However, this does not preclude the existence of a few key biochemical loci involved in the energetic metabolism and reproductive mobilization of the adult. Molecular-genetic methods might ascertain the identity and features of these loci.

Secondly, can we give convincing demonstrations of the independence of characters such as flight duration and ethanol vapour resistance from the lipid/fecundity suite of characters? Obviously selecting for these two characters and examining the responses of starvation resistance and early fecundity would be one method. Laboratory culling would be our preferred technique, but that could prove to be technically difficult with both characters. Flight duration does not directly determine mortality in simple environments. Ethanol vapour levels are difficult to control in cage environments. Perhaps sibling analysis would be an alternative method, such as a design in which fecundity was measured first, then flight duration, and finally time to death in an ethanol vapour environment.

Thirdly, there are those characters that seem to be correlated to some extent with numerous other characters. In at least some experiments, desiccation resistance seems to be pleiotropically connected with all of the aforementioned characters, although in some others it does not appear to be connected. Perhaps some of the loci that determine desiccation resistance contribute pleiotropically to starvation resistance, while others contribute to flight duration. We do not now have any particularly good ideas as to how to sort this out experimentally.

## 2.7 Conclusions

We have discussed the use of various types of selection experiment in the probing of groups of fitness characters in a tractable laboratory insect, *D. melanogaster*. It might be objected that this tells little about procedures that might be used in field systems or with species that are difficult to study directly, in either field or laboratory. Indeed, it could be objected that none of the foregoing tells us much about *Drosophila* in the wild, given the unusual nature of the laboratory environment. In one respect, we would directly concede the validity of all these objections. We have no conclusions to offer that would be specifically relevant to any other system. Rather, the thrust of our research has been to ascertain what the bare possibilities are in *any* system. So long as no investigator had extensively documented antagonistic pleiotropy, in *any* system, doubting that it existed at all remained defensible. Indeed, some still do (e.g. Le Bourg et al. 1988). So long as antagonistic pleiotropy seemed to arise between key combinations of life-history characters, it was a defensible point of view that it was exclusively important (e.g. Williams 1957; Rose 1983). We now suspect that neither of these points of view is tenable (see Kosuda 1985; Mueller 1987; Service et al. 1988), and we are left with the task of unravelling patterns of pleiotropy in particular cases. This conclusion would not have been obtained so quickly if we had not used the *Drosophila* laboratory system, with its characteristic experimental power.

*Acknowledgements*. We are grateful to W. E. Bradshaw and L. MacPhee for their comments on the manuscript. Technical help was provided by S. M. Curtis, P. E. Gallegos, K. K. Grimm, B. Huettner, C. M. Jeong, H. H. T. Ma, J. E. Nelson, Ha T. Nguyen, Huyen T. Nguyen, T. A. Nguyen, A. L. Nohalty, T. B. Tran, K. A. Vu, L. N. Vu and V. H. Vu. This research was supported in part by NIA grant AG-06346 to M.R.R. and a University of California President's Fellowship to J.L.G.

## References

Falconer DS (1981) Introduction to quantitative genetics, 2nd edn. Longman, London

Friedman DB, Johnson TE (1988) A mutation in the age-1 gene in *Caenorhabditis elegans* lengthens life and reduces hermaphrodite fertility. Genetics 118: 75–86

Graves JL, Rose MR (1990) Flight duration in *Drosophila melanogaster* selected for postponed senescence. In: Harrison DE (ed) Genetic effects on aging II. Telford Press, pp 59–65

Graves JL, Luckinbill LS, Nichols A (1988) Flight duration and wing beat frequency in long- and short-lived *Drosophila melanogaster*. J Insect Physiol 34: 1021–1026

Hiraizumi Y (1985) Genetics of factors affecting the life history of *Drosophila melanogaster*. I. Female productivity. Genetics 110: 453–464

Johnson TE, Wood WB (1982) Genetic analysis of life-span in *Caenorhabditis elegans*. Proc Natl Acad Sci USA 79: 6603–6607

Kosuda K (1985) The aging effect on male mating activity in *Drosophila melanogaster*. Behav Genet 15: 297–303

Le Bourg E, Lints FA, Delince J, Lints CV (1988) Reproductive fitness and longevity in *Drosophila melanogaster*. Exp Gerontol 23: 491–500

Lenski RE (1988) Experimental studies of pleiotropy and epistasis in *Escherichia coli*. II. Compensation for maladaptive effects associated with resistance to virus T4. Evolution 42: 433–440

Lints FA, Hoste C (1974) The Lansing effect revisited. I. Lifespan. Exp Gerontol 9: 51–69

Lints FA, Hoste C (1977) The Lansing effect revisited. II. Cumulative and spontaneously reversible parental age effects on fecundity in *Drosophila melanogaster*. Evolution 31: 387–404

Luckinbill LS, Clare MJ (1985) Selection for life span in *Drosophila melanogaster*. Heredity 55: 9–18

Luckinbill LS, Arking R, Clare MJ, Cirocco WC, Buck SA (1984) Selection for delayed senescence in *Drosophila melanogaster*. Evolution 38: 996–1003

Mueller LD (1987) Evolution of accelerated senescence in laboratory populations of *Drosophila melanogaster*. Proc Natl Acad Sci USA 84: 1974–1977

Rose MR (1982) Antagonistic pleiotropy, dominance, and genetic variation. Heredity 48: 63–78

Rose MR (1983) Evolution of aging. Rev Biol Res Aging 1: 19–24

Rose MR (1984a) Genetic covariation in *Drosophila* life history: untangling the data. Am Nat 123: 565–569

Rose MR (1984b) Laboratory evolution of postponed senescence in *Drosophila melanogaster*. Evolution 38: 1004–1010

Rose MR (1984c) Artificial selection on a fitness-component in *Drosophila melanogaster*. Evolution 38: 516–526

Rose MR (1985) Life history evolution with antagonistic pleiotropy and overlapping generations. Theor Popul Biol 28: 342–358
Rose M, Charlesworth B (1980) A test of evolutionary theories of senescence. Nature 287: 141–142
Rose MR, Charlesworth B (1981a) Genetics of life-history in *Drosophila melanogaster*. I. Sib analysis of adult females. Genetics 97: 173–186
Rose MR, Charlesworth B (1981b) Genetics of life-history in *Drosophila melanogaster*. II. Exploratory selection experiments. Genetics 97: 187–196
Rose MR, Dorey ML, Coyle AM, Service PM (1984) The morphology of postponed senescence in *Drosophila melanogaster*. Can J Zool 62: 1576–1580
Sacher GA (1959) Relation of lifespan to brain weight and body weight in mammals. In: Wolstenholme GEW, O'Connor M (eds) Ciba Foundation Colloquia on Ageing. Churchill, London, pp 115–133
Service PM (1987) Physiological mechanisms of increased stress resistance in *Drosophila melanogaster* selected for postponed senescence. Physiol Zool 60: 321–326
Service PM, Rose MR (1985) Genetic covariation among life-history components: the effect of novel environments. Evolution 39: 943–945
Service PM, Hutchinson EW, MacKinley MD, Rose MR (1985) Resistance to environmental stress in *Drosophila melanogaster* selected for postponed senescence. Physiol Zool 58: 380–389
Service PM, Hutchinson EW, Rose MR (1988) Multiple genetic mechanisms for the evolution of senescence. Evolution 42: 708–716
Wattiaux JM (1968a) Cumulative parental age effects in *Drosophila subobscura*. Evolution 22: 406–421
Wattiaux JM (1968b) Parental age effects in *Drosophila pseudoobscura*. Exp Gerontol 3: 55–61
Williams GC (1957) Pleiotropy, natural selection, and the evolution of senescence. Evolution 11: 398–411
Wright S (1968) Evolution and the genetics of populations, vol. 1, Genetic and biometric foundations. University of Chicago Press, Chicago
Wright S (1977) Evolution and the genetics of populations, vol. 3, Experimental results and evolutionary deductions. University of Chicago Press, Chicago

# *Section II*
# Evolution of Life-Cycle Traits

# Introduction

William E. Bradshaw

The previous section reaffirmed the genetic basis of life-history variation and established the potential importance of genetic correlations in the immediate response to selection. The present section examines life-history traits and groups of traits in evolved populations or species. In so doing, there is often a gap between patterns perceived in the present and the sequence of genetic events leading up to these patterns. William Bradshaw and Christina Holzapfel (Chap. 3) propose that life-histories may be separated into groups of traits that covary and function together, thereby forming what Susan Scott and Hugh Dingle (Chap. 4) term adaptive syndromes. Both chapters emphasize that explanations of life-cycle syndromes must rely upon knowledge of both genetic variation and covariation, i.e. upon how genes influence the manner in which traits develop and function together. Yet, both chapters also provide arguments that certain life-cycle syndromes should maintain genetic cohesiveness within themselves but, at the same time, genetic independence from other life-cycle syndromes.

In the third chapter of this section, May Berenbaum (Chap. 5) looks at covariation at a higher level, namely the reciprocal evolution of plants and their herbivores. Berenbaum's own work with the parsnip webworm on wild parsnip has demonstrated genetic trade-offs between plant protection and plant fitness in the absence of herbivory and between webworm chemical defence and webworm fitness on benign plants. Despite the fact that Berenbaum's work establishes the genetic feasibility for coevolution, she also argues convincingly that concluding coevolution from even congruent patterns of host-plant chemical defence and herbivore tolerance may be exceedingly difficult.

In Chap. 6, Francis Gilbert examines life-cycle divergence within a single family of syrphid flies. Gilbert seeks patterns of life-cycle evolution among species within subfamilies and/or genera whose morphometrics, ecology and probable phylogeny are known. In his analyses, he approaches questions about the comparative life-cycle evolution of generalists versus specialists with respect to the trade-off between egg size and egg number, direction in evolution, competition and population stability.

A major component of the life-cycle of any temperate insect is migration and/or dormancy to escape an unfavourable season. Several chapters in this book treat diapause or migration as a given fact characteristic of the particular species in question (Chaps. 3, 4, 11 and 13). In Chap. 7, however, Sinzo Masaki delves into the origin of diapause in a subtropical cricket whose closely related temperate congener enters a typical hibernal diapause. After dissecting the responses to temperature and photoperiod by the subtropical cricket, Masaki is able to project the probable sequence of events in the transition from opportunistic subtropical diapause to determinate temperate diapause.

Most of the chapters in this volume consider plant–insect interaction from the viewpoint of consumer versus herbivore, and only a one-way benefit (Chaps. 5 and 10). John Addicott, Judie Bronstein and Finn Kjellberg (Chap. 8) approach insect–plant interaction from the opposite direction, mutualism. In their chapter, they compare the life-histories of fig wasp and yucca moth host–pollinator mutualisms. Using these examples, they consider the stability of the interactions by showing: (a) how the degree of synchronization between host and symbiont life-cycles affects successful association and, ultimately, the maintenance of mutualism; and (b) under what conditions cheating is favoured over mutualism.

Most models of selection, even selection in seasonal environments, predict evolution of the mean phenotype when confronted with constant, uniform selection. The evolution of insect life-cycles takes place in seasonal environments that vary considerably from year to year. There may be long-term stabilizing selection for some optimum mean phenotype but selection acting on the population may not be uniform either in magnitude or in direction from one year to the next. In the final chapter of this section, Fritz Taylor (Chap. 9) releases the assumption of constancy in selection and examines the approach of the mean phenotype to the optimal when stabilizing (optimizing) selection varies in time with varying degrees of autocorrelation. From his models, he demonstrates that variable environments, especially those that show some degree of correlation from one season to the next, can greatly delay (a) the approach of the selected phenotype to the long-term optimum and (b) the convergence of the standard deviation of the mean on the true standard deviation. Both delays can impede the ability to test theories about the evolution of the mean phenotype in nature.

*Chapter 3*

# Evolution of Phenology and Demography in the Pitcher-Plant Mosquito, *Wyeomyia smithii*

William E. Bradshaw and Christina M. Holzapfel

## 3.1 Introduction

Life-histories may be separated into groups of traits that covary and function together, generally relating to growth and development, reproduction, dormancy and migration (Tauber and Tauber 1981; Tauber et al. 1986; Dingle 1986). Growth, development and reproduction are generally associated with continuous life-cycles and contribute directly to demographies. Dormancy and migration generally disrupt or delay continuous life-cycles. Since a genotype's age at first reproduction and generation time have a large impact on its fitness (Cole 1954; Lewontin 1965; Stearns 1976), developmental or reproductive delays might initially appear to be maladaptive (but see Murphy 1968; Livdahl 1979; Taylor 1980). Organisms in nature do not, however, live in continuous conditions but rather in seasonal environments that are punctuated with periods of unfavourable food, temperature, and/or drought. Dormancy and migration provide "escape in time and space" (Slobodkin 1961) and are thus crucial to long-term fitness despite the concomitant life-cycle delays.

In this paper, we are concerned with evolutionary coordination between two main groups of traits that we define as follows:

*Demographic* traits are those traits involved with direct development, growth, survivorship and reproduction.

*Phenological* traits are those traits involved with migration and dormancy. They are usually invoked seasonally by specific direct (temperature, moisture, food) or indirect (photoperiod) cues (Lees 1955; Danilevskii 1965; Beck 1980, Tauber et al. 1986; Danks 1987).

We envision the distinction between these two types of traits to be fundamental to insect life-cycle strategies. Demographic and phenological traits constitute the major determinants of fitness under favourable and unfavourable conditions, respectively. No insect life-cycle in a seasonal environment can be complete

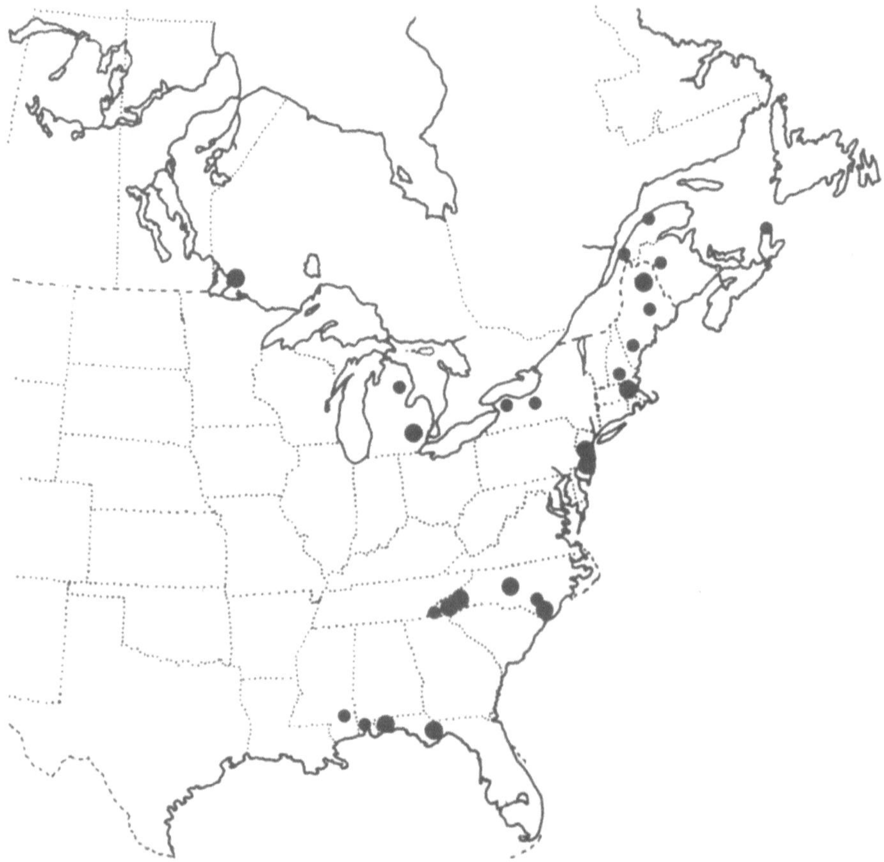

**Fig. 3.1.** Source of populations considered in this paper. *Small circles*, localities contributing to phenological traits; *large circles*, localities contributing to demographic traits.

without tactics to deal with both conditions and without the ability to convert from one to the other. The value of inducible life-cycle conversions is clear: normal growth, development and reproduction may proceed according to one strategy while conditions are favourable, diverting to another strategy when conditions deteriorate or are likely to do so. Since different kinds of selective force are acting upon each of these two groups of traits, normal growth, development and reproduction should be free from constraints imposed by dormancy and migration, i.e. the two groups of traits should be genetically uncorrelated and free to evolve independently (Dingle et al. 1977; Hegmann and Dingle 1982; Dingle 1986; Palmer and Dingle 1986; Lande 1982; Taylor 1986).

Indices of fitness, such as the finite or instantaneous rates of increase, are assembled from components and subcomponents. Variation in the underlying components may be either coordinated or compensatory. Coordinated variation in causal components results in variation of the composite index of fitness. Compensatory variation results in little or no variation of the composite index but may reflect adapted trade-offs among the causal components. In this chapter, we address two questions concerning evolution of the pitcher-plant mosquito, *Wyeomyia smithii*, in eastern North America (Fig. 3.1).

1. Do phenological and demographic traits evolve independently in response to geographical patterns of climate and density-dependent development?
2. Is geographical variation in fitness composed of correlated or independent components and subcomponents?

To answer these questions, we shall examine geographical variation in phenological and demographic traits. These patterns of variation reveal how life-cycle traits have undergone evolutionary divergence during the geographical radiation of *W. smithii*; they do not, however, show whether this divergence has been facilitated, constrained or unaffected by the underlying genetic correlations.

## 3.2 Life-Cycle of *Wyeomyia smithii*

*Wyeomyia smithii* develops (Fig. 3.2) only in the water-filled leaves of pitcher plants in eastern North America. The eggs are dropped on to the water by hovering females and hatch directly without any hatching stimulus. Long days sustain direct development but short days induce a larval diapause in the fourth instar at southern latitudes (≤36° N) or in the third instar at more northern latitudes or higher altitudes (Smith and Brust 1971; Evans and Brust 1972; Bradshaw and Lounibos 1972, 1977). Continuous short days maintain diapause. Larvae respond to photoperiod throughout the winter and into the spring, even after prolonged chilling in nature, so that photoperiod is also probably responsible for the termination of diapause (Evans and Brust 1972; Lounibos and Bradshaw 1975). In the north, diapausing third instars may terminate diapause and enter a second, stable fourth instar diapause that acts as a buffer against unpredictable vernal weather (Lounibos and Bradshaw 1975). Females of all populations obligatorily lay their first batch of eggs without a blood meal (Bradshaw and Lounibos 1977). Northern females never take blood meals but continue to produce repeated, small batches of eggs; southern females require a blood meal for the production of their second and subsequent batches (Bradshaw and Lounibos 1977; Bradshaw and Holzapfel 1983; Bradshaw 1980; O'Meara and Lounibos 1981; O'Meara et al. 1981).

Over its range, *W. smithii* encounters a predictable climatic gradient. In the south, mild winters are followed by long growing seasons, while further north, harsher winters are followed by progressively shorter growing seasons. At the southern end of its range, *W. smithii* may be able to complete up to five generations per year at 30° N (Bradshaw and Holzapfel 1983), declining to two or, rarely, three at 42.5° N (Judd 1959; Istock 1978; Istock et al. 1976), one to two at 45.5° N (Kingsolver 1979), and only one at 49–54° N (Evans 1971). In the south (30° N), generations overlap and blend into each other continuously; in the north (≥42.5° N) generations are relatively discrete.

Resources for developing larvae consist of prey captured by their host leaf.

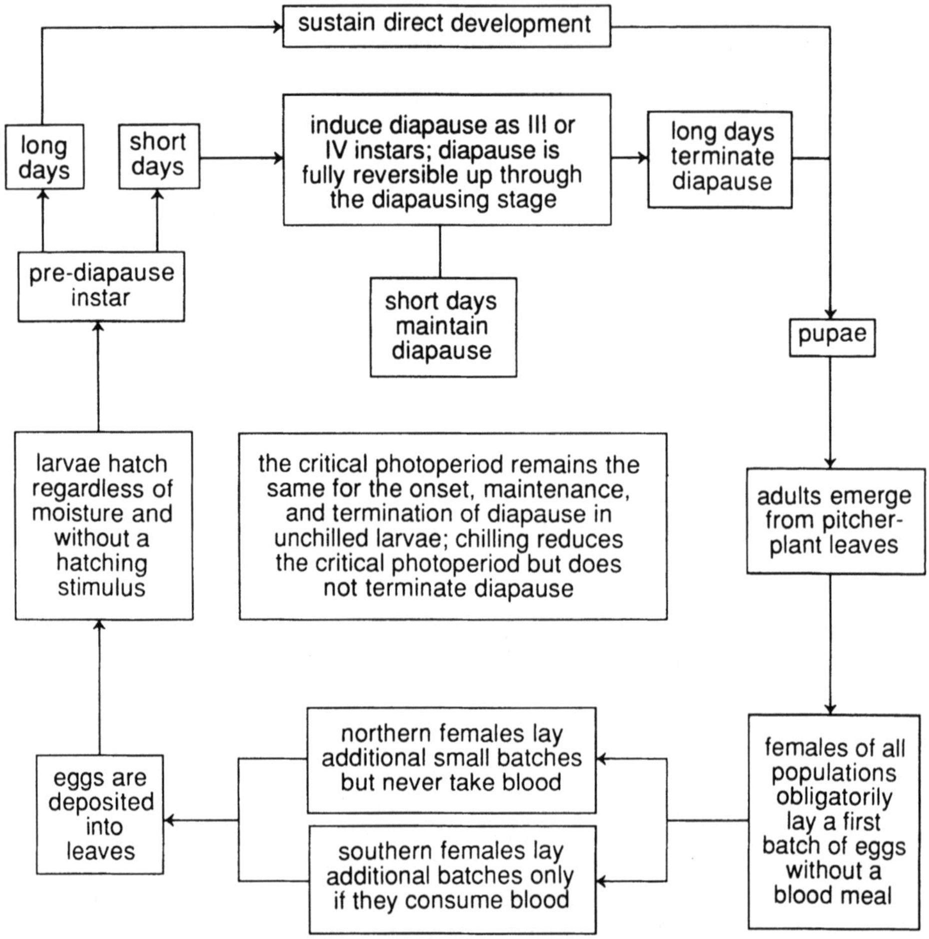


**Fig. 3.2.** Life-cycle of *Wyeomyia smithii*.

Although prey rapidly disarticulate in the leaf, the more heavily chitinized head capsules of victims remain intact and are readily identified as such among the "gut" contents of leaves. Resources per individual *W. smithii* are then measured as larvae per head capsule of prey captured by the host leaf. Along the Gulf of Mexico at the southern end of their range, pupation success of *W. smithii* is limited by larval density (equals *W. smithii* per head capsule of prey: Bradshaw and Holzapfel 1986). Overwintering *W. smithii* begin to eclose as adults 2 weeks before the first leaves open and the first summer generation saturates available leaves; leaves remain saturated for the duration of the growing season (Bradshaw and Holzapfel 1986). Density also determines the age structure of the overwintering population and, subsequently, its pupation success in the spring. Thus, southern *W. smithii* saturate their environment year-round, including the winter, when even diapause does not provide escape in time or space (Bradshaw and Holzapfel 1986). Density-dependent constraints to development then decrease with increasing latitude or altitude (Istock et al. 1976; Bradshaw and Holzapfel 1986). At 42.5° N, *W. smithii* enjoy release from density-dependent constraints during most of the growing season but go through density-dependent bottlenecks in the spring and fall (Istock et al. 1976). The effect of density on vernal development continues to decrease with increasing latitude and, at some northern and high-elevation localities, there is no significant effect of density even on vernal development (Bradshaw and Holzapfel 1986). *Wyeomia smithii* therefore encounter a geographical gradient of density-dependent selection that can be visualized most easily by the crowding of *W. smithii* per head capsule of prey over its entire range from the Gulf of Mexico (30° N) to north-central Canada (Manitoba, 54° N) (Fig. 3.3).

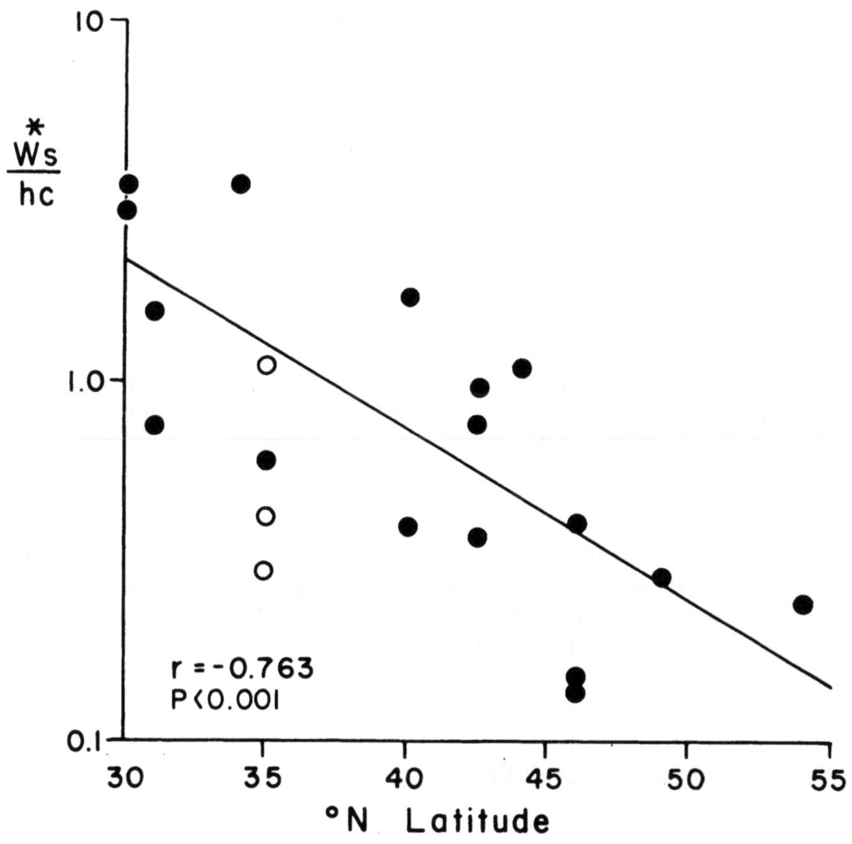


**Fig. 3.3.** Resource availability to the average larva over the range of *W. smithii*. The regression was run on the latitudinal gradient (●); the mean of three censuses at 900 m at 35° N (○) intercepts the regression line at 43° N. Resource availability is measured as mean crowding of *W. smithii* per head capsule (Ws*/hc) of prey in the overwintering generation when 100% of the population is present as larvae in the leaves. (From Bradshaw and Holzapfel 1986.)

## 3.3 Life-Cycle Traits and Geography

On the basis of present-day distribution, physiology, morphology and behaviour, we conclude that the evolution of *W. smithii* has proceeded from south to north.

1. *Present day distribution*: the genus has over 50 tropical and subtropical species (Stone et al. 1959) and one temperate species, *W. smithii*, that extends further north than any other species in the genus or even the whole tribe (Sabethini).
2. *Diapause*: southern populations diapause only in the fourth instar; northern populations diapause as third instars but may enter a second, stable, fourth-instar diapause in the spring (Lounibos and Bradshaw 1975).
3. *Anal papillae*: southern populations possess four long anal papillae as in most mosquitoes; northern populations have but two short anal papillae (Bradshaw and Lounibos 1977).
4. *Blood feeding*: southern populations require a blood meal for the second and subsequent ovarian cycles; northern populations mature repeated egg batches without blood meals (highly unusual among mosquitoes) and are not known to bite under any circumstances (Bradshaw 1983, 1986a; O'Meara et al. 1981; O'Meara and Lounibos 1981).

Life-cycle traits of *W. smithii* should then reflect its evolution along the geographical gradient of climate and density-dependent development.

### 3.3.1 Predictions

Phenological traits, including critical photoperiod, stage of diapause and depth of diapause should respond to the regular geographical change in growing season and seasonal harshness, showing greater conservatism in the north than in the south. Demographic traits, including capacity for increase, its components and their subcomponents should respond to the regular geographical change in density (Fig. 3.3) as well as climate. Theories of r- and K-selection or of stage-specific mortality both predict that northern populations should exhibit lower density-tolerance and higher productivity than southern populations (Stearns 1976; Mueller and Ayala 1981; Mueller 1988). These predictions are reinforced both by the decrease in seasonal harshness (Roughgarden 1971) and by the increase in the growing season (King and Anderson 1971) with the concomitant increase in resource tracking (Bradshaw and Holzapfel 1983) at southern relative to northern latitudes.

### 3.3.2 Observed Phenological Traits

*Critical Photoperiod*

The critical or median photoperiod is the switch-point between long- and short-day responses, i.e. the number of hours of light per day that initiates or maintains 50% diapause in a population. In *W. smithii*, the critical photoperiod is tightly

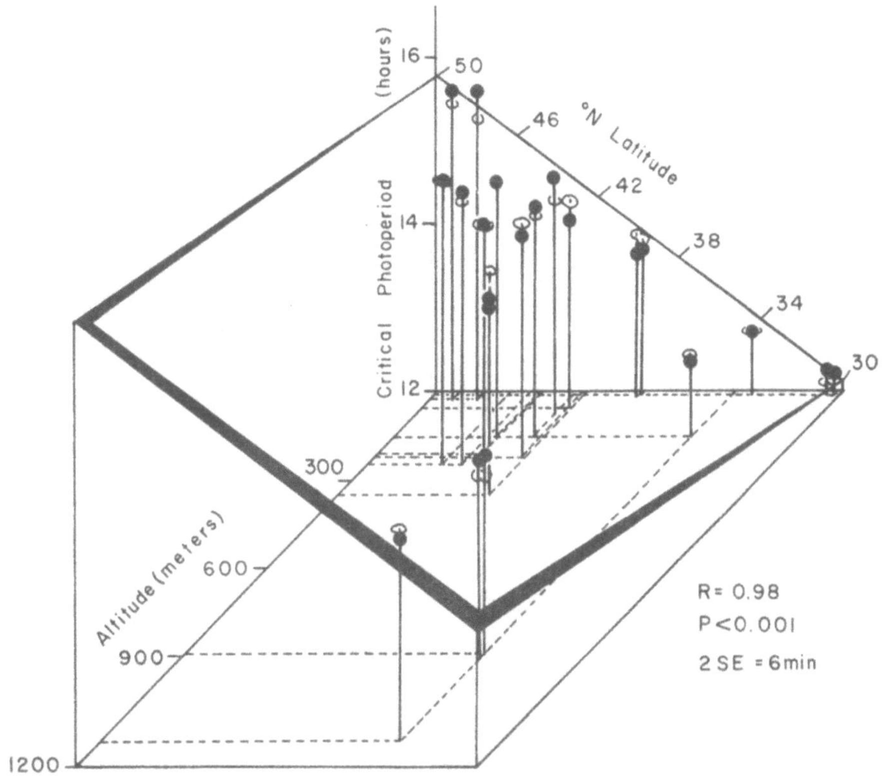


**Fig. 3.4.** Relationship between critical photoperiod, altitude and latitude. Circles on the regression plane indicate where the projections of the data points cross that plane. (From Bradshaw 1976. Reprinted by permission from *Nature* vol. 263, pp. 384–386, copyright © 1976 Macmillan Magazines Ltd.)

correlated with latitude and altitude of origin (Fig. 3.4). Photoperiodic response of *W. smithii* thus tracks the climatic gradient of eastern North America very closely (Bradshaw 1976; Bradshaw and Lounibos 1977).

To determine the genetic basis of critical photoperiod, we subjected laboratory-reared diapausing larvae from Florida (30° N) and Maine (46° N) to naturally increasing photoperiods. We used astronomic time switches (street-lamp timers) that program a sine-function change in photophase over a 365-day period. In our experiments, we used the near-linear increase in photophase (2–3 min increment per day) that occurs in the vicinity of the vernal equinox. At each cumulative 15 min increment in photoperiod, we removed all pupae and allowed the adults from each specific, cumulative 15 min increment to mass swarm in an isolated cage. We reared the resulting $F_1$ progeny on short days, induced diapause, and then subjected them to the same naturally increasing photoperiod as their parents had experienced. From the offspring–parent regression of median (critical) photoperiod (Fig. 3.5, left), we concluded that a substantial amount of genetic variation underlies photoperiodic response within populations of *W. smithii*. However, because of the experimental setup, differences in photoperiod of pupation also

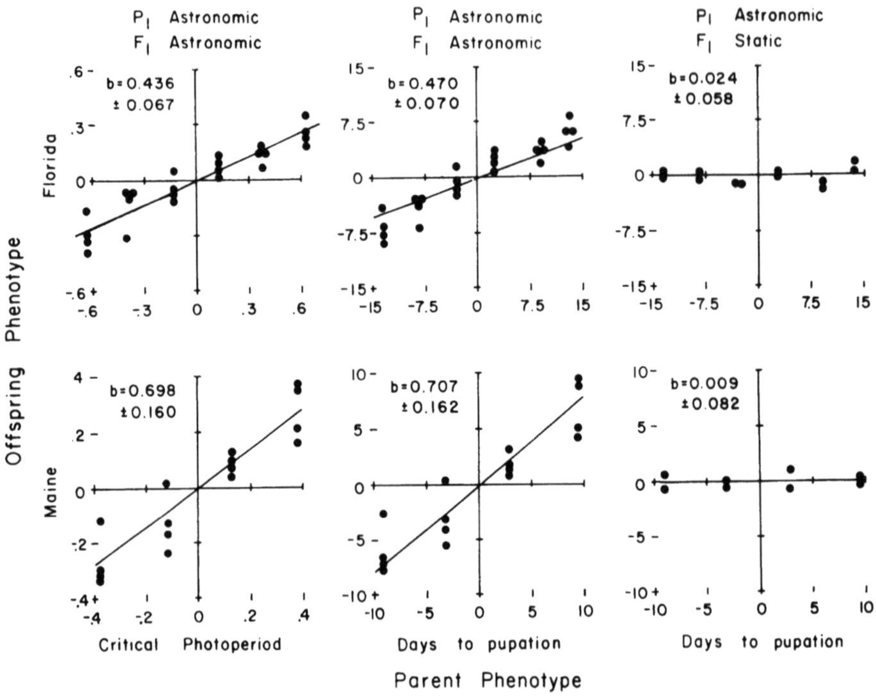


**Fig. 3.5.** Heritability of critical photoperiod within populations from Florida and Maine. ***Left***: critical photoperiod expressed as deviations (in h) from parent mean critical photoperiod (*x*-axis) and from offspring mean critical photoperiod (*y*-axis). ***Middle*** and ***right***: mean days to pupation expressed as deviations (in days) from parent mean days to pupation (*x*-axis) and from offspring mean days to pupation (*y*-axis). In both the left and middle figures, diapausing parents ($P_1$) and diapausing offspring ($F_1$) were exposed to naturally increasing photoperiods that incremented 2–3 min per day. In the right-hand figures, diapausing parents ($P_1$) were exposed to naturally increasing photoperiods that incremented 2–3 min per day and their diapausing offspring ($F_1$) were exposed to static long days. Slopes ±95% confidence limits are shown in the upper left of each graph.

reflect differences in time to develop from the onset of experimental conditions. The offspring–parent regression of days to pupation (Fig. 3.5, middle) indicates a heritability similar to that of critical photoperiod. To see whether differences in response to the astronomic clock may have reflected variation in post-diapause development time rather than in photoperiodic response, we exposed samples of $F_1$ larvae to static, unambiguous long days (Fig. 3.5, right). In this case, we found no correlation between parental and offspring development times. Thus the correlation between offspring and parent development time (Fig. 3.5, middle) reflects the underlying correlation between their critical photoperiods (Fig. 3.5, left) and not inadvertent selection for post-diapause development time. Consequently, within populations of *W. smithii*, variation in photoperiodic response has a large heritable component. Analysis of covariance revealed heterogeneity of slopes for critical photoperiod (Fig. 3.5, left; $F_{1,36} = 12.04$, $P<0.01$), indicating that genetic variation for critical photoperiod is significantly higher in the northern than in the southern population. We therefore conclude that heritable variation for photoperiod response has not been diminished by long-term

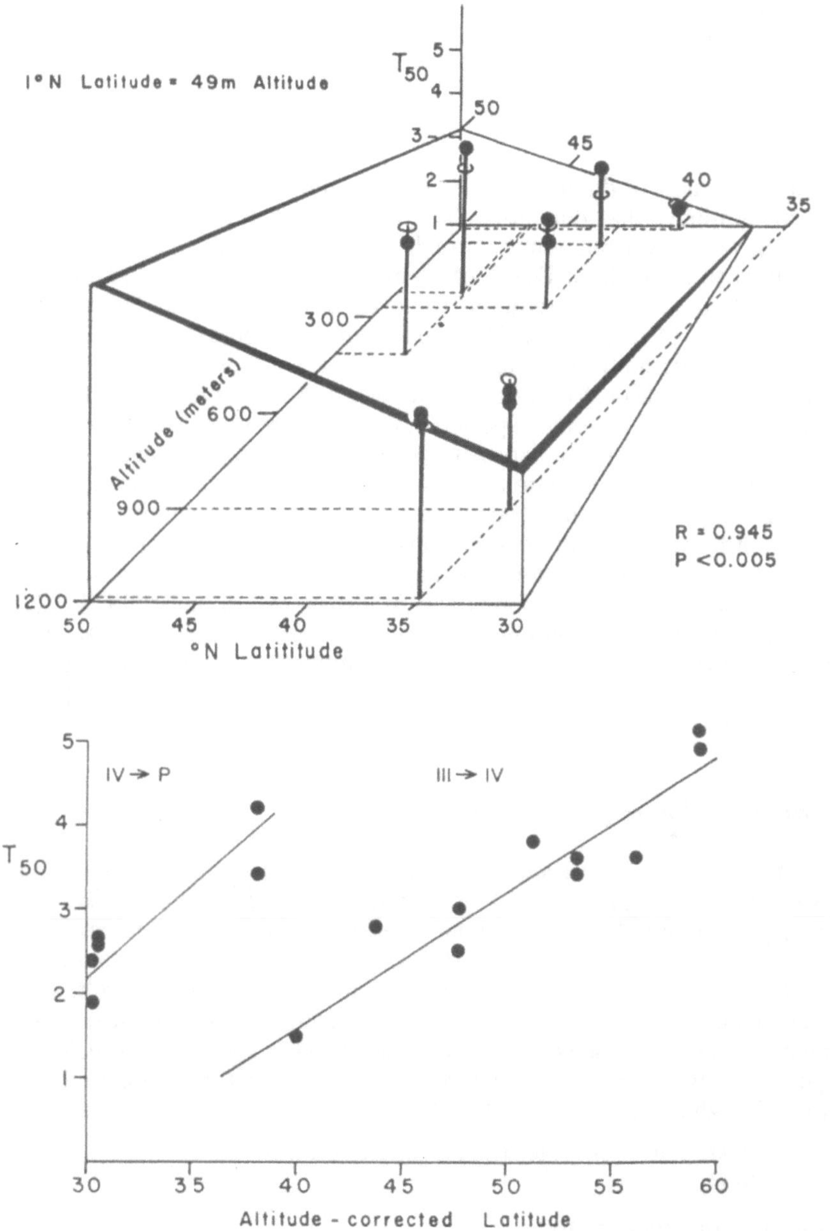


**Fig. 3.6.** Depth of diapause as measured by $T_{50}$, the number of long days required to terminate diapause in 50% of the sample population. *Upper panel*: regression of $T_{50}$ on latitude and altitude for third instar diapausing populations produced an altitude–latitude conversion factor of 49m/°N; this conversion factor was then used to correct for altitude of fourth-instar diapausing populations. *Lower panel*: comparative $T_{50}$ for third-(III) and fourth-(IV) instar diapause plotted on a common axis of altitude-corrected latitude. P, puparium. (From Bradshaw and Lounibos 1977.)

directional selection for increasing critical photoperiod during the northward adaptive radiation of *W. smithii*.

### *Depth of Diapause*

The depth or intensity of diapause can be measured by the number of long days required to terminate that diapause. This number is determined by exposing diapausing larvae to various numbers of long days (0, 1, 2 . . . *n*) and then returning them to short days. Depth of diapause is then measured as $T_{50}$, the number of pulsed long days required to terminate diapause in 50% of the population. Depth of diapause in *W. smithii* generally increases with latitude (Fig. 3.6). but, superficially, appears irregular. At the southern edge of its range, *W. smithii* enters a shallow, fourth-instar diapause and the depth of this diapause increases with latitude. In northern populations, however, diapause occurs in the third instar. Within third-instar diapausing populations, diapause is again shallow among the southernmost populations and increases with latitude or altitude. The shift from a deep to a shallow diapause at 35–40° N is accompanied by a concomitant shift from a fourth- to a third-instar diapause. At the same time, the ability to enter a second, fourth-instar diapause is preserved among northern populations (Lounibos and Bradshaw 1975; Bradshaw and Lounibos 1977). This pattern, then, reflects a continuum of progressive conservatism: at each increment in latitude or altitude, irrevocable commitment to adult emergence becomes increasingly remote. As with critical photoperiod, depth of diapause tracks the climatic gradient of North America very closely.

## 3.3.3 Observed Demographic Traits

To assess the effects of climate and density-dependent selection on demographic traits, we collected populations of *W. smithii* from 12 localities (Fig. 3.1) representing six geographical zones along a latitudinal and altitudinal gradient. After rearing these populations for at least two laboratory generations to eliminate field effects, we reared them at varying densities under controlled but near natural conditions of light, temperature and feeding schedule in the leaves of intact pitcher plants (Bradshaw 1986a). For each locality and density (10, 20, 40 or 60 per leaf) at a constant food level (200 *Drosophila melanogaster*), we determined Laughlin's (1965) capacity for increase:

$$r_c = \ln(\text{replacement rate})/(\text{mean generation time})$$
$$= \ln R_o/T$$

where the components of $r_c$ are:

$$R_o = \Sigma E_x/n_o$$
$$= [\text{eclosion success}]\ [\text{sex ratio}]\ [\text{fecundity}]$$
$$= [\%\ \text{eclosion}]\ [\%\ \text{females eclosing}]\ [\text{eggs/eclosed female}]$$

$$T = \Sigma(xE_x)/\Sigma E_x$$
$$= [\text{development time}] + [\text{maturation time}]$$
$$= [\text{days to median eclosion}] + [\text{days from median eclosion to median egg}]$$

$\Sigma E_x$ = number of eggs oviposited by a uniform aged cohort at time $x$, the number of days since oviposition of the cohort

$n_o$ = initial cohort size

We then performed 2-way analysis of variance (ANOVA) on each demographic trait with treatments geographical zone of origin ($n$=6) and density in the laboratory ($n$=4). Each cell in the ANOVA included two replicates (=independent localities) so that we could test for zone×density interaction.

## *Zone Effects*

Neither $r_c$ nor its components, $R_o$ and $T$, differed among zones (Fig. 3.7). All three traits responded dramatically to density in the laboratory but there was no significant zone×density interaction. Thus, $r_c$, $R_o$, and $T$ did reflect proximal effects of density but not longer-term evolution along a geographical gradient of climate and density-dependent selection. As we have already pointed out (Bradshaw and Holzapfel 1989) and will not belabour here, the theory of r- and

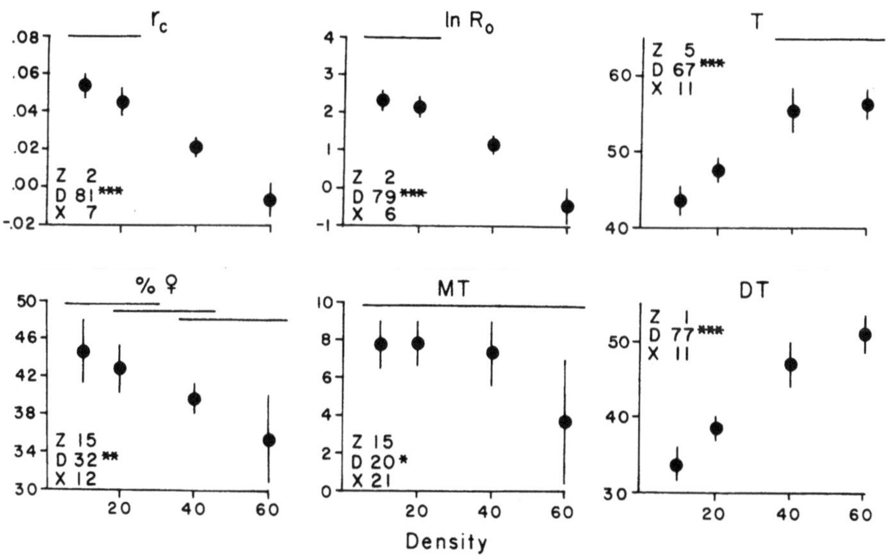


**Fig. 3.7.** Effect of density in the laboratory and geographical zone of origin on mean (±2 SE) capacity for increase ($r_c$), its components replacement rate (ln $R_o$) and mean generation time ($T$), and its subcomponents sex ratio (% female, arcsine transformed), development time ($DT$), and maturation time ($MT$). Within each graph is given reduction in total sum of squares ascribable to zone of origin ($Z$), density in the laboratory ($D$), and their interaction ($X$) from 2-way ANOVA. *$P<0.05$; **$P<0.01$; ***$P<0.001$; otherwise, not significant. *Horizontal lines* at the top of each graph extend over means that are not significantly different ($P>0.05$) from Duncan's multiple range test. ($r_c$, $R_o$ and $T$ from Bradshaw and Holzapfel 1989; © 1989 by The University of Chicago. 0003–0147/89/3306–0007 $02.00. All rights reserved)

K-selection fails completely to predict the pattern of demographic traits in *W. smithii*.

Among the subcomponents of fitness (Fig. 3.7), sex ratio, development time and maturation time did not vary among zones or reflect significant zone×density interaction. Fecundity (Fig. 3.8) varied significantly among zones of origin and among densities in the laboratory but did not exhibit any zone×density interaction. Duncan's multiple range test revealed that fecundity declined steadily with increasing density (Fig. 3.8) but failed to identify a significant difference between any two zones (Fig. 3.8). Inspection of mean fecundities revealed no consistent geographical pattern: far southern populations did not

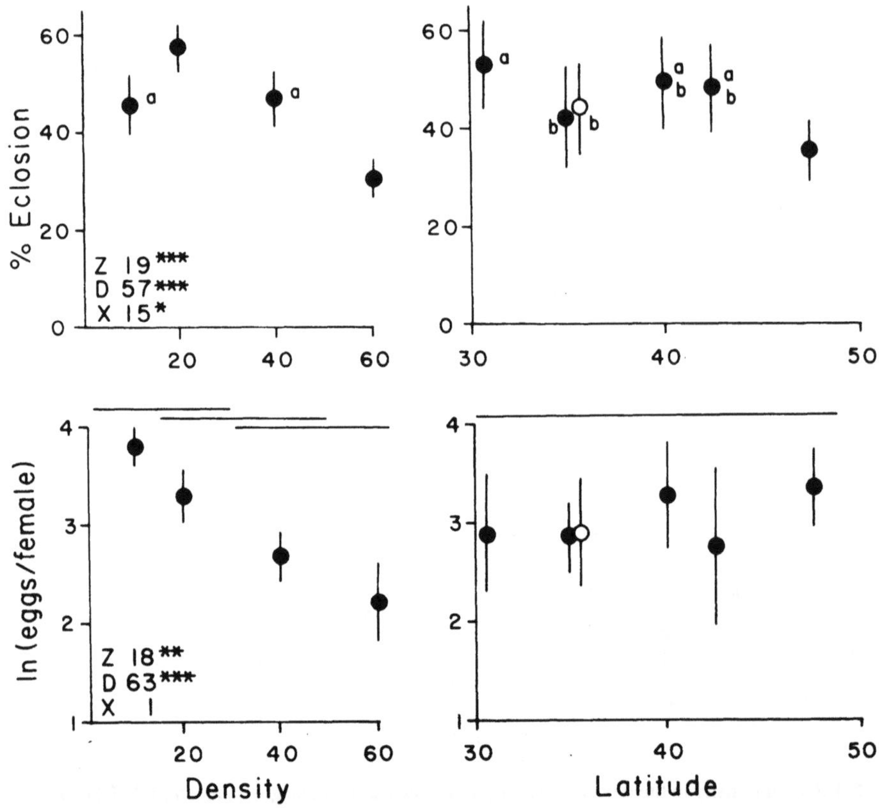


**Fig. 3.8.** Effect on (*top*) eclosion success (mean arcsine transformed percentage eclosion ±2SE) and (*bottom*) fecundity (mean ln (eggs per ecdysed female) ±2 S.E.M.) of density in the laboratory (*left*) and zone of origin plotted as mean latitude of the two contributing localities (*right*). In both the top and bottom plots, the *open circle* denotes the high elevation zone (900 m at 35° N). *Z*, *D* and *X* as in Fig. 3.7. Means accompanied by the same letter or lying underneath the same horizontal line are not significantly different by Duncan's multiple range test ($P$>0.05). (Fecundity data from Bradshaw 1986a.)

differ from far northern populations and the lowest and highest fecundities occurred in adjacent zones. Thus, fecundity, like the other traits described above, reflected primarily the proximal effects of density and not longer term evolution along a geographical gradient of climate and density-dependent selection.

Eclosion success (Fig. 3.8) reflected density, geography, and their interaction, with density being the more important component. Eclosion success was lower at 10 larvae per leaf, peaked at 20 larvae per leaf, and then declined at higher densities. A similar Allee effect was observed in laboratory populations of *Aedes aegypti* (Wada, 1965) and *A. sierrensis* (Hard et al. 1989). In the case of *W. smithii*, at a density of 10/leaf, development time is significantly faster (Fig. 3.7) and there is a non-significant tendency towards greater fecundity (Fig. 3.8) than at a density of 20/leaf so that $R_0$ and, consequently, $r_c$ do not decrease at the lowest density (Fig. 3.7). Eclosion success among zones (Fig. 3.8) appeared to decrease with latitude. Neither zone mean ($r=-0.71$; d.f.=3; $P>0.05$) nor locality mean ($r=-0.61$; d.f.=8; $P>0.05$) was significantly correlated with latitude (excluding the high elevation zone or high elevation localities at 35° N) but these correlations might have been significant with a larger sample size. Further, there was no significant correlation between mean eclosion success and density of *W. smithii* per unit resource at the locality of origin (for all localities except one Gulf Coast population for which we do not have an estimate of field density: $r=0.31$; $n=11$; $P>0.05$). Thus, eclosion success, also, reflected primarily the proximal effects of density and only to a lesser extent, if at all, the longer-term evolution along a geographical gradient of climate and density.

Phenotypic variation of demographic traits in response to variable resources vastly exceeds differences due to locality or zone of origin (Figs. 3.7 and 3.8). Decreased resources result in a linear decline of capacity for increase but the reason for this decline varies with resource level. At modest resource limitations, generation time increases but replacement rate remains much the same. At more severe levels of resource limitation, generation time remains relatively stable, while replacement rate continues to decline. Moeur and Istock (1980) also found that moderately reduced resources in *W. smithii* resulted in the prolongation of pre-adult development, especially in the fourth instar, with no resultant decline in pupal size or adult fecundity. At more severe resource reductions (Fig. 3.7), prolongation of development appears to reach a higher asymptote, whereafter net fecundity is reduced. These observations suggest that there are physiological trade-offs between development time and fecundity in *W. smithii* but that the specific trade-off is density dependent.

The low residual variations in demographic traits (Figs. 3.7 and 3.8) show that homeostatic adjustment greatly exceeds genetic differences in response to density. Slobodkin (1968) has proposed that the optimal strategy for organisms evolving in a heterogeneous environment is to "behave in a homeostatic fashion so as to avoid or counteract the effects of introduced (environmental) variation". On the other hand, homeostatic ability is advantageous over a genetically fixed phenotype only when homeostatic adjustment can take place within the magnitude and time constraints of environmental variation (Bradshaw 1976, 1986b). In the case of mosquitoes, adult size and, consequently, ovariole number and blood-meal capacity, are fixed by the time of pupation (Hawley 1985). The prolongation of larval development and the determination of potential adult fecundity are all developmental alternatives available to larvae within the time frame of reduced resources. Consequently, most of the phenotypic variation in demographic traits of *W. smithii* encountering variable density in the laboratory reflects individual homeostasis rather than genetic divergence among populations.

### *Locality Effects*

The above results indicate that $r_c$, its components and its subcomponents are all refractory to the broader gradient of climate and density-dependent selection experienced by *W. smithii* throughout its range. These results are based on phenotypic differences observed among six geographical zones. The lack of among-zone variation may be due to an impotence of climate- and density-dependent selection to mediate the evolution of demographic traits or to a lack of heritable variation in the traits themselves. However, if there is significant variation among localities regardless of zone, then heritable variation for those traits must exist and phenotypic evolution must have taken place.

To examine differences among localities requires an estimate of within-locality variation. Consequently, we made a second, independent determination of $r_c$, its components, and its subcomponents in nine populations of *W. smithii* (two locations each from 30–31° N, 900 m at 36° N, 40° N and 42.5° N and one locality at 49° N) at two densities (10 and 40 per leaf).

This procedure provided replicated norms of reaction across two densities (Fig. 3.9) for capacity for increase, its components, and its subcomponents. Two-way ANOVA then revealed significant locality and/or locality×density effects for capacity for increase, replacement rate, mean generation time, eclosion success and development time. Eggs per ecdysed female, previously found to vary among zones (Fig. 3.8) did not vary significantly among localities (Fig. 3.9), probably because of the lower discriminating power of the latter ANOVA. Sex ratio, although reflecting effects of density did not vary among zones or localities or exhibit zone or locality×density interaction. Maturation time did not reflect effects of zone, locality, density, or their interaction, in agreement with Moeur and Istock (1980), who likewise found no variation within populations for this trait. Taken together, the results in Figs. 3.7–3.9 indicate that evolutionary divergence has taken place in all the above traits except sex ratio and maturation time. Thus, the lack of among-zone variation in $r_c$, in its components, and in all but two of its subcomponents does not represent a lack of genetic variability underlying the traits but rather a lack of climate or density-dependent development, per se, to select for their evolutionary divergence.

### *Genetic Divergence*

The degree of genetic divergence between two populations may be reflected by the degree of heterosis observed in their hybrids (Endler 1977; Falconer 1981; Dingle et al. 1982).

To examine possible heterotic effects, we performed 13 crosses among the 12 populations of *W. smithii* contributing to Figs. 3.7–3.9. These 13 crosses represented parent populations with median and extreme phenotypes and geographical displacements (Bradshaw and Holzapfel 1989). We considered heterosis to occur if the difference between hybrid and mid-parent phenotypes

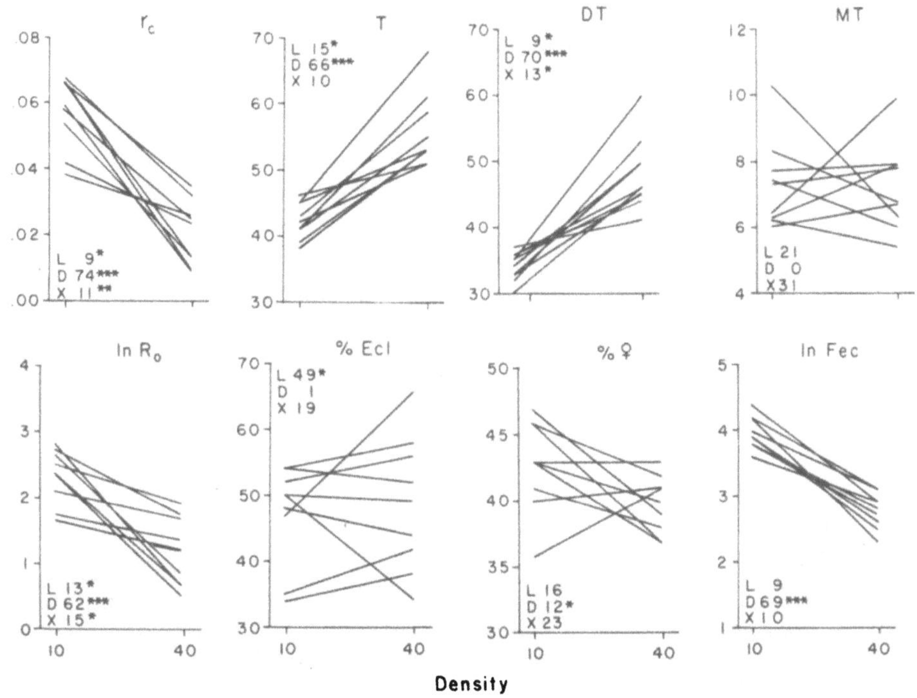


**Fig. 3.9.** Effect of density on capacity for increase, its components, and its subcomponents for each of nine localities of origin regardless of zone. Trait abbreviations are as in Fig. 3.7 except: % Ecl=arcsine transformed percentage eclosion; ln Fec=ln (eggs/female). Each point represents the mean of two independent determinations in the laboratory for each population from a given locality. Within each graph is provided the percentage reduction in total sum of squares from 2-way ANOVA with treatments locality of origin (*L*), density in the laboratory (*D*), and their interaction (*X*). Significance of each treatment is given by $^{*}P<0.05$, $^{**}P<0.01$, $^{***}P<0.001$; otherwise, not significant. A significant effect of locality indicates a significant difference among localities in the trait indicated; a significant effect of *X* indicates that localities differ in response to density by the trait indicated.

was in the direction of increased fitness (increased $r_c$) and tested for significant differences between hybrid and mid-parent values by one-way ANOVA.

Fig. 3.10 shows that capacity for increase, its components, and two of its subcomponents (sex ratio and development time) all exhibited significant heterosis but maturation time, fecundity and pupation success did not. The data in Fig. 3.9 provide evidence that all the demographic traits except sex ratio and maturation time have diverged both genetically and phenotypically among populations of *W. smithii*. The data in Fig. 3.10 show that sex ratio has diverged genetically even though this divergence has not resulted in significant phenotypic differences among populations (Fig. 3.9). Conversely, fecundity that varied significantly among populations (Fig. 3.9) did not produce significant heterosis in hybrid populations (Fig. 3.10). Consistently, maturation time did not differ among populations (Fig.3.9) and did not produce significant heterosis in hybrid populations (Fig. 3.10). The demographic traits in *W. smithii* thus illustrate that within the geographical divergence of a single species, substantial phenotypic divergence can occur through substantial, or minimal, non-detectable genetic

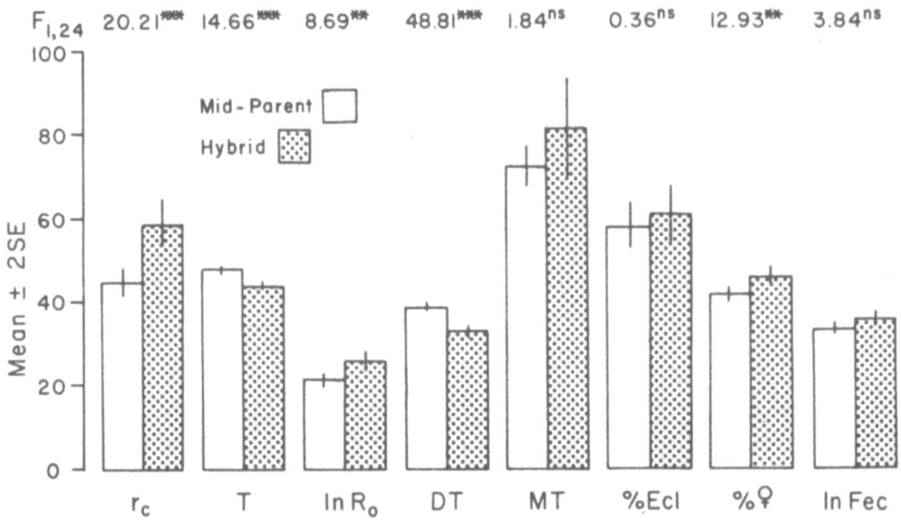


**Fig. 3.10.** Heterosis of demographic traits observed among hybrids between parent populations of *W. smithii* representing median and extreme phenotypic divergence of traits and of geographical separation. At the top of the figure, $F_{1,24}$ is the $F$-value from one-way ANOVA between hybrid and mid-parent phenotypes for each trait. Trait abbreviations as in Fig. 3.9. Note that heterosis is indicated for a significantly lower hybrid than mid-parent value for $T$, $DT$ and $MT$. $^{ns}$not significant; $^{**}P<0.01$; $^{***}P<0.001$.

divergence; and significant genetic divergence can occur with minimal, non-detectable phenotypic divergence.

## *Covariation of Traits*

The above results still leave open the question as to whether geographical variation in capacity for increase is composed of covarying or independently varying components and subcomponents. The independence of demographic traits from broader geographical influence means that evolutionary trade-offs among the traits themselves should be apparent once the proximal effects of density have been factored out. Since density in the laboratory does have such a dramatic influence on demographic traits (Figs. 3.7–3.9), there is no optimal density at which to look for these trade-offs. Consequently, we examined covariation among $r_c$, its components, and its subcomponents at each locality by factoring out density effects and then determined correlations between traits among localities. We corrected for the species-wide effects of density by: (a) finding the mean value of each trait at each density across localities; (b) regressing this mean value on density or, in the case of percentage eclosion, on density and ln (density) to account for the curvilinear response of this trait to density; and (c) calculating the mean deviation from regression across densities for each trait at each locality. Thus, the mean trait at each locality was based on four observations (deviations from regression on density). To look at correlated divergences of traits among localities, we made pair-wise correlations ($n$=12 each) between mean deviation from regression of $r_c$, its components, and its subcomponents.

With a sample size of 12, the least significant correlation is $r=0.57$ ($r^2=0.32$) but the averaging of four independent deviations from regression should reduce sampling variance.

Capacity for increase among localities (Fig. 3.11) was very highly correlated with replacement rate among localities but not mean generation time among localities. Further, replacement rate and mean generation time were not themselves correlated and the non-significant correlation was negative. It would thus appear, firstly, that divergence of capacity for increase is due mainly to divergence of replacement rate and not mean generation time. Secondly, there have been no adaptive (genetic) trade-offs between replacement rate and generation time. That is, among localities, an evolved increase in replacement rate has resulted in a concomitant increase in capacity for increase without a compensatory increase in mean generation time. Although all three traits exhibit evolutionary divergence among localities (Fig. 3.9), only replacement rate covaries with capacity for increase and both vary independently of mean generation time. Among the components of replacement rate, only percentage eclosion and fecundity vary among localities while sex ratio does not (Fig. 3.9) and only fecundity covaries with replacement rate (Fig. 3.11). Among the components of mean generation time, development time varies among localities,

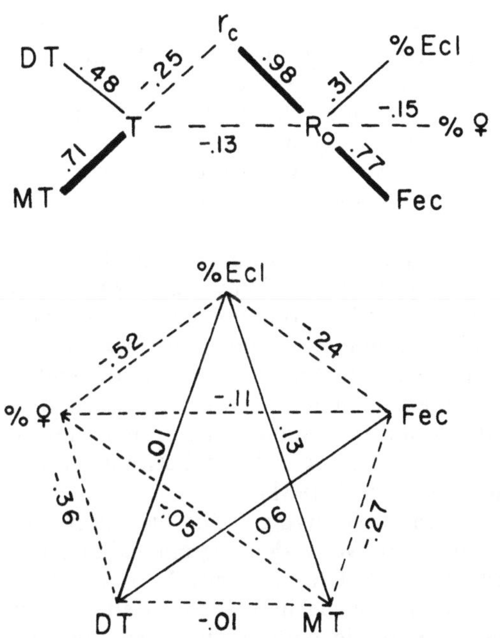


**Fig. 3.11.** Correlation between $r_c$ and its components, between the components and each of their respective subcomponents, and among all the subcomponents from mean values ($n=4$) of each trait at each of 12 localities. Means represent residuals from regression of each trait on density in the laboratory. *Continuous lines* indicate positive correlations; *broken lines* indicate negative correlations; *heavy lines* indicate significant correlations ($P<0.05$). The correlation coefficient ($r$) is given on each line. Note that with a sample size of 12, the least significant $r=0.57$. Abbreviations as for Figs. 3.7 and 3.9.

while maturation time does not (Fig. 3.9), but only the latter covaries with mean generation time (Fig. 3.11). Among the five subcomponents of fitness, three vary among localities while two do not (Fig. 3.9) but none of the five covaries with any other (Fig. 3.11).

From these results, we conclude that among geographically diverse populations of *W. smithii*, subcomponents of fitness have generally evolved independently of one another and may or may not covary with their respective component of fitness. Covariation of components with their subcomponents is not predictable from the magnitude of geographical variation of the individual component or subcomponent. Finally, covariation among capacity for increase and either of its components is not predictable from the magnitude of their separate geographical variation and does not reflect any compensatory response or reflect any genetically based trade-off selection.

## 3.4 Conclusions

We may now answer our original questions. Phenological and demographic traits of *W. smithii* have evolved independently over the geographical range of *W. smithii* in North America from 30–49° N. Phenological traits show strong, consistent trends of progressive conservatism along an eco-climatic gradient. Evolutionary divergence of demographic traits has also taken place but independently of this same gradient. We therefore conclude either that phenological and demographic traits are, as we originally proposed, genetically uncorrelated with each other or that genetic correlations do exist between them but such correlations do not constitute significant barriers to adaptive radiation of this species over geographical distances.

Among demographic traits, there appears to be a coordinated axis of capacity for increase, replacement rate and fecundity. This coordination has not resulted in any substantial compensatory trade-off in other components or subcomponents of fitness. All of the evidence indicates that it is the immediate exigencies of the local microhabitat, not regional differences in climate or density-dependent development, that have selected for divergent demographic traits in *W. smithii*. It is not therefore surprising that capacity for increase, replacement rate and mean generation time all exhibit strong heterosis among hybrids between various populations (Bradshaw and Holzapfel 1989; Istock 1981).

*Wyeomyia smithii* is a single species of mosquito living within the water-filled leaves of a single species of insectivorous plant over wide geographical distances (Fig. 3.1). We have previously (Bradshaw 1976; Bradshaw and Lounibos 1977; Bradshaw and Holzapfel 1983, 1989) cited this unusual habitat consistency among localities as a good reason for considering life-cycle evolution in *W. smithii*, since geographical variation is unlikely to be confounded by local microhabitat variation. The conclusion we have just made above is therefore surprising, but it does suggest that microhabitat variation may be far more important in shaping the demographies of most other organisms that live in more variable habitats than does *W. smithii* (Bradshaw and Holzapfel 1989).

Over the geographical range of *W. smithii*, subcomponents of fitness generally vary independently of each other and may covary with, or be independent of, components of fitness that in turn may covary with, or be independent of, the composite index of fitness. It is therefore very dangerous to extrapolate to fitness from any of its components or correlates without determining actual connection (Livdahl and Sugihara 1984; Hard et al. 1989). The lack of more significant covariation among demographic fitness traits (Fig. 3.11) implies that adaptive trade-offs have not been an important component in the evolutionary divergence of *W. smithii* populations, regardless of the underlying genetic correlations within populations. Specialized adaptations such as facultative iteroparity (Bradshaw, 1986a) or high intraspecific competitive ability (Bradshaw and Holzapfel, 1989) may then evolve and dramatically affect realized fitness in a given local habitat without adversely affecting potential fitness over a broad range of habitats.

*Acknowledgements*. We are grateful to Janet Battista for many months of consistent assistance with long-term experiments. We have profited from discussion and helpful comments from Philip Lounibos, Hugh Dingle, Conrad Istock, William Hawley, Fritz Taylor, Dietrich Neumann, Peter Sauer, Michael Rose and Jeff Hard. We also thank Jeff Hard, Derek Roff and one anonymous reviewer for considering our manuscript. Finally, we are indebted to the National Science Foundation for grants DEB78–22799, DEB81–14953, and BSR–8717151 that supported and made possible this research.

# References


Beck SD (1980) Insect photoperiodism. Academic Press, New York

Bradshaw WE (1976) Geography of photoperiodic response in a diapausing mosquito. Nature 262: 384–386

Bradshaw WE (1980) Blood-feeding and capacity for increase in the pitcher-plant mosquito, *Wyeomyia smithii*. Environ Entomol 9: 86–89

Bradshaw WE (1983) Interaction between the mosquito, *Wyeomyia smithii*, the midge, *Metriocnemus knabi*, and their carnivorous host, *Sarracenia purpurea*. In: Frank JH, Lounibos LP (eds) Phytotelmata: terrestrial plants as hosts of aquatic insect communities. Plexus, Medford, NJ, pp 161–189

Bradshaw WE (1986a) Variable iteroparity as a life-history tactic in the pitcher-plant mosquito, *Wyeomyia smithii*. Evolution 40: 471–478

Bradshaw WE (1986b) Pervasive themes in insect life cycle strategies. In: Taylor F, Karban R (eds) The evolution of insect life cycles. Springer, Berlin Heidelberg New York, pp 261–275

Bradshaw WE, Holzapfel CM (1983) Life cycle strategies in *Wyeomyia smithii*: seasonal and geographic adaptations. In: Brown VK, Hodek I (eds) Diapause and life cycle strategies in insects. Dr W. Junk, The Hague, pp 169–178

Bradshaw WE, Holzapfel CM (1986) Geography of density-dependent selection in pitcher-plant mosquitoes. In: Taylor F, Karban R (eds) The evolution of insect life cycles. Springer, Berlin Heidelberg New York, pp 48–65

Bradshaw WE, and Holzapfel CM (1989) Life-historical consequences of density-dependent selection in the pitcher-plant mosquito, *Wyeomyia smithii*. Am Nat 133: 869–887

Bradshaw WE, Lounibos LP (1972) Photoperiodic control of development in the pitcher-plant mosquito, *Wyeomyia smithii*. Can J Zool 50: 713–719

Bradshaw WE, Lounibos LP (1977) Evolution of dormancy and its photoperiodic control in pitcher-plant mosquitoes. Evolution 31: 546–567

Cole LC (1954) The population consequences of life history phenomena. Q Rev Biol 29: 103–137

Danilevskii AS (1965) Photoperiodism and seasonal development of insects. Oliver and Boyd, Edinburgh

Danks HV (1987) Insect dormancy: an ecological perspective. Biological Survey of Canada Monograph Series no 1. Biological Survey of Canada, Ottawa.

Dingle H (1986) The evolution of insect life cycle syndromes. In: Taylor F, Karban R (eds) The evolution of insect life cycles. Springer, Berlin Heidelberg New York, pp 187–203

Dingle H, Brown CK, Hegmann JP (1977) The nature of genetic variance influencing photoperiodic diapause in a migrant insect, *Oncopeltus fasciatus*. Am Nat 111: 1047–1059

Dingle H, Blau WS, Brown CK, Hegmann JP (1982) Population crosses and genetic structure of milkweed bug life histories. In: Dingle H, Hegmann JP (eds) Evolution and genetics of life histories. Springer, Berlin Heidelberg New York, pp 209–229

Endler JA (1977) Geographic variation, speciation, and clines. Princeton University Press, Princeton, NJ

Evans KW (1971) The bionomics of *Wyeomyia smithii* (Coquillett), the pitcher plant mosquito (Diptera: Culicidae: Sabethini). MSc thesis, University of Manitoba, Winnipeg.

Evans KW, Brust RA (1972) Induction and termination of diapause in *Wyeomyia smithii* (Diptera: Culicidae), and larval survival studies at low and subzero temperatures. Can Entomol 104: 1937–1950

Falconer DS (1981) Introduction to quantitative genetics, 2nd edn. Longman, London

Hard JJ, Bradshaw WE, Malarkey DJ (1989) Resource- and density-dependent development in tree-hole mosquitoes. Oikos 54: 137–144

Hawley WA (1985) A high fecundity aedine: factors affecting egg production of the western treehole mosquito, *Aedes sierrensis* (Diptera: Culicidae). J Med Entomol 22: 220–225

Hegmann JP, Dingle H (1982) Phenotypic and genetic covariance structure in milkweed bug life history traits. In: Dingle H, Hegmann JP (eds) Evolution and genetics of life histories. Springer, Berlin Heidelberg New York, pp 177–185

Istock CA (1978) Fitness variation in a natural population. In: Dingle H (ed) Evolution of insect migration and diapause. Springer, Berlin Heidelberg New York, pp 171–190

Istock CA (1981) Natural selection and life history variation: theory plus lessons from a mosquito. In: Denno RF, Dingle H (eds) Insect life history patterns: habitat and geographic variation. Springer, Berlin Heidelberg New York, pp 113–127

Istock CA, Vavra KJ, Zimmer H (1976) Ecology and evolution of the pitcher-plant mosquito: 3. Resource tracking by a natural population. Evolution 30: 548–557

Judd WW (1959) Studies on Byron Bog in southwestern Ontario. X. Inquilines and victims of the pitcher plant, *Sarracenia purpurea* L. Can Entomol 91: 171–180

King CE, Anderson WW (1971) Age-specific selection. II. The interaction between r and K during population growth. Am Nat 105: 137–156

Kingsolver JG (1979) Thermal and hydric aspects of environmental heterogeneity in the pitcher plant mosquito. Ecol Monogr 49: 357–376

Lande R (1982) Elements of a quantitative genetic model of life history evolution. In: Dingle H, Hegmann JP (eds) Evolution and genetics of life histories. Springer, Berlin Heidelberg New York, pp 21–29

Laughlin R (1965) Capacity for increase: a useful population statistic. J Anim Ecol 34: 77–91

Lees AD (1955) The physiology of diapause in arthropods. Cambridge Monogr Exp Biol 4: 1–151

Lewontin RC (1965) Selection for colonizing ability. In: Baker HG, Stebbins GL (eds) The genetics of colonizing species. Academic Press, New York, pp 77–94

Livdahl TP (1979) Environmental uncertainty and selection for life-cycle delays in opportunistic species. Am Nat 113: 835–842

Livdahl TP, Sugihara G (1984) Non-linear interactions of populations and the importance of estimating per capita rates of change. J Anim Ecol 53: 573–580

Lounibos LP, Bradshaw WE (1975) A second diapause in *Wyeomyia smithii*: seasonal incidence and maintenance by photoperiod. Can J Zool 53: 215–221

Moeur JE, Istock CA (1980) Ecology and evolution of the pitcher-plant mosquito. IV. Larval influence over adult reproductive performance and longevity. J Anim Ecol 49: 775–792

Mueller LD (1988) Density-dependent population growth and natural selection in food-limited environments: the *Drosophila* model. Am Nat 132: 786–809

Mueller LD, Ayala FJ (1981) Trade-offs between *r*-selection and *K*-selection in *Drosophila*. Proc Natl Acad Sci USA 78: 1303–1305

Murphy GI (1968) Patterns in life history and the environment. Am Nat 102: 391–403

O'Meara GF, Lounibos LP (1981) Reproductive maturation in the pitcher-plant mosquito, *Wyeomyia smithii*. Physiol Entomol 6: 437–443

O'Meara GF, Lounibos LP, Brust RA (1981) Repeated egg clutches without blood in the pitcher-plant mosquito. Ann Entomol Soc Am 74: 68–72
Palmer JO, Dingle H (1986) Direct and correlated responses to selection among life-history traits in milkweed bugs (*Oncopeltus fasciatus*). Evolution 40: 767–777
Roughgarden J (1971) Density-dependent natural selection. Ecology 52: 453–468
Slobodkin LB (1961) Growth and regulation of animal populations. Holt, Rinehart and Winston, New York
Slobodkin LB (1968) Toward a predictive theory of evolution. In: Lewontin RC (ed) Population biology and evolution. Syracuse University Press, Syracuse, NY, pp 187–205
Smith SM, Brust RA (1971) Photoperiodic control of maintenance and termination of larval diapause in *Wyeomyia smithii* (Coq.) (Diptera: Culicidae) with notes on oogenesis in the adult female. Can J Zool 49: 1065–1073
Stearns SC (1976) Life-history tactics: a review of the ideas. Q Rev Biol 51: 3–47
Stone A, Knight KL, Starke H (1959) A synoptic catalog of the mosquitoes of the world (Diptera, Culicidae). Entomological Society of America, Washington, DC
Tauber CA, Tauber MJ (1981) Insect seasonal cycles: genetics and evolution. Annu Rev Ecol Syst 12: 281–308
Tauber MJ, Tauber CA, Masaki S (1986) Seasonal adaptations of insects. Oxford University Press, New York
Taylor F (1980) Timing in the life histories of insects. Theor Popul Biol 18: 112–124
Taylor F (1986) Toward a theory for the evolution of the timing of hibernal diapause. In: Taylor F, Karban R (eds) The evolution of insect life cycles. Springer, Berlin Heidelberg New York, pp 236–257
Wada Y (1965) Effect of larval density on the development of *Aedes aegypti* (L.) and the size of adults. Quaest Entomol 1: 223–249

*Chapter 4*

# Developmental Programmes and Adaptive Syndromes in Insect Life-Cycles

Susan M. Scott* and Hugh Dingle

## 4.1 Introduction

Insect life-cycles are examples of "complex adaptations" (Frazzetta 1975) that involve complementary interactions among traits to achieve a common adaptive result, and the different stages provide enormous potential for adaptation to varying environments (Fig. 4.1). The ultimate understanding of complex adaptations requires knowledge of how the genome is *organized*, in particular how genes may influence how traits develop and function together (Dingle et al. 1982; Dingle 1986). Central to this issue are not only genetic variances for single traits, but also genetic correlations among traits, especially *which* combinations are correlated or uncorrelated. Any evolutionary explanation of life-cycles depends on knowledge of both genetic variances and covariances, since natural selection acts on variation but may be constrained or enhanced by covariation, at least in the short term, with respect to both magnitude and direction (Istock 1982; Via and Lande 1985; Groeters and Dingle 1987, 1988). Traits influenced by the same genes are interdependent, while traits influenced by independent genes are freed from a "cost of correlation" (Dingle 1984; Palmer and Dingle 1986). Complex adaptations thus depend on the interplay of genetic flexibility and genetic (and developmental) constraint. Permutations in single life-cycle traits will, as a consequence, have ramifications beyond their own immediate short-term effects.

One such life-cycle permutation is diapause, which can occur at any stage of the life-cycle (Fig. 4.1), and in some species may occur more than once. The flexibility achieved by the division of the developmental cycle into active and diapause phases has been crucial to the success of many temperate-zone insects (Tauber et al. 1986; Danks 1987). Further, geographical variation in diapause is closely linked to patterns of voltinism. In univoltine populations diapause may be essentially obligate, either because the trait is genetically fixed or because

---

* Order of authorship decided by coin toss.

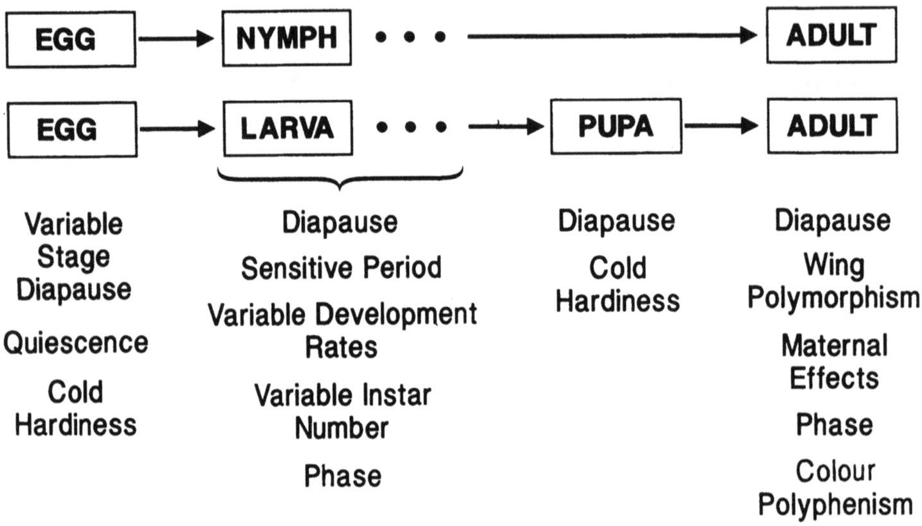


**Fig. 4.1.** Some of the variation possible in insect life-cycles. Hemimetabolous (*top line*) and holometabolous pathways are indicated.

environmental conditions always exceed the threshold for induction. In multivoltine populations the induction of diapause may result from the influence of photoperiod either acting directly or acting on the females to produce diapause offspring (a maternal effect) (see e.g. Dean 1982b). Diapause intensity may vary, often being greatest where the risk of warm days before or after overwintering is high (Masaki 1978). In univoltine species, eggs laid early in the season, for example, may undergo a more intense diapause than later eggs, preventing premature hatching, while late-season eggs develop more rapidly (Tanaka 1984). Patterns of voltinism and diapause display several correlated trends involving life-history characters and form coadapted systems with development, fecundity and migration (Tauber et al. 1986; Danks 1987). In migratory milkweed bugs, diapause facilitates migration (Dingle 1978), and as we shall see below, migratory flight is genetically correlated with fecundity (Palmer and Dingle 1986; Dingle 1988). Diapause can also be crucial in range expansion through the establishment of new immigrants in novel environments, as has apparently occurred in both milkweed bugs and monarch butterflies (Dingle 1978). The existence of interactions among life-cycle traits poses interesting questions concerning the evolution of underlying genetic and developmental programmes.

In this chapter we outline some attempts that we are making toward understanding these programmes, including the role of the genes and the interactions between the genes and varying environments. We present two examples, the grasshopper *Melanoplus sanguinipes* (Orthoptera) and the water strider *Gerris remigis* (Heteroptera), of life-cycles we have studied in the field, to indicate the possibilities for adaptation to varying environments. We then discuss briefly genetic correlations in the life-history of the milkweed bug, *Oncopeltus fasciatus*, to examine the question of constraints. It would be especially interesting to know if genetic correlations among life-history traits occur across the boundaries of stadia; for example, do genes influencing, say, age at first reproduction in the adult also influence diapause in, say, the egg? Understanding of trade-offs within or between stages requires assessment of both genetic and phenotypic correlations. Unfortunately, on present information, we can do little more than pose questions. Finally, we consider in our closing discussion whether there are likely to be rules governing the partitioning of physiological variation among life-cycle stages, and if so, what those rules might be.

## 4.2 *Melanoplus sanguinipes*

Our first example of life-cycle adaptations to varying environments is the lesser migratory grasshopper, *Melanoplus sanguinipes*. This species is a major agricultural and rangeland pest (Hewitt 1979; Onsager 1983; Shotwell 1930) and contributes significantly to the recurring grasshopper outbreaks in North America. It is distributed from Alaska through Canada and the United States, excluding peninsular Florida (Gurney and Brooks 1959), and from sea-level to over 3000 m (Alexander 1964; Alexander and Hilliard 1969). In California, *M. sanguinipes* is found from sea-level to above 2700 m. Along with this broad distribution is a diversity of life-history patterns involving voltinism and diapause. Every stage of development which we and others have examined exhibits variation, and we shall examine this variation in each stage (egg, nymph and adult) in turn.

Over most of its range *M. sanguinipes* is univoltine; it is, however, bivoltine in places with a long relatively humid summer (for example in Kansas; Dean 1982a, b) and possibly trivoltine in parts of Arizona (J. E. Henry, personal communication). It may require 2 years to complete its life-cycle in upland British Columbia (Gurney and Brooks 1959). There may be a partial second generation in parts of the Imperial Valley of southern California where populations are known to produce both diapause and non-diapause eggs. These patterns of voltinism covary with the incidence of egg diapause, but, even in univoltine field populations, there is variation in the proportion of diapausing eggs (at least in the laboratory), in the stage of embryonic development in which the eggs overwinter (apparently in diapause) and in temperature effects on their subsequent emergence (Hilbert et al. 1985).

Our own observations of California *M. sanguinipes* confirm the presence of many of the above responses and reveal further interesting variation. We have examined six populations along an altitudinal gradient from a coastal, nearly sea-level habitat at 90 m to one at 2700 m in the Sierra Nevada. All of these populations appear to be exclusively univoltine, but they nevertheless display considerable life-cycle variation across altitudes, suggesting adaptation to local climates and season lengths.

During four summers we collected female grasshoppers from the six sites along the altitudinal gradient and allowed these females to oviposit in the laboratory for about 2 weeks. The proportion of eggs that did not diapause (i.e. hatched within 45 days at 27 °C, hereafter called non-diapause eggs) varied among the six populations from less than 1% to more than 60% (Fig. 4.2). The highest- and

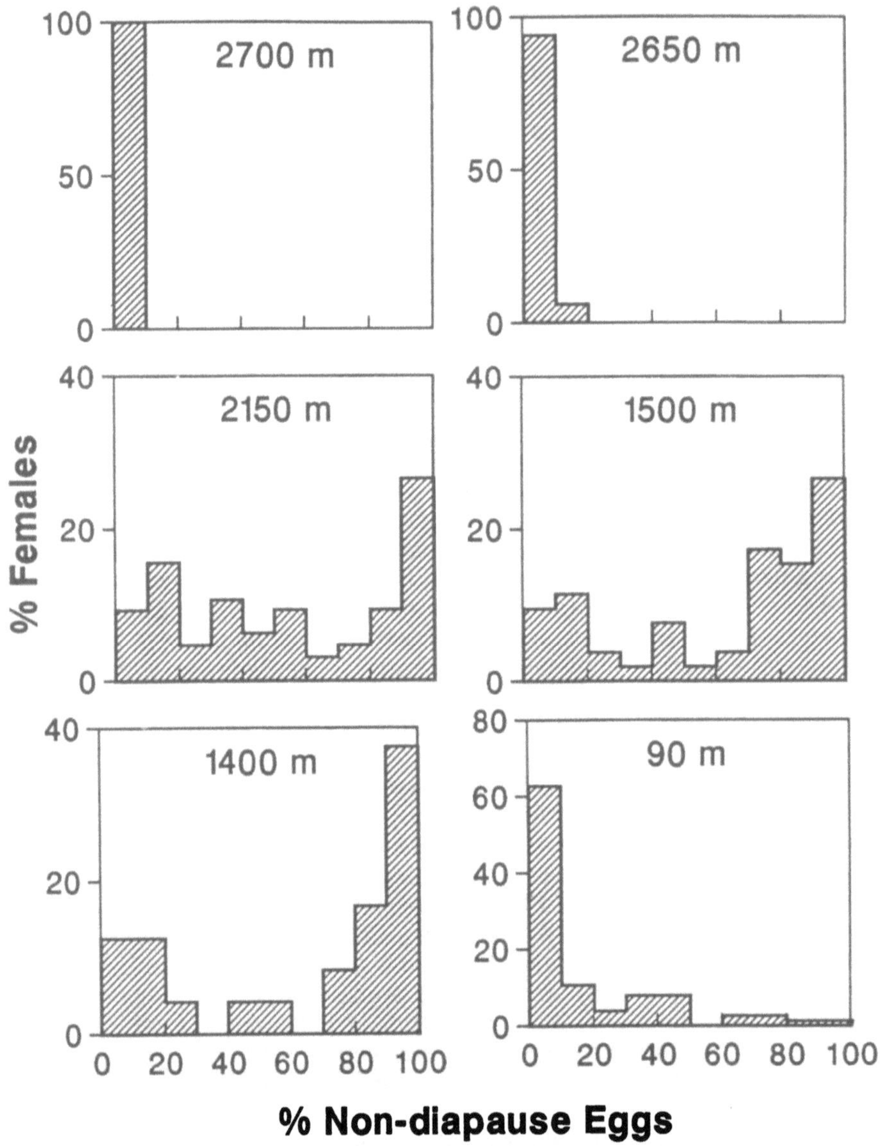


**Fig. 4.2.** Proportions of female *Melanoplus sanguinipes* producing varying proportions of non-diapause eggs in populations along an altitudinal gradient in central California (circa 38° N). The highest (2700 m, 2650 m) and lowest (90 m) altitude populations displayed the lowest non-diapausing proportions. Note changes of scale on the ordinate in different panels. *N*=22 to 56 females.

lowest-altitude populations produced the highest frequency of diapause eggs, while the three populations from mid-altitude sites produced less than 50% diapause eggs. This pattern did not change from mid- to late-summer nor from one year to the next. In addition to variation between populations, there were clear and consistent (over four summers) differences in the amount of variation among females within populations, with populations from the middle altitudes showing much more within-population variability. Populations from the two high-altitude sites showed very little variation; virtually all females laid exclusively diapause eggs (Fig. 4.2), and there were no significant differences among females (analysis of variance (ANOVA) $P>0.05$). In contrast, there was great variation in the types of egg laid by grasshoppers from the mid-altitude populations.

In these intermediate populations, the frequency distributions of numbers of females laying various proportions of diapause and non-diapause eggs revealed three types of female (Fig. 4.2). Pods laid by a single female contained similar proportions of diapause and non-diapause eggs (i.e. there was little within-female variance) so the population distributions result largely from differences among females. Thus, these mid-altitude populations appear to include females that lay diapause eggs, other females that lay non-diapause eggs, and some females that produce varying proportions of each type of egg. Because collections made one month apart in the late summer and early autumn of 1986 (the year with the largest sample sizes) had similar distributions (i.e. no marked shift from one class to another), female age and temperature–photoperiod regime, both of which naturally changed during the interval, probably do not account for these distributions. Thus, it seems likely that the groups reflect different genotypes. Indeed in the first laboratory generation of the high-altitude Frog Lake population, non-diapause eggs were laid by only five of 52 females, and three of the five were sisters. The distribution of diapause types in the mid-altitude populations is similar to that found by Dean (1982a) in his study of *M. sanguinipes* in Iowa. He reported a U-shaped distribution with about 35% of the population laying mostly non-diapause eggs (bivoltine), 27% laying mainly diapause eggs (univoltine) and 38% laying varying mixtures of non-diapause and diapause eggs ("varivoltine").

In addition to variation in the incidence of egg diapause in *M. sanguinipes*, there is also variation in the stage of development within the egg at the time of diapause. Moore (1948) described and categorized the embryonic development of *M. sanguinipes* based on two key features of grasshopper development that can be observed through the chorion of intact eggs: the size and pigmentation of the eyes and their position along the longitudinal axis of the egg. Using these characters, we assigned diapause eggs (i.e. eggs that had not hatched after 45 days at 27 °C) to early, mid or late stage of development and found significant differences among populations in the stage of embryogenesis reached before development was interrupted by diapause (Table 4.1). The majority of eggs from all of the sites except the coastal location diapaused at an advanced stage of development and required only 6 to 8 days to hatch when re-exposed to 27 °C after exposure to 2 °C for more than 60 days. Eggs from the coastal population, however, generally diapaused at a much earlier stage of development (Table 4.1). In addition there was greater variation in rate of development in this population. Indeed individual pods often contained eggs in more than one stage. This variation resulted in a more scattered pattern of hatching, with some eggs hatching after 6 to 7 days and others after 14 to 18 days. Presumably this reflected their degree of pre-diapause development. Most interesting is the contrast between the highest- and lowest-altitude populations in the way variation in incidence of diapause reflects variation in stage at diapause. These two populations both display high proportions of egg diapause (Fig. 4.2). The coastal sea-

**Table 4.1.** Percentage of diapausing eggs in various stages of embryonic development. Values are averages of means for each female (1986 only)

| Altitude (m) | Stage of embryonic development | | | No. of females |
|---|---|---|---|---|
| | Early[a] (%) | Mid[b] (%) | Late[c] (%) | |
| 2700 | 4.2 | 0.4 | 95.5 | 10 |
| 2650 | 2.1 | 1.8 | 96.1 | 19 |
| 2150 | 14.3 | 2.6 | 83.1 | 34 |
| 1500 | 12.5 | 3.7 | 83.7 | 29 |
| 1400 | 24.9 | 17.2 | 52.6 | 24 |
| 90 | 44.0 | 45.3 | 8.8 | 26 |

[a] Embryo small, eye crescents not visible, <50% developed (Moore 1948).
[b] Embryo turned, eye crescents small but visible, 60%–70% developed (Moore 1948).
[c] Embryo almost fully developed, head at anterior end of egg, eyes large, >80% developed (Moore 1948).

level population, however, is somewhat variable, and this is the population with the greater variation in embryogenesis.

The coastal population experiences a very different climate from that to which the other *M. sanguinipes* populations are exposed, in that there is very little seasonal variation in temperature. However, there is probably considerable uncertainty associated with winter and spring rain. Rain is a major source of early nymphal mortality in *M. sanguinipes* (Pickford 1960, 1966), and the variation in hatching dates as a consequence of variation in stage of development during diapause may help to "spread the risk" by not having all of a clutch or cohort hatch just before a wet period.

There is still further variation in the nature of egg diapause. The proportion of diapause eggs that hatch after various lengths of exposure to cold is a standard measure of diapause intensity (Tauber et al. 1986). The more intense the diapause, the longer the stimulus required for diapause termination. The coastal population had the least-intense diapause, requiring less than 40 days of exposure to 2° C to produce a 76% hatch (Table 4.2). Given the mild winters along the northern coast of California (the daily minimum temperatures rarely fall below freezing) this result is not surprising. The 2700 m population required between 40 and 60 days of cold temperatures to produce 72% hatch, a value certainly achieved before the onset of spring at such altitudes. The mid-altitude populations are more difficult to analyse because most females produced few diapause eggs. With such small numbers of diapause eggs, variances are high, but it appears that between 60 and 80 days of cold are required at 1500 m and perhaps more than 100 days at 2200 m to achieve the same hatching success as in the other populations (Table 4.2). A more intense diapause may be important to mid-altitudes if there are brief periods of warm weather during the winter and spring.

Because all the eggs have been treated identically, differences in egg diapause must reflect either maternal effects or different genotypes of the eggs. The role of maternal effects has been shown in bivoltine populations by Dean (1982a, b). Our preliminary data, however, suggest that, at least for our highest- and lowest-altitude populations, the pattern of diapause incidence and stage of development persist in first-generation laboratory populations. Because these laboratory populations were reared under identical conditions, the results imply a genetic basis to the population differences.

**Table 4.2.** Intensity of diapause as measured by the proportion of diapause eggs hatching after various exposures to cold (2 °C)

| Altitude (m) | Number of days at 2 °C | | | |
|---|---|---|---|---|
| | 100 | 80 | 60 | 40 |
| 2700 | 0.79[a] | 0.67 | 0.72 | 0.28 |
| | 0.61–0.92[b] | 0.44–0.86 | 0.52–0.88 | 0.06–0.58 |
| 2200 | 0.67 | 0.47 | 0.39 | 0.50 |
| | 0.13–1.00 | 0.17–0.78 | 0.00–0.97 | 0.00–0.95 |
| 1500 | 0.93 | 0.98 | 0.46 | 0.13 |
| | 0.72–1.00 | 0.73–0.92 | 0.04–0.93 | 0.00–0.57 |
| 90 | 0.92 | 0.88 | 0.81 | 0.76 |
| | 0.80–0.99 | 0.69–0.98 | 0.64–0.93 | 0.60–0.88 |

[a] Underlined sets of means within a population do not differ at $P<0.05$ (Student–Neumann–Keuls multiple range test).
[b] 95% confidence interval.

Variability is not confined to the egg stage in *M. sanguinipes*. The two populations that we have studied in greatest depth show differences in nymphal development rates, with the sea-level population developing more slowly (Fig. 4.3). More importantly they show different responses to environmental variables. Neither the coastal sea-level population nor the high-altitude population showed any photoperiodic variation in response to day length at 33 °C, but at 27 °C the coastal population developed significantly faster in short days than it did in long days (Fig. 4.3).

Are there any patterns between variation in the egg, nymph and adult that

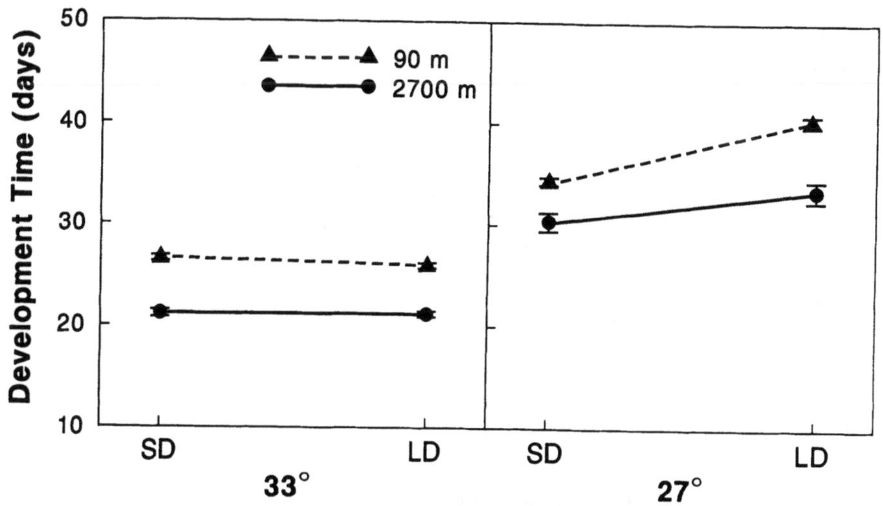


**Fig. 4.3.** Nymphal development rates of California populations of *Melanoplus sanguinipes* at two photoperiods, a long day at LD 14 : 10 h (LD) and a short day at 11 : 13 h (SD), and two temperatures (33 °C and 27 °C). The low-altitude population develops more slowly under all conditions.

suggest that variation in the subunits, or tactics, of which a life-history is made results in a coordinated whole, or strategy, in response to environmental conditions? Sawtooth geographical clines in body size, egg size and age at first reproduction (Fig. 4.4) occur latitudinally in *M. sanguinipes* in conjunction with patterns of voltinism (egg diapause) (Dean 1982a). The more rapid nymphal development of northern univoltine individuals results in smaller-sized adults that reproduce earlier despite lower ambient temperatures. Size increases to the

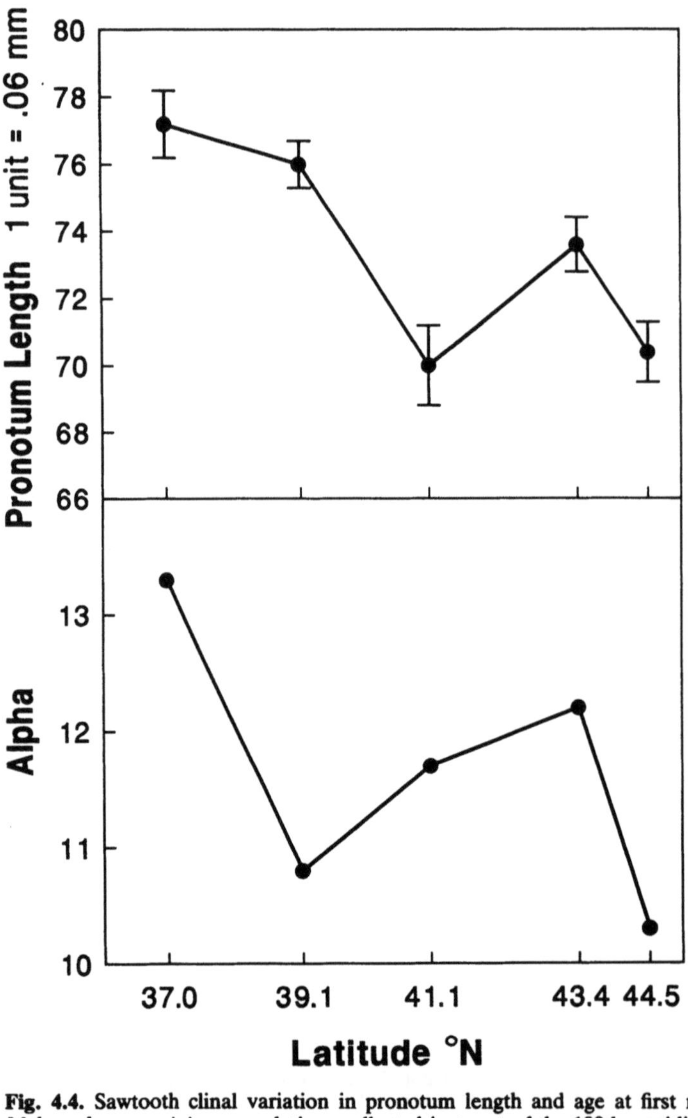


**Fig. 4.4.** Sawtooth clinal variation in pronotum length and age at first reproduction (*Alpha*) in *Melanoplus sanguinipes* populations collected just east of the 100th meridian in North America. A break in the cline indicates a shift from univoltine to bivoltine populations. (After Dean 1982a.)

south until a transition zone is reached where two generations are often possible. In this transition zone, the incidence of egg diapause is variable and influenced by maternal photoperiod. Selection will favour shortening of development to accommodate the extra generation (Masaki 1978; Roff 1983). As season length and mid-summer temperatures increase farther south, size increases even as age at first reproduction declines. Here the populations have more than sufficient time to complete two generations, and selection probably favours increased size and fecundity. Variation in the maternal effects responding to environmental inputs provide the mechanism for producing the sawtooth pattern (Dean 1982a, b).

*Melanoplus sanguinipes* is also one of a number of species of insect displaying a density-dependent polyphenism known as phase polymorphism. The phenomenon was originally described in African migratory locusts (*Locusta migratoria* and *Schistocerca gregaria*) by Uvarov (1921, 1966) as a tendency for morphological and behavioural differentiation to occur as a function of population densities. In both locusts and armyworm moths, *Spodoptera exempta* (Gatehouse 1986), for example, immature stages produced at high densities are more melanic and more gregarious, and adults are more migratory. In the armyworm moth this greater migratory tendency in the crowded form is largely due to the expression of a migratory genotype under high densities (a genotype×environment interaction). In both species the migratory adults colonize new habitats made available by rainfall.

The phase polyphenism of *M. sanguinipes* parallels that of *M. differentialis* studied by Dingle and Haskell (1967). Isolated nymphs are pale brown with little melanin at low densities and often become green at high humidities; crowded nymphs, on the other hand, show extensive black pigmentation due to more extensive melanin in the cuticle (Fig. 4.5). The isolated forms frequently insert an extra instar between the fourth and fifth stadia and in any event produce larger adults. The smaller, darker, crowded adults also display maturational colour changes going from yellow to dark olive green or brown (although the ventral surface may remain yellow). Unlike African locusts no behavioural differences in the crowded form have been found by us or by McAnelly and Rankin (1986),but an explicit test for the environment-specific expression of a migratory genotype (as occurs in crowded armyworm moths) has yet to be made.

## 4.3 *Gerris remigis*

Our second example of the adaptations of insect life-cycles to varying environments illustrates the point that the patterns observed can sometimes derive in part from unexpected sources, in this case sexual selection. The species in question is the widely distributed North American water strider, *Gerris remigis* (Hemiptera: Gerridae). This species is primarily a denizen of small streams. In eastern North America these streams are usually permanent, and the water striders are primarily wingless. Even when winged forms occur, they may not display much flight (Fairbairn and Desranleau 1986; Fairbairn 1986). In California, in contrast, many populations inhabit foothill streams which are either completely dry or

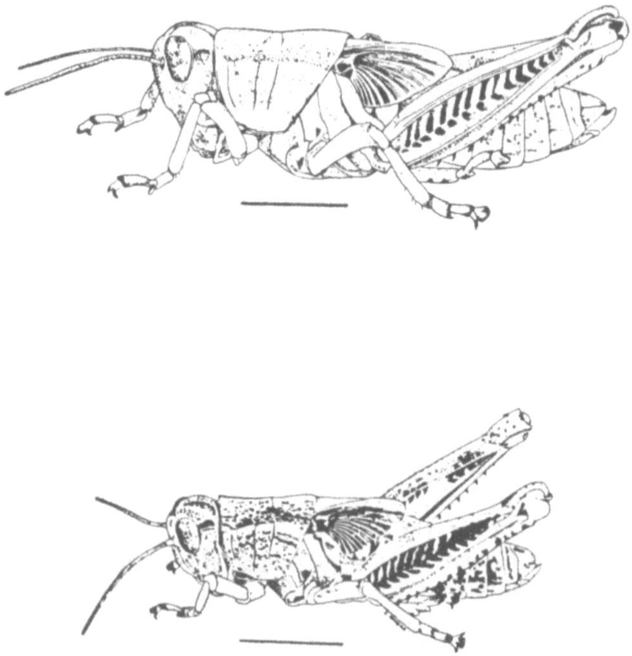

**Fig. 4.5.** Isolated (*top*) and crowded (*bottom*) female siblings from an Iowa population of *Melanoplus differentialis*. Scale bars represent 1 cm.

reduced to a few pools during the summer dry season. Breeding takes place when the streams run during the winter rains. Here populations are highly wing-polymorphic, and, as might be expected, the seasonal cycle involves considerable migration by flight (A. Kaitala, unpublished data).

During the summer, winged individuals take refuge in large permanent streams where they may form rafts along vegetated banks, but insect prey are scarce, and the water striders are in a state of suspended reproduction. A few wingless individuals also take refuge in the few remaining pools of temporary foothill streams, but those populations are at high risk of extinction. By the end of the summer only a few of these populations of wingless individuals remain. With the onset of the rains in the autumn, winged individuals migrate to the breeding streams and colonize both the permanent pools remaining from summer and the first new flows. The first pairs are seen in tandem at this time, and reproduction starts soon after. When the streams are flowing again, females establish feeding territories. Here the larger, winged females are at an advantage and establish their territories at the base of riffles which bring floating insect food, much as described by Rubenstein (1984) for an eastern population.

In the face of the obvious advantages that winged individuals have in being able to escape from drying streams and hold more favourable territories, what maintains the wingless form in the population? One factor appears to be post-diapause age at first reproduction (Fairbairn 1988a). The wingless females begin reproducing sooner following diapause than the larger winged females; earlier reproduction by wingless females appears to be true of many polymorphic insects (Dingle 1985). This would give wingless females that did survive the summer dry season (apparently in a starvation-induced diapause) a head start over their winged counterparts when the fall and winter rains again make streams into suitable breeding habitats. A second advantage is that wingless females produce more eggs than the winged forms (Fairbairn 1988a). A third unexpected advantage accrues to wingless males which, in spite of their small size, have a mating advantage relative to winged males, as indicated in Fig. 4.6. The largest of the wingless males overlap the winged males only at the low end of the size distribution, yet in these field observations wingless males display greater mating success than even the largest winged males. Within a morph there is size-assortative mating, as noted also by Fairbairn (1988b), but small wingless males do mate with much larger winged females. The greater success of wingless males is probably due to the larger femora of the forelegs, which are used to grasp the female when pairs are in tandem with the male riding the female's back. Rubenstein (1984) has postulated that, since females usually resist mountings by males, they may be "de facto" choosing the males with the strongest forelegs as partners. Sexual selection thus favours wingless males and consequently contributes to the maintenance of wing polymorphism in this population; in doing so it also contributes coincidentally to the flexibility of the *G. remigis* life-cycle. An interesting parallel occurs in some thrips where the larger forelegs of wingless males also assure greater mating (and territorial) success (Crespi 1986).

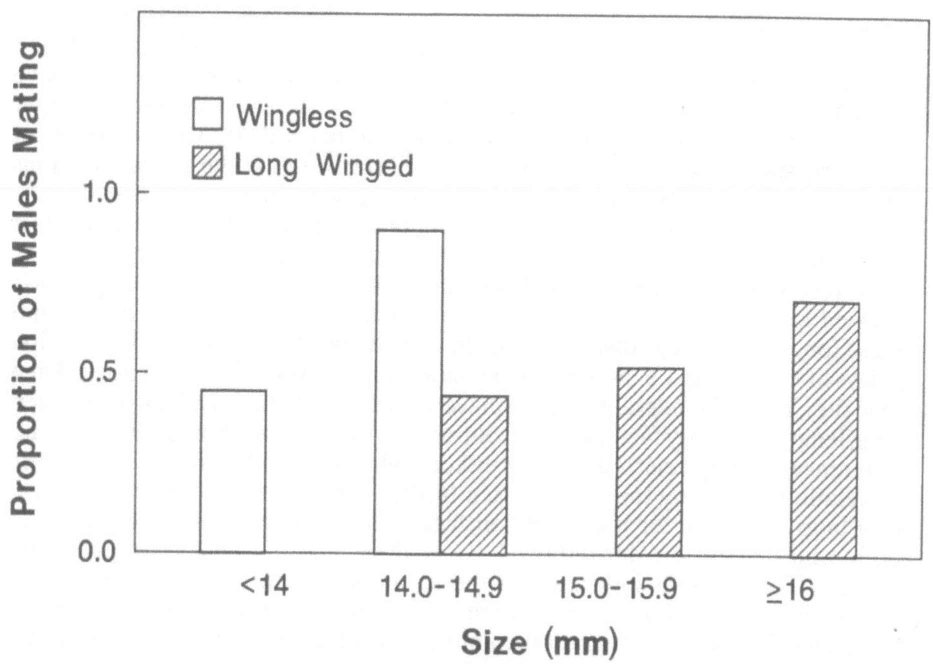


**Fig. 4.6.** Mating success in the field of males of the water strider *Gerris remigis*, as a function of size and wing morph. Note greater success of smaller wingless males.

## 4.4 *Oncopeltus fasciatus*

The results from grasshoppers and water striders discussed above strongly suggest interactions among different stages and functions of life-cycles. One way to analyse these interactions in greater depth is to consider genetic correlations among the relevant traits. Additive genetic correlations indicate the extent to which traits share genes in common and are important in evolution because selection on any one trait of a correlated complex will influence other traits as well, with response to selection being constrained or enhanced depending on the relative strengths and directions (i.e. whether positive or negative) of the correlations (Via and Lande 1985). The relation between genetic and phenotypic correlations can be expressed by the relation:

$$r_P = h_x h_y r_A + e_x e_y r_E$$

where $r_P$ is the phenotypic correlation between two traits, $x$ and $y$, $r_A$ and $r_E$ are the additive genetic and environmental correlations respectively, $h_x$ and $h_y$ are the square roots of the heritabilities ($h^2 = V_A/V_P$ where $V_A$ is the additive genetic variance and $V_P$ is the phenotypic variance), and $e^2 = 1 - h^2$ (Falconer (1981) gives a detailed discussion).

Although we have not examined genetic correlations among different stages of an insect life-cycle, we have computed them for long-duration (migratory) flight and life-history traits in the milkweed bug, *Oncopeltus fasciatus*. These are discussed in detail elsewhere (Dingle 1988, 1990 and references therein) so a brief summary will suffice here to illustrate the major conclusions. Comparisons were made between a migratory population (Iowa) and a non-migratory population (Puerto Rico). The Iowa bugs were also larger and had longer wings. Extensive comparative studies have suggested an association between migration, larger size, and longer wings (Dingle 1985; Roff and Fairbairn 1990), so an experiment was designed to select for both longer and shorter wings in the two populations and to assess correlated responses after several generations of selection.

In both populations there was a rapid reponse to selection for long and short wings and a correlated response in body size. The two populations differed, however, in their flight and fecundities after selection. The migratory population showed much greater long-duration flight activity in the long-winged line, and these long-winged bugs displayed significantly higher fecundities during the first 5 days of egg-laying, i.e. at the time a migrant female would have just arrived at a new site. Thus, the migrant bugs displayed positive genetic correlations between wing length and early fecundity. In contrast the non-migratory bugs displayed no differences among selected lines in either flight or fecundity and so no genetic correlations involving these traits. In a further experiment involving the migratory bugs, selection for long duration flight resulted in correlated responses in both wing length and fecundity (J. O. Palmer and H. Dingle, unpublished results). These results indicate that the life-cycle syndromes of the two populations differ. In migratory bugs there are positive genetic correlations among traits contributing to a migration-colonization syndrome, while in at least this population of non-migratory bugs there are not. Natural selection has thus acted to produce a specific syndrome of correlated traits in the one population, but not in the other.

## 4.5 Discussion

Our studies of *O. fasciatus* indicate that populations of the same insect species may differ in the genetic correlation structures uniting life-cycle traits in complex adaptations. Natural selection may act on the individual traits, on the correlations among traits, and on the strength and direction (positive or negative) of the correlations themselves (Bradshaw 1986). Genetic variation and especially covariation thus provide the framework for life-cycles which are then modified by developmental and environmental variation. The genetic correlation structure described for *O. fasciatus* includes a component of nymphal development rates (Palmer and Dingle 1986), but it is largely constructed of adult characters. If we are to formulate rules for life-cycles, we shall need to know the correlations across stadia.

There is some evidence to suggest the existence of such correlations in *Melanoplus sanguinipes*. Hollingsworth and Capinera (1983), for example, developed a diapause strain with slower nymphal development and a sixth instar as compared to a non-diapause strain with five instars. Our high- and low-altitude populations differ in stage of diapause and development rate (Fig. 4.3) even when raised in a common environment; diapause intensities also differ. We are currently selecting on diapause in *M. sanguinipes* and a comparison with the results of Hollingsworth and Capinera (1983) will be most interesting. Further, Dean (1982a, b) found that development rates, body size, age at first reproduction, and production of diapause eggs all responded to changes in photoperiod and showed similar patterns of geographical variation. It was not possible in Dean's experiments to separate genetic and environmental contributions, but his results and the fact that diapause is highly heritable in this grasshopper (Pickford and Randell 1969; and suggested by observations of family differences in our data) indicate at least the possibility of interesting genetic correlations. With our selection experiments on diapause, we should be able to assess these correlations with respect to diapause and other aspects of the life-cycle and to begin defining rules concerning the organization of the complexity we have observed (Fig. 4.1). It is also worth noting that correlations between development rate and diapause have been found in other insects, for example the pitcher-plant mosquito *Wyeomyia smithii* (Istock 1978; and see Chap. 3).

Correlations across stadia may also occur in *Gerris remigis*. The wingless forms of this species reproduce earlier following diapause than do the winged individuals and produce more eggs as noted above. Wing polymorphism has a genetic component in some gerrids (Vepsäläinen 1974; Zera et al. 1983), although it appears likely that it is largely environmentally influenced in *G. remigis*, at least in populations from eastern Canada (Fairbairn 1988a). To the extent that, across species, wing polymorphisms include components of developmental rates and are also genetically influenced, genetic correlations across stadia and their contributions to phenotypic correlations may be important components of migration syndromes and provide at least some of the rules for life-cycle organization in migratory species. These notions are highly speculative, but analysis of wing polymorphisms from this perspective would seem a potentially fruitful area of future investigation. Certainly it would provide further insight into the evolution of wing polymorphism and the costs and benefits involved (Roff 1986a) and probably into more general aspects of life-cycles as well.

Because the balance of evidence suggests important correlations among the elements of insect life-cycles, it is natural to ask how these might be coordinated. Roff (1986a) suggests an interaction between a hormone and the threshold for wing production or suppression in the evolution of wing polymorphisms, and a similar interaction seems likely for life-cycle variation as well. Juvenile hormone is a probable candidate because of the multitude of effects it produces, but brain hormone and/or a diapause hormone as found in *Bombyx mori* (Sonobe and Ohnishi 1971) could also be involved in at least some species. Titre (see e.g. Rankin et al. 1986), timing of production and tissue responsiveness can all influence the effect a hormone has on an insect life-cycle, and various combinations of these factors provide the potential for considerable variation and adaptation.

The phylogeny of the developmental programme may also be a factor in determining life-cycle organization. In holometabolous insects, for example, the imaginal disc system provides a whole new level of organization in the developmental process. Cowley and Atchley (1988) have found that the magnitude of sexual dimorphism in *Drosophila melanogaster* is similar for traits derived from the same imaginal disc, but the level of dimorphism varies across discs. Furthermore, heritability and genetic correlation estimates are more similar for traits from the same discs, which suggest similar genetic effects within discs, although there is no evidence to suggest that traits from different discs are genetically independent, since genetic correlations for such traits were generally positive (Cowley and Atchley 1990). These results concern morphological characters, but to the extent that timing or tissue sensitivities are involved, life-cycle traits could be similarly influenced. Programme effects across stadia or in hemimetabolous insects remain to be examined, but once again would seem to be a fruitful line of enquiry (see also Cowley and Atchley 1990).

Regulation of the development programme is the key to providing the observed flexibility of insect life-cycles. If the environment is predictable, switches triggered by external cues such as photoperiod can produce the alternative forms best suited for the prevailing conditions. Because environments almost always vary in some unpredictable fashion, polygenic modes of inheritance will be more common because they can produce any frequency of offspring type rather than the limited combinations possible with one or few locus systems (Roff 1986b). Indeed it may be possible to evolve plasticity itself with adaptive responses in each environment encountered (Groeters and Dingle 1987, 1988 and references therein), although additive genetic variance for plasticity is not always observed (Groeters and Dingle 1988). Evolution toward a joint optimum may in fact be greatly retarded depending on across-environment genetic correlations (Via and Lande 1985).

One way the constraining effects of genetic correlations across environments can be circumvented is by selection for environmentally cued maternal control of the characteristics of offspring (Groeters and Dingle 1987, 1988). Maternal effects can thereby provide considerable flexibility in the life-cycle because mothers can influence future performance of their offspring in temporally heterogeneous environments and override offspring genotypes. These maternal effects can potentially affect whole suites of life-history traits, since such effects appear to influence the programme of juvenile hormone production throughout offspring lifespan (Nijhout and Wheeler 1982). The environmentally cued maternal effects noted for diapause in *Melanoplus sanguinipes* (Dean 1982a, b) and *Oncopeltus fasciatus* and similar effects in other insects (for references, see Groeters and Dingle 1988) thus assume added significance for the understanding of life-cycle syndromes.

Amidst speculation a final, cautionary note is also necessary. As Roff (1986b) has pointed out, it is not a requirement that there be a selective advantage for a pattern to exist in a life-cycle. Selection may be acting to produce optima, but there will inevitably be variation as a consequence of available genotypes. This variation may produce "chance" patterns, and these, too, must be sorted out if we are to understand the true nature of adaptive syndromes. The importance of comparative studies in searching for patterns common among species and populations and in determining how these patterns relate to fitness in the respective environments thus becomes further evident.

*Acknowledgements.* This work is supported by grants from the U.S. National Science Foundation to H. Dingle and a National Science and Engineering Research Council of Canada postdoctoral fellowship to S. M. Scott. Our thanks to Bill Bradshaw and Valerie Brown for organizing the symposium of which this contribution was a part, to Arja Kaitala for the data in Fig. 4.6, to Elizabeth Owens for the drawings in Fig. 4.5, to David Cowley for a copy of his manuscript in press, and to Daphne Fairbairn, Timothy Mousseau, and Derek Roff for comments on the manuscript.

## References

Alexander G (1964) Occurrence of grasshoppers as accidentals in the Rocky Mountains of northern Colorado. Ecology 45: 77–86

Alexander G, Hilliard JR Jr (1969) Altitudinal and seasonal distribution of Orthoptera in the Rocky Mountains of northern Colorado. Ecol Monogr 39: 385–431

Bradshaw WE (1986) Pervasive themes in insect life cycle strategies. In: Taylor F, Karban R (eds) The evolution of insect life cycles. Springer, Berlin Heidelberg New York, pp 261–275

Cowley DE, Atchley WR (1988) Quantitative genetics of *Drosophila melanogaster*. II. Heritabilities and genetic correlations between the sexes for head and thorax traits. Genetics 119: 421–433

Cowley DE, Atchley WR (1990) Developmental quantitative genetic analysis of correlation structure among body parts of *Drosophila melanogaster*. Am Nat 135: 242–268

Crespi BJ (1986) Territoriality and fighting in a colonial thrips, *Hoplothrips pedicularis*, and sexual dimorphism in Thysanoptera. Ecol Entomol 11: 119–130

Danks HV (1987) Insect dormancy: an ecological perspective. Biological Survey of Canada Monograph Series no. 1. Biological Survey of Canada, Ottawa

Dean JM (1982a) Latitudinal variation in voltinism and life history traits in the lesser migratory grasshopper. PhD. thesis, University of Iowa

Dean JM (1982b) Control of diapause induction by a change in photoperiod in *Melanoplus sanguinipes*. J Insect Physiol 28: 1035–1040

Dingle H. (1978) Migration and diapause in tropical, temperate, and island milkweed bugs. In: Dingle H (ed) Evolution of insect migration and diapause. Springer, Berlin Heidelberg New York, pp 254–276

Dingle H (1984) Behavior, genes, and life histories: complex adaptations in uncertain environments. In: Price PW, Slobodchikoff CH, Gaud WS (eds) A new ecology: novel approaches to interactive systems. Wiley, New York, pp 169–194

Dingle H (1985) Migration. In: Kerkut GA, Gilbert LT (eds) Comprehensive insect physiology, biochemistry, and pharmacology, vol. 9, Behaviour. Pergamon Press, London, pp 375–415

Dingle H (1986) Evolution and genetics of insect migration. In: Danthanarayana W (ed), Insect flight: dispersal and migration. Springer, Berlin Heidelberg New York, pp 11–26

Dingle H (1988) Quantitative genetics of life history evolution in a migrant insect. In: DeJong G (ed) Population genetics and evolution. Springer, Berlin Heidelberg New York, pp 83–93

Dingle H (1990) The evolutionary genetics of animal migration. Am Zool (in press)

Dingle H, Haskell JB (1967) Phase polymorphism in the grasshopper, *Melanoplus differentialis*. Science 155: 590–592

Dingle H, Blau WS, Brown CK, Hegmann JP (1982) Population crosses and the genetic structure of milkweed bug life histories. In: Dingle H, Hegmann JP (eds) Evolution and genetics of life histories. Springer, Berlin Heidelberg New York, pp 209–229

Fairbairn DJ (1986) Does alary dimorphism imply dispersal dimorphism in the waterstrider, *Gerris remigis*? Ecol Entomol 11: 355–368

Fairbairn DJ (1988a) Adaptive significance of wing dimorphism in the absence of dispersal: a comparative study of wing morphs in the waterstrider, *Gerris remigis*. Ecol Entomol 13: 273–281

Fairbairn DJ (1988b) Sexual selection for homogamy in the Gerridae: an extension of Ridley's comparative approach. Evolution 42: 1212–1222

Fairbairn D, Desranleau L (1986) Flight threshold, wing muscle histolysis, and alary polymorphism: correlated traits for dispersal tendency in the Gerridae. Ecol Entomol 11: 13–24

Falconer DS (1981) Introduction to quantitative genetics, 2nd edn. Longman, London

Frazzetta TH (1975) Complex adaptations in evolving populations. Sinauer, Sunderland, MA

Gatehouse AG (1986) Migration in the African armyworm, *Spodoptera exempta*: genetic determination of migratory capacity and a new synthesis. In: Danthanarayana W (ed) Insect flight: dispersal and migration. Springer, Berlin Heidelberg New York, pp 128–144

Groeters FR, Dingle H (1987) Genetic and maternal influences on life history plasticity in response to photoperiod by milkweed bugs (*Oncopeltus fasciatus*). Am Nat 129: 332–346

Groeters FR, Dingle H (1988) Genetic and maternal influences on life history plasticity in milkweed bugs (*Oncopeltus*): response to temperature. J Evol Biol 1:317–333

Gurney AB, Brooks AR (1959) Grasshoppers of the *mexicanus* group, genus *Melanoplus* (Orthoptera: Acrididae). Proc US Natl Mus 110: 1–93

Hewitt GB (1979) Hatching and development of rangeland grasshoppers in relation to forage growth, temperature and precipitation. Environ Entomol 8: 24–29

Hilbert DW, Logan JA, Swift DM (1985) A unifying hypothesis of temperature effects on egg development and diapause of the migratory grasshopper, *Melanoplus sanguinipes* (Orthoptera: Acrididae). J Theor Biol 112: 827–838

Hollingsworth CS, Capinera JL (1983) Comparison of nymphal development times in diapausing and nondiapausing strains of *Melanoplus sanguinipes* (F.) (Orthoptera: Acrididae). J Kans Entomol Soc 56: 612–616

Istock CA (1978) Fitness variation in a natural population. In: Dingle H (ed) Evolution of insect migration and diapause. Springer, Berlin Heidelberg New York, pp 171–190

Istock CA (1982) The extent and consequences of heritable variation in fitness characters. In: King CR, Dawson PS (eds) Population biology: retrospect and prospect. Columbia University Press, New York, pp 61–96

Masaki S (1978) Seasonal and latitudinal adaptations in the life cycles of crickets. In: Dingle H (ed) Evolution of insect migration and diapause. Springer, Berlin Heidelberg New York, pp 72–100

McAnelly ML, Rankin MA (1986) Migration in the grasshopper *Melanoplus sanguinipes* (Fab.). I. The capacity for flight in non-swarming populations. Biol Bull 170: 368–377

Moore HW (1948) Variations in fall embryonic development in three grasshopper species. Can Entomol 80: 83–88

Nijhout HF, Wheeler DE (1982) Juvenile hormone and the physiological basis of insect polymorphisms. Q Rev Biol 57: 109–133

Onsager JA (1983) Relationships between survival rate, density, population trends, and forage destruction by instars of grasshoppers (Orthoptera: Acrididae). Environ Entomol 11: 127–133

Palmer JO, Dingle H (1986) Direct and correlated responses to selection among life history traits in milkweed bugs (*Oncopeltus fasciatus*). Evolution 40: 767–777

Pickford R (1960) Survival, fecundity, and population growth of *Melanoplus bilituratus* (Wlk.) (Orthoptera: Acrididae) in relation to date of hatching. Can Entomol 92: 1–10

Pickford R (1966) Development, survival and reproduction of *Camnula pellucida* (Scudder) (Orthoptera: Acrididae) in relation to climatic conditions. Can Entomol 98: 158–169

Pickford R, Randell RL (1969) A non-diapause strain of the migratory grasshopper, *Melanoplus sanguinipes* (Orthoptera: Acrididae). Can Entomol 101: 894–896

Rankin MA, McAnelly ML, Bodenhamer JE (1986) The oogenesis-flight syndrome revisited. In: Danthanarayana W (ed) Insect flight: dispersal and migration. Springer, Berlin Heidelberg New York, pp 27–48

Roff D (1983) Phenological adaptation in a seasonal environment: a theoretical perspective. In: Brown VK, Hodek I (eds) Diapause and life cycle strategies in insects. Dr W. Junk, the Hague, pp 253–270

Roff DA (1986a) The evolution of wing dimorphism in insects. Evolution 40: 1009–1020
Roff DA (1986b) Evolution of wing polymorphism and its impact on life cycle adaptation in insects. In: Taylor F, Karban R (eds) The evolution of insect life cycles. Springer, Berlin Heidelberg New York, pp 204–221
Roff DA, Fairbairn DJ (1990) Wing dimorphisms and the evolution of migratory polymorphisms among the Insecta. Am Zool (in press)
Rubenstein DI (1984) Resource acquisition and alternative mating strategies in water striders. Am Zool 24: 345–353
Shotwell RL (1930) A study of the lesser migratory grasshopper. Tech Bull US Dept Agric, 190
Sonobe H, Ohnishi E (1971) Silkworm *Bombyx mori* L.: nature of diapause factor. Science 174: 835–838
Tanaka S (1984) Seasonal variation in embryonic diapause of the striped ground cricket, *Allonemobius fasciatus*. Physiol Entomol 9: 97–105
Tauber MJ, Tauber CA, Masaki S (1986) Seasonal adaptations of insects. Oxford University Press, New York
Uvarov BP (1921) A revision of the genus *Locusta* L. (*Pachytylus* Fab.) with a new theory as to the periodicity and migrations of locusts. Bull Entomol Res 12: 135–163
Uvarov B (1966) Grasshoppers and locusts, vol. 1. Cambridge University Press, Cambridge
Vepsäläinen K (1974) Determination of wing length and diapause in water-striders (*Gerris* Fabr., Heteroptera). Hereditas 77: 163–175
Via S, Lande R (1985) Genotype–environment interaction and the evolution of phenotypic plasticity. Evolution 39: 505–523
Zera AJ, Innes DT, Saks ME (1983) Genetic and environmental determinants of wing polymorphism in the water strider, *Limnoporus canaliculatus*. Evolution 37: 513–522

*Chapter 5*

# Coevolution Between Herbivorous Insects and Plants: Tempo and Orchestration

May R. Berenbaum

For a herbivorous insect, a plant is more than just a meal – it is a way of life. For some species, just about every aspect of the life-cycle involves the plant host – eating, escape from predators, overwintering and mating, to name a few. Particularly for insects with narrow host ranges, life-cycles must coordinate or the insect misses out not only on a meal but on all those other things that make insect life worth while; thus, there must be tremendous selective pressure on insects to adapt to the peculiarities of their hosts. Walsh (1864, 1865) was among the earliest to recognize the fact that specialization by insect species on different host species is heritable and results from natural selection. A passionate early advocate of Darwinism, Walsh (1867) observed in his studies on *Rhagoletis pomonella*, the apple-maggot fly, that several populations of the native North American species fed not on the ancestral and typical hawthorn host (*Crataegus*) but rather on the introduced novel host, apple (*Malus*). He attributed the existence of intraspecific groups with different host usage patterns to evolution resulting from isolation due to "attachment" to different food plants. The apple-maggot fly is a species that does perform many vital functions – including mating, oviposition, and larval development – in or around its hosts. Walsh offered no explanation, however, for the motive forces behind host shifts.

Even before Walsh made the point that plants can affect insect life-cycles, the effect of insects on plant life-cycles had been observed. Probably the earliest recognition of the phenomenon dates back to 1788, when Isaac Chapman observed that, eight years after the introduction of the Hessian fly into the United States in 1776, certain wheat populations, particularly those with an "early growth habit," escaped damage consistently (Snelling 1941). He was among the first to advocate plant breeding for resistance. That insects, or the damage they cause, are in some way responsible for the manifestation of heritable resistance traits, is implicit in the time-honoured techniques of selection, according to which plants are exposed to heavy infestations of insects and only the survivors are used for subsequent seed production. In this way, Harlan (1917) selected for resistance in Seredo cotton to black scale. The genetic basis for this superior survival in the face of insect attack has been demonstrated repeatedly in hybridization studies (e.g. Collins and Kempton 1917).

There is therefore general recognition that insects can and do under certain circumstances adapt to particular hosts and that plants can and do under certain circumstances adapt to particular insect enemies. Where arguments arise is in discussions of how these two phenomena interrelate. There is no shortage of ideas on the subject. One early idea was advocated by Brues (1924), who suggested that insects and their food plants evolve in parallel without any coordination. In support of his idea, he cites (Brues 1924) the case of the Pieridae:

> The Pierid butterflies are essentially feeders on Cruciferae, but a few genera affect Leguminosae. The family is widely distributed and has produced many forms restricted to crucifers, some living on Leguminosae and a few . . . restricted to the single genus *Cassia* in widely separated parts of the world. We can not be far amiss if we assume that the acquisition of a leguminous foodplant was a mutation or at least a sudden shift and that diversification of the insect has proceeded not necessarily with any reference to that of the plants, but that the two have gone on side by side, most of the insects gradually acquiring additional, although not necessarily very close related, leguminous food plants.

Thus, after some form of mutation, insects finding themselves on a new host plant proceed to differentiate more or less independently of their host plant, which also proceeds to differentiate independently of its insect associates.

Yet another point of view, discussed and more or less dismissed by Brues (1924), was that of Vassiliev (1913). Examining beet-feeding insects in Russia, he found *Bothynoderus punctiventris* attacking two species of *Chenopodium*, one species of *Atriplex*, *Beta vulgaris* and *Polygonum vulgare*. The first four hosts are members of the Chenopodiaceae, while the last belongs to Polygonaceae. Since the two families are closely related, Vassiliev argued that the weevil presumably originated on an intermediate form and acquired hosts according to the pattern of plant differentiation and evolution. In other words, plants evolve and/or speciate and insects, of necessity, follow. Plants, therefore, in the words of Hodkinson (1988), are "rafts transporting the insects down the river of evolutionary time. As the original raft began to break up (plant diversification) each section drifted apart, carrying its own fauna, which then began to diversify or adapt to its changed environment." Parallel evolution, in this view, is driven by plant adaptation and diversification. Arguments of this nature are fuelled by the conviction that insects are not a major selective force on plants but that plants do exert substantial selection pressure on insects (Jermy 1976).

A third alternative viewpoint is that insects and plants provide reciprocal selection pressure. In his inimitable fashion, Darwin (1859; as cited by Thompson (1989)) succinctly described the concept of reciprocal evolution (which he called coadaptation) in the context of plant–pollinator interactions: "Thus I can understand how a flower and a bee might slowly become, either simultaneously or one after the other, modified and adapted in the most perfect way to each other by continued preservation of individuals presenting mutual and slightly favourable deviations of structure." Generally, this alternative process is now called coevolution. The term coevolution first appeared not in the context of plant–insect interactions but rather in the context of plant–pathogen interactions. Mode (1958) designed a "mathematical model for the co-evolution of obligate parasites and their hosts," (note hyphen) based on work published earlier by Flor on flax and flax rust (Flor 1955, 1956). In this system, host and parasite have "complementary genetic systems: . . . a gene for resistance acts if and only if there is a corresponding gene in the pathogen for avirulence".

This reciprocity was applied by Fraenkel (1959) to insect–plant interactions. "Reciprocal adaptive evolution" according to Fraenkel is mediated by plant secondary chemicals, elaborated by plants as a defence against insects much in the same manner that thorns were thought to have arisen as a defence against mammalian herbivores. Insects in turn adapt to chemicals in their host such that they can be used for host location and recognition. Ehrlich and Raven (1964) catapulted this proposed process into public awareness by using it to account for patterns of host-plant use among families of butterflies – that "plants and phytophagous insects have evolved in part in response to one another . . . in a stepwise manner". Plant secondary chemicals convincingly coincided with otherwise incomprehensible host shifts and, according to Ehrlich and Raven, "play the leading role in determining patterns of utilization". Novel defensive compounds permit adaptive radiation in plant taxa until such time as insects adapt and undergo their own independent radiation. Ehrlich and Raven stopped short at testing phylogenetic predictions, stating that "the reconstruction of phylogenies on the basis of this sort of information would seem an unwarranted imposition on the data".

Although Ehrlich and Raven (1964) provided evidence for the pattern, they provided no evolutionary mechanism for reciprocal radiations. Janzen (1980) in a criticism of the promiscuous use of the term coevolution (introduced into the literature by Ehrlich and Raven without the hyphen) suggested that coevolution was fundamentally a microevolutionary process involving reciprocal genetic changes in interacting populations.

The nature of the processes underlying plant–insect evolution remains controversial. All schools of thought on the subject still have passionate advocates. As is usually the case in scientific controversies, in all probability everyone is right at least part of the time. Whether the primary evolutionary process behind insect–plant interactions is colonization based on pre-adaptation, association by descent, or mutual descent may depend to a large extent on both the taxon in question and the level of the taxonomic hierarchy.

For example, what has been variously described as coevolution in the narrow sense, classical coevolution, stepwise coevolution, or "the co-evolutionary arms race" (e.g. see Whittaker and Feeny 1971) seems an unlikely scenario for most insect–plant interactions for several reasons. First, there are likely to be important inequities in the frequency of interaction, i.e., the "rare enemy effect" (Dawkins and Krebs 1979). In a two-species interaction, the interaction may be critical to one species which is none the less so rare or unimportant to the interacting species that it at the same time has no appreciable effect on the evolution of the interacting species. There are also probably inequities in the impact of interaction, i.e. the "life–dinner principle" (Dawkins and Krebs 1979). Inequities in selection pressures (e.g. "the fox is running for its dinner while the rabbit is running for its life") result in asymmetries in evolutionary responses. In an interaction between a folivorous insect monophagous on a single plant species, for example, consuming host-plant tissues is absolutely vital for the survival of the insect but results only in the loss of non-essential body parts of the plant. Finally, there are probably inequities in evolutionary rates. Generation times can differ dramatically between herbivorous insects and their host plants and consequently evolutionary rates can differ dramatically as well. A tree-feeding insect species, for example, can undergo several generations a year, whereas its host may require several decades to undergo a single generation.

These sorts of objection have been raised repeatedly, notably by Jermy (1976, 1984), to argue against reciprocal responses at any level of organization. However, these objections serve a useful purpose in pointing toward those interactions in which reciprocal responses are more likely to occur. Such interactions should:

1. Be predictable and constant year to year
2. Involve species with equivalent generation times – e.g. a univoltine insect on an annual or biennial plant
3. Entail equivalent selective impact on plant and herbivore – e.g. an oligophagous insect that consumes indispensible body parts, such as reproductive organs, of its plant host

At the level of population interactions, methods of quantitative genetics are appropriate for use in the estimation and quantification of reciprocal genetic responses.

Berenbaum et al. (1986) examined a candidate plant–insect interaction that appears to meet the minimum criteria for reciprocal interactions. *Depressaria pastinacella*, the parsnip webworm, is a univoltine oecophorid caterpillar introduced from Europe that feeds on *Pastinaca sativa*, wild parsnip, an introduced biennial herb from Europe. Although it is known to feed on two other closely related genera in the Umbelliferae (Hodges 1974), in certain parts of its range, such as east-central Illinois, populations are effectively monophagous on parsnip. Webworms predictably infest from 50% to 100% of plant individuals in a given population (A. R. Zangerl and M. R. Berenbaum, unpublished results). Since larvae feed on buds, flowers and developing fruits on a monocarpic plant, they stand to have an effect on fitness. Moreover, life-cycles are nearly congruent – the insect is univoltine and the plant biennial.

As is typical for apioid umbellifers, *P. sativa* produces furanocoumarins, photoactive chemicals that are toxic and/or antifeedant to many insects (Berenbaum 1978; Klocke et al. 1989 and references therein). In the interaction between wild parsnip and the parsnip webworm, the furanocoumarin phenotype of the plant determines its susceptibility to damage by parsnip webworm; moreover, furanocoumarin variation in the wild parsnip population is genetically based and thus subject to selection. Different plant genotypes experience differential fitness in the presence of the herbivore and, as a result of selection pressure by parsnip webworms, furanocoumarin phenotypes (and hence genotypes) are capable of changing in response to insect herbivory. A significant selection differential indicates that bergapten, the principal furanocoumarin associated with resistance, would increase in concentration in parsnip populations as a result of insect herbivory (Berenbaum et al. 1986).

Bergapten, the furanocoumarin most closely associated with resistance in wild parsnip to parsnip webworm, has an adverse effect on growth rate, a fitness character, of the parsnip webworm (Berenbaum et al. 1989). Since, in the parsnip webworm, growth rate is a character with measurable amounts of additive genetic variance and thus subject to selection (M. R. Berenbaum and A. R. Zangerl, unpublished results), a selection response in the parsnip webworm population to chemical phenotype is theoretically possible. Thus, in the interaction between wild parsnip and the parsnip webworm, herbivory can effect genetically based chemical changes in a plant population which in turn will decrease insect fitness. The process is effectively simultaneous – an increase in fitness in a plant (e.g. in the form of increased amounts of bergapten) decreases the fitness of the herbivore. As Flor (1955) described for the flax/flax rust interaction:

> The criterion for degree of rust resistance or susceptibility is the pustule type that develops on the host as a result of infection by a race of the parasite. The criterion for degree of avirulence or virulence of a race of the rust fungus is the pustule type that develops on the host as a result of infection by the parasite. Thus, a single phenomenon is the measure of both pathogenicity of the parasite and reaction of the host. Pustule type is the expression of the interaction of the genic and cytoplasmic complex of the host with the genic and cytoplasmic complex of the parasite as affected by the environment.

At the population level, then, reciprocal genetic changes between insects and their host plants may occur. However, just as it is difficult to extrapolate from microevolutionary to macroevolutionary phenomena, it is difficult to extrapolate from microcoevolutionary to macrocoevolutionary phenomena (Brooks 1988). Although Ehrlich and Raven (1964) refrained from phylogenetic inference, their description of reciprocal radiations suggests that phylogenies should reflect coevolutionary interactions. Indeed, the relative rarity of groups of closely related insect species on closely related plant species has been one point made repeatedly by Jermy (1976, 1984) as evidence that reciprocity does not have much evolutionary significance.

Evaluating insect–plant interactions above the level of population has proved operationally challenging. Early attempts, borrowing from the host–parasite literature, sought to demonstrate that, in associations above the species level, primitive herbivore taxa associate with primitive host taxa and advanced herbivore taxa with advanced host taxa ("Farenholz's rule"). These early evaluations (e.g. Benson et al. 1975; Nault and DeLong 1980) were essentially descriptive and non-quantitative (Mitter and Brooks 1983). Moreover, establishing such a pattern does not allow for ascertaining the evolutionary mechanism underlying the pattern (Wiebes 1979); an association between primitive hosts and herbivores and derived hosts and herbivores can originate through both sequential evolution and reciprocal coevolution.

Several recent methodological innovations have increased the information value and overall utility of phylogenetic assessments of evolutionary patterns. Phylogeny analysis using cladistic methodology allows for a quantitative estimate of congruence in sequence of phylogenetic events (Mitter and Brooks 1983). In cladistic analysis, organisms are grouped according to shared derived characters, which are used in the construction of a branching diagram that depicts evolutionary relationships among monophyletic assemblages (Eldredge and Cracraft 1980). These diagrams can be compared quantitatively and statistically for similarities with a variety of methods. Cladistic analyses of plant–insect interactions have been applied at various hierarchical taxonomic levels with various degrees of success (Table 5.1). In virtually none of the these studies, however, is any analysis undertaken of the importance of phytochemical "defensive novelties" (Mitter and Brooks 1983) in determining patterns of host association.

In some ways, even a rigorous cladistic analysis of plant and insect taxa may fail to yield a definite answer to coevolutionary questions. The extremely erratic

**Table 5.1.** Tests for insect–plant coevolution above the species level

| Insect | Plant | Authors |
|---|---|---|
| Heliconiine butterflies | Passifloraceae | Benson et al. 1975 |
| Agaonidae | *Ficus* | Ramirez 1974 |
| *Dalbulus* | *Zea, Tripsacum* | Nault and DeLong 1980 |
| Papilionidae | Angiosperms | Richard and Guedes 1983 |
| *Semudobia* | *Betula* | Roskam 1985[a] |
| *Eriococcus* | *Nothofagus* | Humphries et al. 1986[a] |
| Psylloidea | Angiosperms | Hodkinson 1988 |
| Graphiine swallowtails | Magnoliidae | Miller 1987[a] |
| *Phylloxera* | *Vitis* | Wapshere and Helm 1987 |
| *Euphilotes* | *Eriogonum* | Shields and Reveal 1988 |
| *Larinus* | Cynaroideae | Zwölfer and Herbst 1988[a] |

[a] Cladistic analysis was used.

nature of host records, particularly for crucial intermediate or unusual forms, complicates the analysis of host-plant relationships; host records, for example, cannot be used with any degree of certainty to distinguish between consistent patterns of evolutionary association, opportunistic intermittent colonization, or transient occupation of a non-host immediately prior to collection by an entomologist. There is also a general lack of consensus concerning both insect and host phylogenies and rarely does a single investigator have the expertise to evaluate both host and herbivore phylogenies. For example, when Miller (1987) posed the question "Host–plant relationships in the Papilionidae (Lepidoptera): parallel cladogenesis or colonization?" he reached the conclusion that colonization, rather than coevolution, was the major form of host acquisition in the Graphiini. He reached his conclusion, however, without any cladistic analysis of host plants (e.g. an examination of the phylogeny of the Annonaceae, host for over half the taxa in his cladogram).

Parallel cladogenesis per se (as evidenced by congruent phylogenies) does not in and of itself demonstrate reciprocal evolutionary interaction. Association by descent may result from simple host tracking by the herbivore. Precise estimates of evolutionary age would be helpful (albeit difficult to obtain) in order to determine whether speciation events were successive or synchronous. Since coevolution does not leave behind an abundance of fossil evidence, other estimates of taxon age must be used in most cases. Biogeographical evidence has been used to demonstrate that the taxa in question were probably associated over evolutionary time (Humphries et al. 1986). One promising innovation in phylogenetic analyses that can potentially distinguish between sequential and simultaneous radiations is the use of gel electrophoresis for estimation of genetic distances among taxa. Amounts of protein change (assuming a molecular clock hypothesis) should be comparable if host and insect taxa are of comparable evolutionary age. This technique has been used to examine host–parasite evolution with success (Hafner and Nadler 1988) and holds much promise for evolutionary studies of insect–plant interaction.

Again, the relatively low rate of success in matching host and herbivore phylogenies suggests that coevolution at the macro scale is more probably the exception rather than the rule. Both logic and empirical evidence suggest that the same characteristics predisposing populations to reciprocal genetic interactions would result in parallel phylogenesis. Such associations are typically predictable and constant in time, with equivalent fitness impacts on herbivore and host. Herbivores that are totally dependent upon a narrow range of hosts, characterized perhaps by physiological dependence upon host chemistry (e.g. a host recognition kairomone that elicits oviposition or feeding, a pheromone derived from plant precursors, or, in the case of gallmakers, a critical interaction between insect and plant chemistry to induce gall formation), are likely to demonstrate differential fitness in response to changes in host chemistry, ecology and distribution. By the same token, plants that suffer major and consistent fitness losses to a particular herbivore are most likely to evolve in response to herbivory. Those few cases in which congruence of phylogenies is suggested (Table 5.1; Roskam 1985; Zwölfer and Herbst 1988; Humphries et al. 1986) tend to meet this requirement (are gall formers, flower feeders, and sedentary scale insects, respectively). Taxa must also be of equivalent age, such that there is at least a possibility that major radiations in both taxa were synchronous over time.

In view of the a priori evidence of reciprocal genetic selection in *Depressaria pastinacella* and its umbelliferous host, it seems a propitious starting-point to examine patterns of evolutionary association above the species level in these and related taxa. Cladistic analyses are currently underway (S. Passoa, unpublished results) but are as yet unavailable. However, phenetic studies may offer some preliminary evidence. *Depressaria* is a genus of approximately 100 species, primarily Palaearctic in distribution but with both Nearctic and Holarctic representatives. Hannemann (1953) recognized six species groups among Old World *Depressaria*. In a phenetic analysis based on morphological chracters, he proposed a hierarchical ranking of the groups from ancestral to advanced (Fig. 5.1). In general, umbellifer-feeding appears to be ancestral to composite-feeding

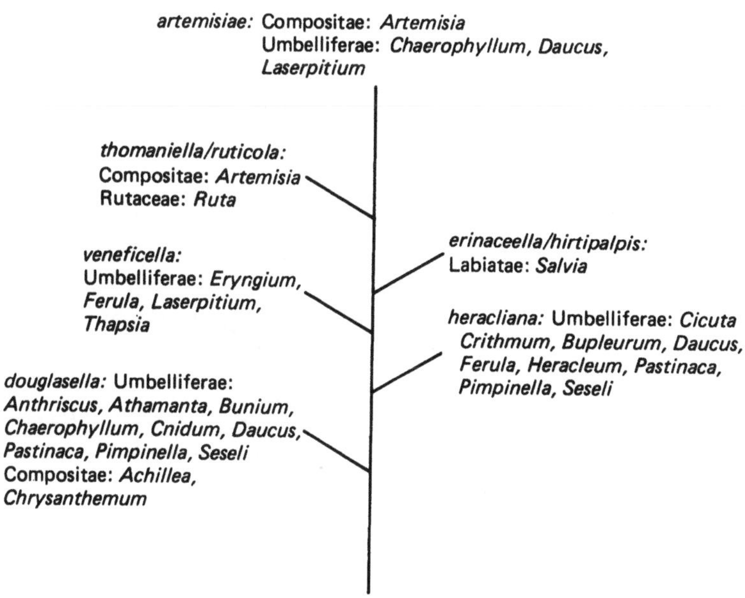


**Fig. 5.1.** Phenetic analysis of species groups in the genus *Depressaria*. (After Hannemann 1953.)

in this genus. Hannemann (1953) also provided a phenetic analysis of the genera in the tribe Depressariinae, to which ***Depressaria*** belongs (Fig. 5.2). ***Horridopalpus***, which could be considered a sister group to ***Depressaria*** if cladistic methodologies are applied, feeds on Rutaceae. In the remainder of the tribe, ***Agonopteryx*** feeds primarily on Umbelliferae and Compositae, although three species feed on Rutaceae, and ***Martyrhilda***, proposed to be the most advanced taxon along this branch, is associated with the Compositae.

The association between purportedly primitive, intermediate and advanced hosts with Rutaceae, Umbelliferae, and Compositae, respectively, is intriguing from both a chemical and an evolutionary perspective. Chemical characters can be used to account very well for host shifts within or among these taxa. These three plant families share a number of unique chemical characters (Table 5.2). The extraordinary concatenation of chemicals among host families suggests phylogenetic relationship; in the absence of such a relationship, these groups of chemicals must have evolved independently in the same precise combination, a seemingly unlikely scenario if random mutation is indeed responsible for the evolution of defensive novelties (although, biosynthetically, there are links among several classes – hemiterpenoid constituents, for example, are characteris-

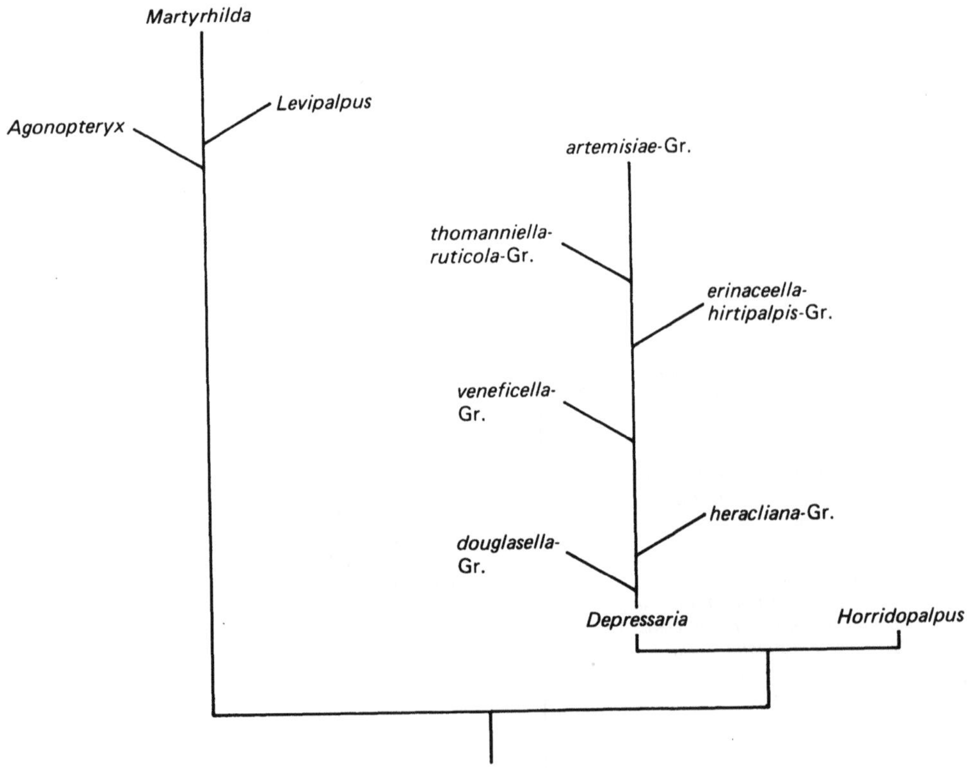


**Fig. 5.2.** Phenetic analysis of depressariine genera (Lepidoptera: Oecophoridae). (After Hannemann 1953.)

**Table 5.2.** Chemical similarities among depressariine oecophorid host families (from de Emerenciano et al. 1987; Fernandes da Silva et al. 1988; Holub et al. 1987; Berenbaum 1990)

| | Rutaceae | Umbelliferae | Compositae |
|---|---|---|---|
| Chromenes | x | Apioideae | Asteroideae |
| Coumarins | x | x | Asteroideae |
| Essential oils | x | x | x |
| Isobutyl amides | x | — | x |
| Polyacetylenes | x | x | Asteroideae |
| Sesquiterpene lactones | — | Apioideae | x |

x indicates presence in all subfamilies.

tic of coumarins, essential oils and chromenes). Phytochemical similarities, in combination with other evidence, have prompted Hegnauer (1973) to propose a phylogenetic relationship among these families; Cronquist (1968), principally on morphological grounds, derives Umbellales from the Sapindales–Rutales as well.

Thus, within the depressariine Oecophoridae, ancestral genera are associated with Rutaceae, intermediate genera with Umbelliferae, and derived genera with Compositae; phylogenetic status, albeit loosely determined, is more or less coincident between host and herbivore. Interestingly, the same patterns of relationship and host use are evident in an unrelated group of herbivorous insects. Hancock (1983) presented a cladistic analysis of genera in the tribe Papilionini (subfamily Papilioninae, family Papilionidae) (Fig. 5.3). In this group of genera, feeding on Rutaceae is regarded as ancestral. Umbellifer-feeding appears to have derived from Rutaceae-feeding at least three times in the tribe. Compositae-feeding in turn is thought to have arisen from umbellifer-feeding in the *Papilio machaon* complex (Sperling 1987).

Identical patterns of relationship and host use in unrelated taxa provides at least provisional evidence for diffuse coevolution (Janzen 1980; Thompson 1989). Diffuse coevolution probably results not only from selection by a large assortment of herbivorous taxa with similar feeding habits but also from mediation through an assemblage of chemicals, rather than a single class or group of chemicals. Although at the population level single compounds or classes of compounds may suffice to account for reciprocal interactions, at higher taxonomic levels chemical complexity must enter into any discussion of reciprocal radiations. In order to confirm these patterns, an estimate of the age of both lepidopteran groups (relative to the estimated time of origin of Umbelliferae and Compositae) would be extremely useful in evaluating this scenario, among other things to overcome a seemingly insuperable difficulty. If biochemical syntheses link related taxa and if biosynthetic innovations are the basis for host shifts, then there is really no way to separate out colonization from mutual descent as a mechanism for determining the evolution of host use. Both processes may generate identical cladograms. In this situation, the distribution of phytochemical "defensive novelties" may provide useful information. As predicted by Mitter and Brooks (1983) exceptions to phylogenetically consistent host use patterns can be accounted for by convergence in plant taxa in defensive novelties. The genus *Psoralea*, in the Leguminosae, for example, is host for both *Agonopteryx* and *Papilio* species. *Psoralea* is among relatively few leguminous genera that produce furanocoumarins. Since both swallowtails (Berenbaum 1981) and oecophorids

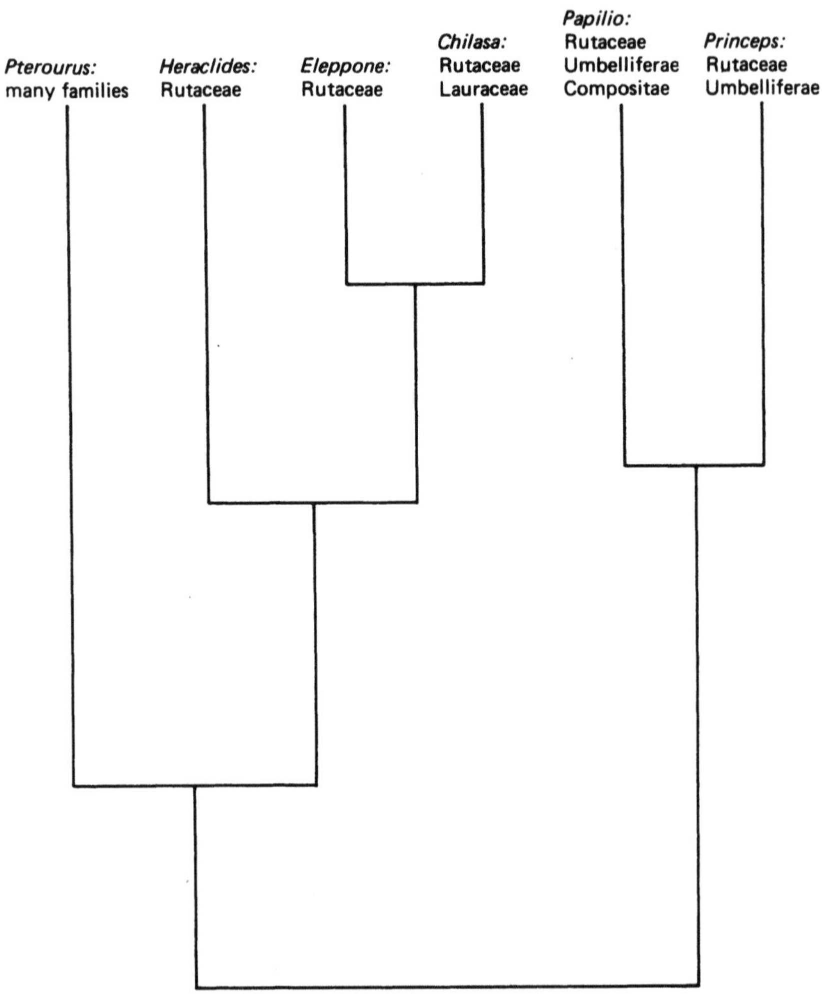


**Fig. 5.3.** Cladistic analysis of papilionine genera. (After Hancock 1983.)

(Berenbaum et al. 1989) are known to display a feeding response to furanocoumarins, *Psoralea* feeding may represent a host transfer by colonization in an otherwise coevolutionary pattern.

Thus, at taxonomic levels higher than the species, demonstrating the process of coevolution – as well as failing to demonstrate the process – may prove to be extremely difficult. These difficulties are not at all surprising, since they are essentially inherent in the study of any macroevolutionary phenomenon. At the level of the population, it is far simpler to identify those genes involved in reciprocal genetic changes; it is in fact possible to record and estimate selection in progress. The macroevolutionary consequences of such selection are not intuitively obvious, nor are they are easily observable or measurable. It is therefore also not surprising that there has been considerable disillusionment with the concept of coevolution (Bernays and Graham 1988); it certainly has not become a singular explanation for all manner of insect/host-plant associations. It has not, however, been evaluated thoroughly enough to merit its outright retirement either. If mutual descent is a mode of evolutionary association, it is likely to occur in those interactions in which insects do have a major selective impact on their host plants and the host plant plays a central role in virtually all aspects of the life-cycle of the insect (see e.g. Roskam 1975; Zwölfer and Herbst 1988). Efforts to evaluate the importance of coevolution as an evolutionary mechanism should be concentrated on these sorts of interaction.

However, at hierarchical levels above the species, a definitive test for coevolution "sensu stricto" has yet to be devised. Evidence of congruent phylogenies, even with supporting biogeographical or molecular genetic data, is open to more than one interpretation. Undoubtedly, controversy will continue to rage, as it has for over a century, not only about the extent to which host plants act as selective agents in insect lives and insects as selective agents in the lives of plants but also about the appropriate ways in which to investigate these issues, regardless of the amount of evidence, or its quality. As Walsh (1864) bemoaned when these discussions were in their nascent stage, "the constitution of the human mind is such that the same evidence carries with it very different degrees of weight, when presented to different intellects".

*Acknowledgements*. I thank Steve Passoa for helpful discussions. This work was supported by NSF Grant BSR 8818–205.

## References

Benson WW, Brown KS, Gilbert LE (1975) Coevolution of plants and herbivores: passion flower butterflies. Evolution 29: 659–680

Berenbaum MR (1978) Toxicity of a furanocoumarin to armyworms: a case of biosynthetic escape from insect herbivores. Science 201: 532–534

Berenbaum M (1981) Effects of linear furanocoumarins on an adapted specialist insect (*Papilio polyxenes*). Ecol Entomol 6: 345–351

Berenbaum MR (1990) Evolution of specialization in insect-umbellifer associations. Ann Rev Entomol 35: 319–343

Berenbaum MR, Zangerl AR, Nitao JK (1986) Constraints on chemical coevolution: wild parsnips and the parsnip webworm. Evolution 40: 1215–1228

Berenbaum MR, Zangerl AR, Lee K (1989) Chemical barriers to adaptation by a specialist herbivore. Oecologia 80: 501–506

Bernays EA, Graham M (1988) On the evolution of host specificity in phytophagous arthropods. Ecology 69: 886–892

Brooks DR (1988) Macroevolutionary comparisons of host and parasite phylogenies. Annu Rev Ecol Syst 19: 235–259

Brues CT (1924) The specificity of food-plants in the evolution of phytophagous insects. Am Nat 58: 127–144

Chapman I (1826) Some observations on the hessian fly; written in the year 1797. Mem Philos Soc Prom Agric 5: 143–153

Collins GN, Kempton JH (1917) Breeding sweet corn resistant to the corn earworm. J Agric Res 11: 549–572

Cronquist A (1968) The evolution and classification of flowering plants. Houghton Mifflin Co, Boston, MA

Dawkins R, Krebs JR (1979) Arms races within and between species. Proc R Soc Lond [Biol] 205: 489–511

Ehrlich PR, Raven PH (1964) Butterflies and plants: a study in coevolution. Evolution 18: 586–608

Eldredge N, Cracraft J (1980) Phylogenetic patterns and the evolutionary process: method and theory in comparative biology. Columbia University Press, New York

de Emerenciano V, Ferreira ZS, Kaplan MAC, Gottlieb OR (1987) A chemosystematic analysis of tribes of Asteraceae involving sesquiterpene lactones and flavonoids. Phytochemistry 26: 3103–3115
Fernandes da Silva MF, Gottlieb OR, Ehrendorfer F (1988) Chemosystematics of the Rutaceae: suggestions for a more natural taxonomy and evolutionary interpretation of the family. Pl Syst Evol 161: 97–134
Flor HH (1955) Host–parasite interaction in flax rust – its genetics and other implications. Phytopathology 45: 680–685
Flor HH (1956) The complementary genic systems in flax and flax rust. Adv Gen 3: 29–54
Fraenkel GS (1959) The raison d'être of secondary plant substances. Science 12: 1466–1470
Hafner MS, Nadler SA (1988) Phylogenetic trees support the coevolution of parasites and their hosts. Nature 332: 258–259
Hancock DL (1983) Classification of the Papilionidae (Lepidoptera): a phylogenetic approach. Smithersia 2: 1–48
Hannemann HJ (1953) Natürliche Gruppierung der Europäischen Arten der Gattung *Depressaria*, s.1. (Lep. Oecoph.). Mitt Zool Mus Berlin 29: 269–373
Harlan SC (1917) A note on resistance to black scale in cotton. West Indian Bull 16: 255–256
Hegnauer R (1973) Chemical patterns and relationships of Umbelliferae. In: Heywood VH (ed) The biology and chemistry of the Umbelliferae, pp. 267–278 (Bot J Linn Soc Suppl. 1)
Hodges R (1974) The moths of America North of Mexico, Fas 6.2, Gelechioidea Oecophoridae. E. W. Classey, London
Hodkinson ID (1988) Coevolution between psyllids (Homoptera: Psylloidea) and rain-forest trees: the first 120 million years. Tropical Rain Forest: The Leeds Symp. pp 187–194
Holub M, Toman J, Herout V (1987) The phylogenetic relationships of the Asteraceae and Apiaceae based on phytochemical characters. Biochem Syst Ecol 15: 321–326
Humphries CJ, Cox JM, Nielsen ES (1986) *Nothofagus* and its parasites: a cladistic approach to coevolution. In: Stone AR, Hawksworth DL (eds) Systematics Association special vol. 32 Coevolution and Systematics, Clarendon Press, Oxford, pp 55–76
Janzen DH (1980) When is it coevolution? Evolution 34: 611–612
Jermy T (1976) Insect–host-plant relationship – co-evolution or sequential evolution? Symp Biol Hung 16: 109–113
Jermy T (1984) Evolution of insect/host plant relationships. Am Nat 124: 609–630
Klocke JA, Balandrin MF, Barnby MA, Yamasaki RB (1989) Limonoids phenolics and furanocoumarins as insect antifeedants, repellents, and growth inhibitory compounds. In: Arnason JT, Philogene BJR, Morand P (eds) Insecticides of plant origin. American Chemical Society, Washington DC, pp 136–149 (ACS Symp. Series 387)
Miller JS (1987) Host-plant relationships in the Papilionidae (Lepidoptera): parallel cladogenesis or colonization? Cladistics 3: 105–120
Mitter C, Brooks DR (1983) Phylogenetic aspects of coevolution. In: Futuyma DJ, Slatkin M (eds) Coevolution. Sinauer, Sunderland, MA, pp 65–98
Mode CJ (1958) A mathematical model for the co-evolution of obligate parasites and their hosts. Evolution 12: 158–165
Nault LR, DeLong DM (1980) Evidence for co-evolution of leafhoppers in the genus *Dalbulus* (Cicadellidae: Homoptera) with maize and its ancestors. Ann Entomol Soc Am 73: 349–353
Ramirez RB (1974) Coevolution of *Ficus* (Moraceae) and Agaonidae (Hymenoptera). Ann MO Bot Gard 61: 770–780
Richard D, Guedes M (1983) The Papilionidae (Lepidoptera): co-evolution with the angiosperms. Phyton 23: 117–126
Roskam JC (1985) Evolutionary patterns in gall midge-host plant associations (Diptera, Cecidomyiidae). Tijdschr Entomol 128: 193–213
Shields O, Reveal JL (1988) Sequential evolution of *Euphilotes* (Lycaenidae: Scolitantidini) on their plant host *Eriogonum* (Polygonaceae: Eriogonoideae). Biol J Linn Soc 33: 51–93
Snelling RO (1941) The place and methods of breeding for insect resistance in cultivated plants. J Econ Entomol 34: 335–367
Sperling FAH (1987) Evolution of the *Papilio machaon* species group in western Canada (Lepidoptera: Papilionidae). Quaest Entomol 23: 198–315
Thompson JN (1989) Concepts of coevolution. Trends Ecol Evol 4: 179–183
Vassiliev EM (1913) Plants serving as food for some herbivorous insects and the causes for their selection. Stud Exp Ent Stn All-Russian Soc Sugar-Refiners 1912: 63–66
Walsh BD (1864) On phytophagic varieties and phytophagous species. Proc Entomol Soc Phila 3: 403–430

Walsh BD (1865) On the phytophagic varieties of phytophagous species, with remarks on the unity of coloration of insects. Proc Entomol Soc Phila 5: 194–216
Walsh BD (1867) The apple worm and the apple maggot. Am J Hort 2: 338–343
Wapshere AJ, Helm KF (1987) Phylloxera and *Vitis*: an experimentally testable coevolutionary hypothesis. Am J Enol Vitic 38: 216–222
Whittaker RH, Feeny PP (1971) Allelochemics: chemical interactions between species. Science 171: 757–770
Wiebes JT (1979) Coevolution of figs and their insect pollinators. Annu Rev Ecol Syst 10: 1–12
Zwölfer H, Herbst J (1988) Präadaptation, Wirtskreiserweiterung und Parallel-Cladogenese in der Evolution von phytophagen Insekten. Z Zool Syst Evolutionsforsch 26: 320–340

*Chapter 6*

# Size, Phylogeny and Life-History in the Evolution of Feeding Specialization in Insect Predators

Francis Gilbert

## 6.1 Introduction

The existence of a vast array of animals that are feeding specialists presents evolutionary theory with a problem. How are we to explain their existence? Either we accept that there are adaptive reasons why many animals, particularly insects, should specialize (and then we must understand the selective forces operating to promote specialization in some circumstances, and generalization in others: Scriber 1983; Rausher 1983; Futuyma and Moreno 1988), or we must explain specialists as "dead ends" in evolution (Simpson 1944; Moran 1988).

Feeding specialization is a major ecological feature of animals, with many ramifications. The vast bulk of the literature is concerned with understanding the feeding strategies of herbivorous insects on their host plants, and understanding the ecology of host preference (for a review, see Scriber 1983; Rausher 1983; Futuyma and Peterson 1985; Thompson 1988a). Much of our understanding of the evolution of feeding patterns among herbivorous insects involves a consideration of plant chemistry. Recently, Tauber and Tauber (1987) pointed out that virtually nothing is known about the evolution of feeding specialization in carnivorous insects, particularly species that are not parasitoids. This lack of attention is all the more surprising in view of the fact that predators and parasitoids are responsible for nearly all biocontrol programmes, and make up the majority of insect species. Bristowe (1988) remarked that the "conventional wisdom" about predators is that they tend to be generalists precisely because animal tissues are supposed to be much more nutritious, providing a balanced diet and not differing greatly from species to species. Despite this, all entomologists and ecologists are well aware of the high degree of specificity shown by many if not most carnivorous insects, especially parasitoids but also many predators. The field of predator specialization is thus largely unexplored.

In this chapter I consider some predictions about specialization, garnered largely from the literature concerned with the relationships between herbivorous insects and their host plants. I test these predictions using data from predatory and non-predatory species from an insect group particularly suitable for this purpose, namely the hoverflies (Diptera, Syrphidae). Adult feeding ecology is reasonably uniform in this family, since the adult flies all visit flowers for pollen and nectar (Gilbert 1981, 1985a, 1986), whilst the larval feeding biology varies widely, and includes predators and non-predators.

In testing these predictions, I categorize the degree of specialization as a species characteristic: I am aware that the concept of a generalist species is a difficult one to maintain, since populations can be local specialists and regional generalists (Fox and Morrow 1981), and even within populations some individuals are more specialized than others (Via 1986; Rossiter 1987; Ng 1988). Here, hoverfly species are assigned to categories of specialization based upon within-habitat sampling of food types. Arguments could be made against all of my predictions; I am, however, attempting to test them against the data.

## 6.2 Predictions

I make three predictions about the evolution of specialization in the diet range of larval insects. Specialization is defined here as the narrowing of diet range, and often involves a narrowing of the range of habitats used by the larvae, e.g. in saprophagous species. The three predictions are concerned with the trade-off between egg size and egg number, the direction of evolution of diet range, and the reasons why selection might favour a decreased range of food types.

### *The Trade-off Between Egg Size and Egg Number*

The basic postulate is that females of insect specialists on average encounter suitable oviposition sites at a lower rate than generalists, and hence on average ("ceteris paribus") lay fewer eggs in their lifetime. This allows them to increase the allocation of nutrients to each egg.

**Prediction 1:** *Specialists will lay fewer, larger eggs than generalists.*

A trade-off in nutrient allocation to various activities including reproduction has been a constant feature of theories of life-histories for many years. A recent review (Reznick 1985) concluded that, while some phenotypic studies have failed to detect any trade-offs, genetic studies have nearly always shown negative genetic correlations between fitness components (but see Rollo (1986) for counter arguments). However, some more recent genetic studies have shown positive, not negative, genetic correlations between stocks (e.g. in *Drosophila*; Giesel 1986). van Noordwijk and de Jong (1986) suggested that studies of interspecific differences often show trade-offs between fitness components, whereas studies of individual or population variation often do not: J. Stevens and P. Calow (personal communication) and van Noodwijk and de Jong (1986) independently suggested that a likely cause of this apparent paradox lies in individual differences in resource gathering abilities. Some individuals merely have more resources to allocate, generating positive genetic correlations. In this study, I assume that the principle of allocation is valid for syrphids in interspecific comparisons, giving data that support this contention.

The prediction that specialists will tend to lay fewer, larger eggs seems very simplistic. However, there is a good deal of direct and indirect support for it, and I cite three examples, below.

Lack of oviposition opportunities may play an important role in determining the size of butterfly eggs. Wiklund et al. (1987) suggest that sun-loving satyrid and pierid butterflies have plenty of time available for oviposition, since they spend much of their time in the sun; instead they are limited in fecundity by egg size, and therefore have reduced egg sizes to a physiological minimum. In shade-loving species of the same groups, there is nothing to be gained by decreasing egg size, since this will not lead to larger numbers of eggs being laid: lack of ovipositional opportunities limits reproductive success in these species. Egg size is related to body size in shade-loving species, but is a constant independent of body size in sun-loving species. Thus, where opportunities for oviposition are limited, eggs are larger in size.

Tauber and Tauber's (1987) study of two *Chrysopa* predators is also consistent with the prediction. The egg volume of the specialist *C. slossonae* is 42% larger than that of the generalist *C. quadripunctata.*

Finally, I have used data from the Drosophilidae (from Kambysellis and Heed 1971; Atkinson 1979) to produce Fig. 6.1, similar to, but not the same as, the

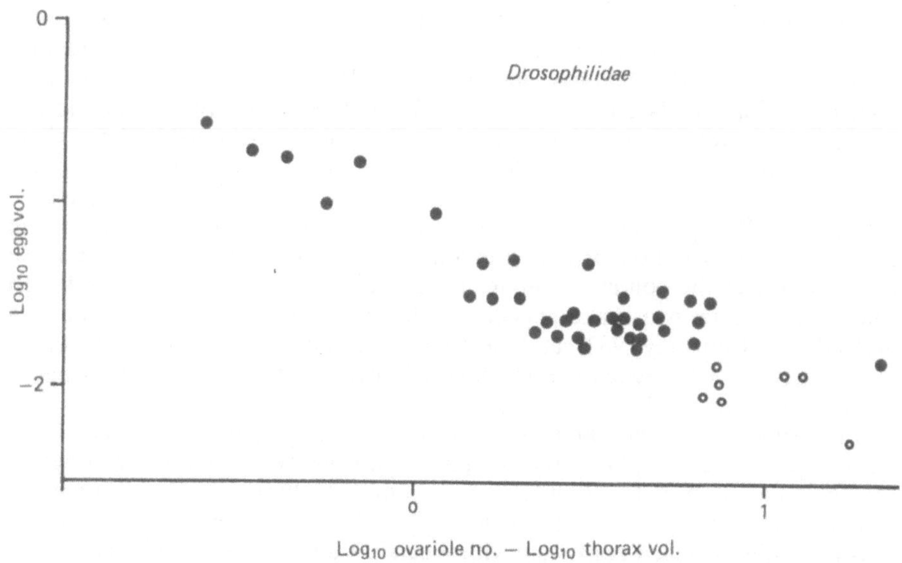


**Fig. 6.1.** Trade-off between egg size and egg number for the Drosophilidae. Egg numbers are represented by the number of ovarioles in the ovaries, and are adjusted for body size (= thorax length$^3$) as indicated. Data are taken from Kambysellis and Heed (1971: *solid circles*) and Atkinson (1979: *open circles*).

representation of the same data by Montague et al. (1981). The graph shows egg size plotted against the number of eggs per unit of body size, and clearly illustrates the trade-off between egg size and egg number in this group. The generalists such as *D. melanogaster* are at the low end of this spectrum, i.e. having many small eggs, whilst the specialized flower-breeding species are at the opposite end, with a few, very large eggs.

However, note that Courtney (1986) challenges the assumption that fecundity is a function of host encounter rate. Some insects fail to lay on many potential host plants even when encountered. This phenomenon is also encountered in parasitoids, and constitutes in part the evidence for "spreading of risk" theories of population regulation (see Hengeveld 1989)

### *The Direction of the Evolution of Prey Range*

Do specialized species always develop from more generalized ancestors, or vice versa; or is there no general direction of the evolution of prey range?

**Prediction 2a:** *Specialization is a derived or apomorphic trait.*
**Prediction 2b:** *Specialists tend to be larger in size than generalists.*

Evidence from studies of the host range of phytophagous insects is unclear on whether prediction 2a is reasonable. "It is commonly presumed that specialists evolve from generalists, and this is surely true in many instances. . . . Many, perhaps most, specialists arise from other specialists" (Futuyma and Moreno 1988).

Strong et al. (1984 p 176) suggested that monophagy is neither primitive nor derived, because the evidence shows that diet breadth can expand or contract within a taxon, according to the prevailing selective pressures. The two references that they cited in support of this statement were Benson's (1950) work on British sawflies and Holloway and Hebert's (1979) study of trends in host selection within the Macrolepidoptera. Neither of these studies used a phylogeny, or provided quantitative evidence for their viewpoints: presumably these viewpoints are beliefs derived from their extensive experience with these groups. While such feelings should be taken seriously, they clearly need to be tested.

Courtney (1986 p 97) considered five theories of diet width in some detail in the context of the evolution of diet breadth in pierid butterflies. Three of these theories predict the conditions under which polyphagy is favoured or selected against; the other two make specific predictions, one that specialization is a derived trait (Courtney 1982), and the other that, while high host densities favour monophagy, monophagous species generally evolve from other monophagous species (Futuyma 1983).

An alternative and venerable view is seen in the recent paper by Moran (1988) on host-plant use in aphids. By comparing morphs with identical genotypes, Moran found support for Simpson's (1944) opinion that specialization was an evolutionary dead end, and that it limited further evolutionary opportunities. This viewpoint is clearly consistent with specialization as a derived characteristic. In contrast, however, Thompson (1988b) recently looked for, and found, genetic variation in host choice in a monophagous swallowtail butterfly (*Papilio oregonius*), implying that the local monophagy of this species masks genetic variation for host choice that could allow host shifts if opportunities became available.

Tauber and Tauber (1987) suggested that in the predatory Chrysopidae the evidence indicates that prey specificity is a derived trait. I follow Futuyma and Moreno (1988) and take as an initial working hypothesis the prediction that specialization is a derived or apomorphic trait.

An additional prediction concerns the average size of specialists as against generalists. Most phyletic sequences have involved size increases rather than decreases (Hayami 1978). Also, older life-history theories such as r- and K-selection predict that larger species are more likely to be specialized (e.g. Horn 1978). I do not regard this prediction as a particularly strong one, but it is supported by data from the Drosophilidae used in Fig. 6.1: in this case, specialists are decidedly larger than generalists.

### *Selective Forces Involved in Specialization*

**Prediction 3a:** *The selective pressure for specialization involves at least in part an escape from competitive forces.*
**Prediction 3b:** *Populations of specialists should be less stable.*

We know very little about the selective forces that promote specialization. In the theory of host-plant selection, efforts have concentrated upon differences in the suitability of hosts, in part because of our knowledge of plant toxins. Berenbaum et al. (1986) provided clear evidence for a selective effect of herbivores on wild parsnips *Pastinaca sativa* (see Chap. 5); in the absence of herbivory, the heritable traits for resistance via toxin production are disadvantageous. Insect oviposition preferences can also be adapted to the suitability of the host as a larval food source (Via 1986; Thompson 1988a). This agrees with the idea that coevolution has occurred in insect–plant relationships.

However, in other instances adult oviposition preferences appear to be unrelated to larval performance (e.g. Courtney 1981; Wiklund 1975). In the *Dacus* fruit flies studied by Fitt (1986), larvae could develop successfully on most of the fruits offered as food, which included hosts not normally used in the wild. Diet breadth of these species in the wild, which ranged from a broadly generalized diet to monophagy, was a product of adult oviposition preferences and not larval physiological specialization.

In this paper I am concerned with diet breadth in predators and saprophagous species, where arguments about diet suitability may have less force. It has always been argued that the populations of generalists should be more stable than those of specialists because of the buffering effects of alternatives to fluctuating prey abundances (see Owen and Gilbert 1989). The populations of specialists are tied to fluctuations in the abundance of their food resources. If this is true, then generalists should be much more likely to compete with one another than with specialists, whilst specialists should rarely compete with each other. Part of the evolutionary pressure for specialization (if specialization is to be regarded as a derived trait) can then be seen as an escape from competition with generalists. Competition is not seen as an important general force in communities of phytophagous insects (Lawton and Strong 1981; Shorrocks et al. 1984), but here I am predicting that among predators and saprophages competition may be important only to part of the community, perhaps only a very small part.

Predators and parasitoids can also have a major impact on the evolution of host characteristics, and the concept of "predator-free space" (Jeffries and Lawton 1984) is likely to be important in the evolution of insect life-cycles (see e.g. Zwölfer 1975). Although not considered here, parasitoid-induced selective pressures are likely to have been important in the evolution of predatory syrphids (G. E. Rotheray & F. S. Gilbert, unpublished results).

## 6.3 The Hoverflies (Diptera, Syrphidae)

The hoverflies are a particularly suitable group for testing these predictions because of the wide variety of larval feeding habits found among substantial sections of the family. Most dipteran families have one very common larval feeding habit, usually with a few small groups of mavericks (e.g. amongst the Drosophilidae; Ashburner 1981). In contrast, larvae from large proportions of the 5378 recognized syrphid species (data on numbers of species are taken from Thompson (1989)) belong to the different major feeding groups of predators (40%), phytophages (16%), and a diverse array of saprophages (44%); the saprophage group contains species with a wide variety of feeding mechanisms, from dung-feeders to aquatic filter feeders (see Gilbert 1986). Larval habitats in the saprophage group can be divided roughly into generalized aquatic species (45% of saprophage species), species associated with trees and tree-holes (34%), or terrestrial rotting organic matter (19%) or inquilines in the nests of Aculeata (2%). Larvae belonging to each major group can be specialized or generalized: for example, in the predators some species are monophagous while others take a wide range of different aphids (Rotheray and Gilbert 1989); aquatic species can be specialized to relatively rare habitats (e.g. elm sap flows).

There are three subfamilies: the Syrphinae (all with predatory larvae), the Eristalinae (=Milesiinae) (with phytophagous or saprophagous larvae), and the Microdontinae (whose larvae are traditionally regarded as saprophages (Donisthorpe 1927) but are probably predators (Garnett et al. 1985) in the nests of ants). Microdons are relatively rare, and there are not enough data on them for any conclusions to be drawn: nearly all analysis and discussion will therefore be restricted to species of the other two subfamilies.

### 6.3.1 Species Considered

*Predators.* I consider here only data from the Syrphinae, whose 1644 recognized members are all predatory, as far as is known. I include in the Syrphinae the tribe Pipizini, a predatory group previously placed in the Eristalinae. The predatory habit of feeding on aphids has almost certainly evolved only once, and the Syrphinae probably form a monophyletic group (Rotheray and Gilbert 1989). There is a range of specialization in larval habitat, and also in prey range (these are of course not independent classifications). Prey range was scored on the basis of the experience of G. E. Rotheray's systematic sampling of aphid colonies for syrphid predators (see Owen and Gilbert 1989). Species from North America were assigned to categories where known (some are Holarctic, and I assume that feeding habits are the same across the entire range). Larval habitats were scored as follows: occurring in leaf litter; widely on herbs, shrubs and trees; only on herbs; only on trees; and occurring only in specialized enclosed habitats (in galls or underground). This order is assumed to represent increasing specialization of habitat, since it undoubtedly represents diminishing proportions of the available prey.

*Saprophages.* I use data from saprophages of any type, from the generalized aquatic filter feeders such as *Eristalis tenax* to the specialized scavengers in wasp and bee nests such as *Volucella*. In this case specialization of feeding habits is equivalent to a narrowing of larval habitat, and this is the basis upon which decisions were made about positions on the spectrum of specialization. Categories were as follows, in order of increasing specialization: generalized aquatic or semi-aquatic detritivores, tree-hole species, and inquilines.

*Phytophages.* All *Cheilosia* species were assumed to belong to this group, even though the larvae of some species feed in rotting fungi, presumably as saprophages. Not enough is known to be able to differentiate them without specific rearings, but the vast majority of the 386 species are almost certainly phytophagous (cf. Rotheray 1988). Species of *Merodon* and *Eumerus* from the Eumerini were also assumed to be phytophagous, although experiments with *Eumerus* have shown that fungal decay of plants is necessary before larvae can develop successfully (Creager and Spruijt 1935). Nothing adequate is known about specialization in phytophagous syrphids, and therefore the group is not subdivided further.

## 6.3.2 Data Used

The data are of three types: phenotypic characteristics, population densities, and a cladogram of generic evolution.

### *Size, and the Trade-off Between Egg Size and Number*

As part of a larger study on ecomorphological relationships within communities (Gilbert 1985a,b), I have measured a standard set of 25 variables on more than 250 species of hoverfly from Cambridge (UK), Maine, Florida, Arizona and Oregon (USA). Only three of these measures are used here to assess body size and reproductive strategy; complete data are available for 184 species. These data will shortly be published in full, so the raw values are not presented here. Of the 184 species, 67 are predators, for many of which the degree of larval specificity can be assessed.

Thorax volume is used as an index of body size. Three variables were actually measured on each fly: thorax width between the wing bases, thorax length from the extreme front edge of the prothorax (antepronotum) to the tip of the scutellum, and thorax height from the lowermost tip of the katepisternum (sternopleuron) (see Speight 1987). The three measures are multiplied together to give an index of thorax volume, and averaged over individuals to give a mean value for each species.

Egg sizes were measured on mature eggs dissected from gravid females; egg maturity is recognized by complete chorionic development. Maximum lengths ($L$) and widths ($W$) of ten eggs were measured for each female where possible, and egg volumes calculated from the equation for an ellipsoid:

$$\text{vol} = (4/3)\ \pi\ (L/2)\ (W/2)^2$$

The number of eggs laid during the lifetime of an individual female is difficult and time-consuming to measure on a large number of species. In parasitoids, there is a good correlation between lifetime fecundity and the number of ovarioles in the ovaries (Price 1975), and ovariole number is widely recognized as an index of potential fecundity. I therefore decided to use the numbers of ovarioles in females as an index of egg number.

In saprophagous species that often lay their eggs in a large batch, the number of ovarioles sets a maximum on batch size, since no mature eggs are retained in the oviducts. Predatory Syrphinae usually lay only a single egg at a time, and it is not known whether they lay all their mature eggs during a single day. Unlike the synchronous development of ovarioles in saprophagous species, ovarioles of syrphines develop asynchronously and only a few eggs are mature at any one time (F. S. Gilbert, unpublished results).

There is a clear allometry of reproductive effort, here taken to be the maximum clutch volume (ovariole number × egg volume). Fig. 6.2 shows the relationship

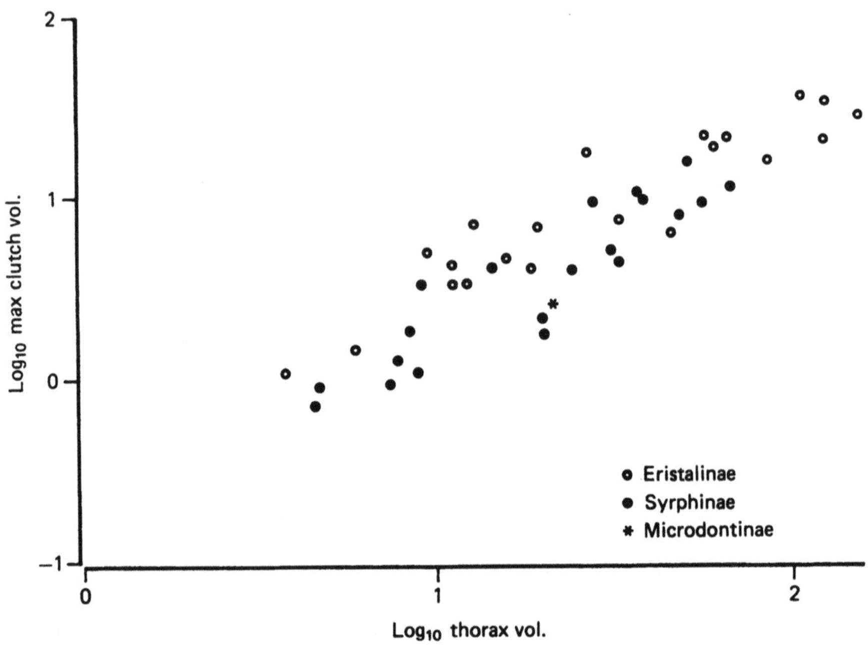


**Fig. 6.2.** Relationship between the maximum volume of eggs that could be produced at one time (= number of ovarioles × egg volume) and body size. There is no significant difference between the slopes of the regression lines of the Syrphinae and Eristalinae, but the elevations are significantly different (see the text).

between clutch volume and body size for data from one site (Maine), and it is clear that size exerts a strong influence. There is no difference between the slopes of the regression lines of the Eristalinae and the Syrphinae ($F_{1,123} = 1.0$, n.s.), but the elevations are significantly different ($F_{1,124} = 17.1$, $P << 0.001$). The adjusted mean clutch volumes differ by 0.16, and since this is a logarithmic scale, this implies that the mean clutch volume of the Syrphinae is only 69% of that of the Eristalinae. This is interesting, since it implies little difference in reproductive allocation between the synchronous batch layers (Eristalinae) and those that lay only a single egg at a time, maturing eggs asynchronously (Syrphinae); I conclude from this that both subfamilies have a similar basic allocation to reproduction, which is to mature one batch of eggs synchronously, even though many do not. Why should this be? It is probably because adults do not live long in the wild, and I have only rarely found evidence for more than one egg being laid from each ovariole (F. S. Gilbert, unpublished results).

The trade-off between egg size and number is most clearly seen by plotting data from syrphids in the same way as for the Drosophilidae. Fig. 6.3 shows the number of ovarioles per unit thorax volume plotted against egg volume (in a log–log plot). Clearly larger eggs imply smaller numbers of ovarioles for a given body size. The extremes of this line run from the huge (2.5 mm long) eggs of *Volucella* species to the numerous and tiny eggs (0.6 mm) of the Xylotini (e.g. *Xylota*, *Criorhina*). The predatory Syrphinae encompass a narrower range of size-adjusted egg numbers and egg sizes. Once again there is no difference between the slopes of the regression lines for the Eristalinae and the Syrphinae ($F_{1,123} = 2.65$, n.s.), but the elevations are significantly different ($F_{1,124} = 16.5$, $P << 0.001$): the different elevations imply that the Eristalinae lay eggs that are

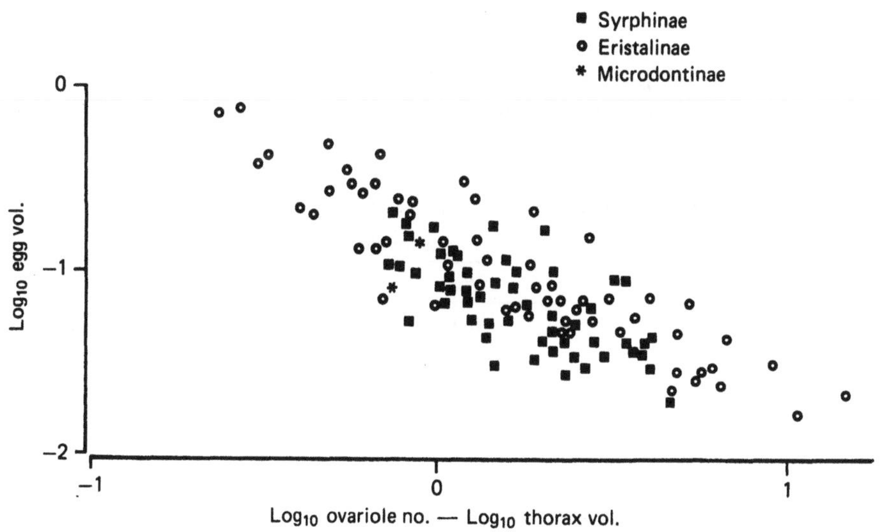


**Fig. 6.3.** Trade-off between egg size and egg number for adult Syrphidae. Egg numbers are denoted by the number of ovarioles in the ovaries, adjusted for body size in the way indicated. Regression lines for the Syrphinae and Eristalinae do not differ in slopes, but differ significantly in elevation (see the text).

41% larger for a standardized fecundity. These regression lines are very similar to that of the Drosophilidae (Fig. 6.1), with an identical slope ($F_{2,164}$ = 1.86, n.s.), but different intercept ($F_{2,166}$ = 30.1, $P << 0.001$). Thus the trade-off between fecundity and egg size is very similar in these different dipteran families.

With these data, body size and the trade-off between egg size and number are confounded. It is not adequate merely to divide one variable by body size to remove the effects of body size. Using multiple regression one can look at the relationship between the two predictor variables with relatively low intraspecific variance (body size and egg size) and ovariole number, which varies much more between individuals of one species (F. S. Gilbert, unpublished results). This is very successful ($R$ = 0.82, $F_{2,127}$ = 131.6, $P << 0.001$), and the relationship is:

$$\log_{10} (\text{ovariole number}) = 0.83 \log_{10} (\text{size}) - 0.70 \log_{10} (\text{egg size}) - 0.34$$

However, principal components analysis (PCA) is a much better way of separating the two influences of body size and the egg size–egg number trade-off. PCA is a method of taking intercorrelated, and producing uncorrelated, variables. Table 6.1 shows the results of a PCA of the three log-transformed variables. The first axis explains 60% of the variation in the data, and has high positive loadings on egg volume and number, and very high with thorax volume: it is clearly a body size axis. The second axis explains a further 36% of the variation, and has a very low loading on thorax volume, but loads positively with egg volume, and negatively with ovariole number. Thus, from three intercorrelated variables, PCA produces two uncorrelated ones that explain 96% of the original variance: scores along the first axis are a measure of body size, whereas scores along the second axis measure the trade-off between egg size and number, independent of body size. I therefore use these PCA scores to test predictions about the effects of specialization, and I label scores along axis 1 (PCA 1) as measuring "size", and scores along axis 2 (PCA 2) as measuring the "reproductive strategy", the position of the species along the egg size–number trade-off. To test for differences between groups,I use a one-way analysis of variance using program P7D of the BMDP package of programs (Dixon 1983): this uses Levene's test to check whether variances are homogeneous, and if not, calculates Welch's $F$-ratio in place of the usual $F$-ratio.

**Table 6.1.** Results of a principal components analysis of size and reproductive variables

| Original variables | Factor loadings | |
|---|---|---|
| | Axis 1 | Axis 2 |
| Thorax volume | 0.97 | 0.00[a] |
| Egg volume | 0.65 | 0.74 |
| Number of ovarioles | 0.65 | −0.74 |
| **Order** | **Eigenvalues** | **% variance** |
| 1 | 1.7888 | 60 |
| 2 | 1.1044 | 36 |
| 3 | 0.1067 | 4 |

[a] <0.25, and set to 0.0 by the BMDP program.

Fig. 6.4 uses these PCA scores to look for differences in size and reproductive strategy between species in the three major larval feeding categories. There are

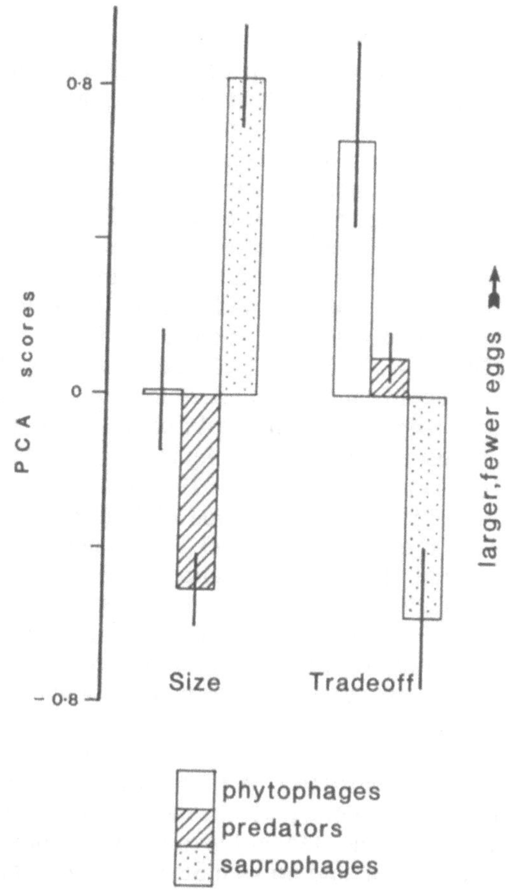


**Fig. 6.4.** Mean values for "size" (scores along PCA1) and "reproductive strategy" (scores along PCA2) for the major feeding groups of larval Syrphidae. For details, see the text.

very clear and significant differences between the categories in both size ($F_{2,124}$ = 32.1, $P<<0.001$) and reproductive strategy (Welch $F_{2,42}$ = 8.6, $P<0.001$). Phytophagous species are of average size, but have relatively few, large eggs; predators are small with an average number of average-sized eggs; saprophagous species are large, with a large number of small eggs.

## *Phylogeny of the Syrphidae*

To test quantitatively whether specialization is a derived trait, a well-differentiated phylogeny is needed, ideally at the species level. This is not available for any syrphid group, and therefore we cannot make a quantitative test of the prediction at the species level. However, my colleague Graham Rotheray and I have developed a generic-level phylogeny for the predatory Syrphinae (Rotheray and Gilbert 1989), which does allow a certain amount of analysis to be done.

The phylogeny is based upon larval morphological characters, such as details of the posterior respiratory process, that are constant across genera. We used a numerical cladistic technique (parsimony) implemented by the computer package PAUP (Swofford 1985) to generate the most likely cladogram from the characters that were scored. Fig. 6.5 shows the resulting cladogram, which incidentally confirms in many details opinions about the phylogeny of those who have worked on adult or larval Syrphinae.

I regard Fig. 6.5 as a reasonably accurate phylogeny, for two reasons. Firstly, the constancy of characters within genera also confirms nearly all the generic boundaries that taxonomists working with adults have spent decades reorganizing, refining and finally agreeing. This engenders confidence in the use of larval characters to indicate relationships. Secondly, I believe that larval characters clearly contain a great deal more phylogenetic information than do adult characters (Rotheray and Gilbert 1989). We are currently working on generic-level phylogeny of the whole family from larval characters (G. E. Rotheray, F. S. Gilbert, C. T. Maier, F. C. Thompson and J. C. Hartley, unpublished results).

I identify the quantitative influence of phylogeny on size and reproductive traits using the technique of Cheverud et al. (1985). There has been a good deal of interest in exploring techniques for assessing the role of phylogeny (Stearns 1983; Harvey and Clutton-Brock 1985; Felsenstein 1985; Derrickson and Ricklefs 1988; Bell 1989), but the method of Cheverud et al. (1985) is the only one with an explicit quantitative model. It uses network autocorrelations and an attempt to describe the genetics of ancestor–descendant relationships in the same way as Mendel's laws describe the genetics of parent–offspring relationships.

The method requires a species × species matrix of the degree of relatedness. With a reasonably well-differentiated phylogeny, one can estimate this relatedness matrix by counting the number of hypothetical ancestors ("nodes") between pairs of species. This is the approach adopted here. The technique is only valid if the evolution of taxa occurs at a constant rate.

I am indebted to J. M. Cheverud for sending me the program for calculating the relevant parameters for the model, but note that I have been unable to reproduce accurately the standard errors (SE) of phylogenetic autocorrelations using his sample data, and no such SE values are reported here.

## *Population density and stability*

A third type of data used in testing the predictions concerns population fluctuations and competition between species. Manipulative experiments to test for the occurrence of competitive forces among adults are very difficut if not impossible to do with a highly mobile group such as the hoverflies. The large number of species involved make it impractical to do manipulations of larval densities, even if we knew where they all were in the habitat. An indirect method is required that analyses long-term population data to look for reciprocal fluctuations in density between competitors. As a by-product of such data, we can assess the stability of populations and thereby test another of the predictions (i.e. that generalists will have more stable populations).

Jennifer Owen has trapped insects in her suburban garden in Leicester (UK) with a Malaise trap every week for the past 17 years, and the Syrphidae are one of the groups that she has already identified to species level. This constitutes one of the finest and most detailed sets of long-term population data for an insect community.

The analysis of stability of these populations involves using an index of

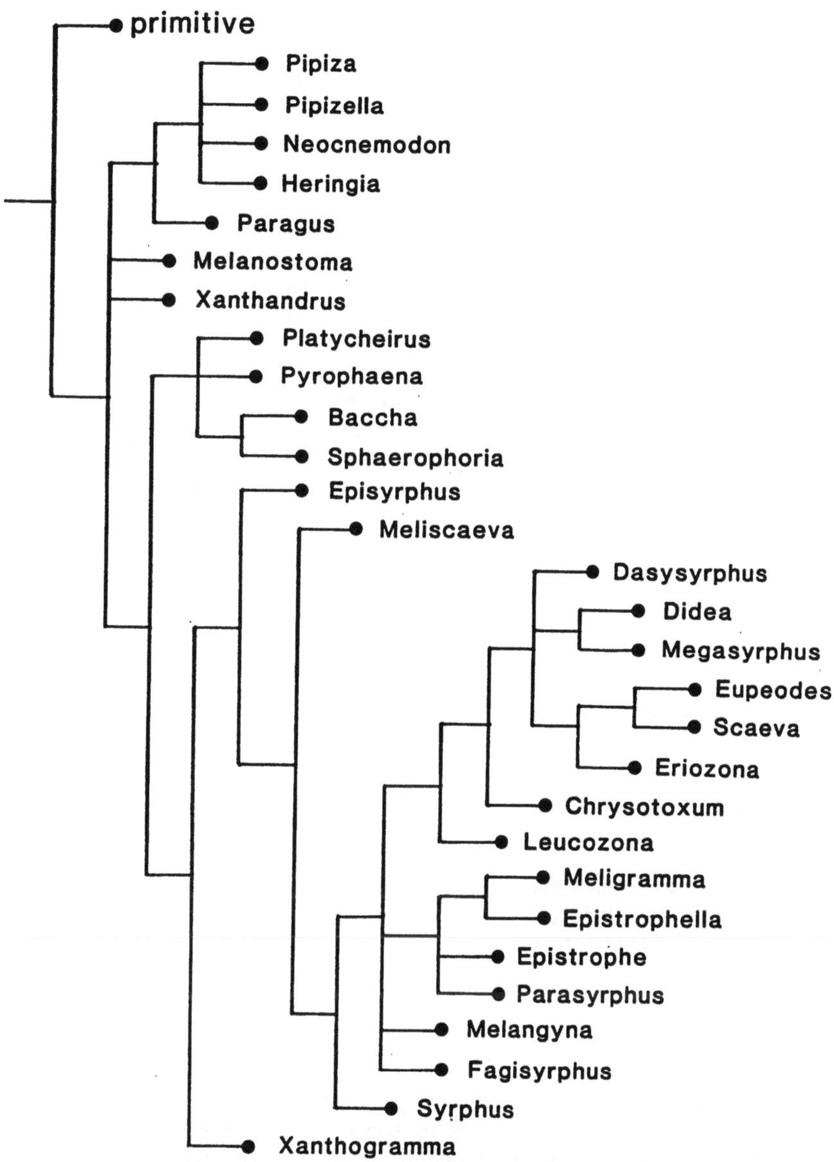


**Fig. 6.5.** Cladogram of the genera of the Syrphinae (redrawn from Rotheray and Gilbert 1989). The cladogram results from a consensus tree of the multiple equally parsimonious trees from the PAUP package (for details, see Rotheray and Gilbert 1989).

stability, the standard deviation of log-transformed annual totals; the detailed analysis has recently been published (Owen and Gilbert 1989). Here I test whether groups of differing specificity have different population stabilities, using a one-way analysis of variance (ANOVA).

I test for competition using the similarity measure from correspondence analysis, $\chi^2$ similarity. Correspondence analysis is a multivariate technique for analysing contingency-table data, and is ideal for this purpose since it compares the profiles of annual abundances between pairs of species using what is very similar to a calculation of $\chi^2$. It thus automatically allows for differences in the relative abundances of species, and in the overall suitability of different years for syrphids. The test assumes that no time delays are involved in the competition. A large $\chi^2$ similarity between two species implies reciprocal deviations from the expected abundances generated from the marginal totals of years and species.

To test for an unusually large $\chi^2$ similarity, the mean of $\chi^2$ similarities between all pairs of species of a subgroup of the data (e.g. generalists) is compared to 100 means of the same subgroup generated from a random model, one that randomizes the differences of the observed numbers from the "expected" numbers generated from the marginal totals. This randomization of residuals will disrupt any putative reciprocal fluctuations in density. If the observed mean is in the top five of the random means, I take it as being a significantly large mean value for $\chi^2$, and therefore as an indicator of the reciprocal fluctuations in density that would indicate the action of competitive forces. Full details of this approach are given by Gilbert and Owen (1990). It remains possible that groups consistently respond differently to environmental variables, such that resulting covariation has nothing to do with competitive effects. The fact that these are very closely related species taxonomically should minimize this confounding effect.

## 6.4 Testing the Predictions

**Prediction 1:** *Specialists should lay fewer, larger eggs.*
**Prediction 2b:** *Specialists should tend to be larger.*

I test these predictions separately on saprophages and predators. Species in both groups were assigned to different categories according to larval habitat (saprophages) or prey range (predators).

### *Saprophages*

Species were assigned to three groups: generalized detritivores, tree-hole detritivores, and inquilines in the nests of social Hymenoptera. Undoubtedly among the "generalized detritivore" group, there are species with narrower niches than others. Because of the lack of detailed data for the majority of species, I am unable to divide into a finer classification. I ordered these groups a priori into a sequence of increasing specialization, from generalized detritivores to inquilines.

Fig. 6.6 compares the sizes and reproductive strategies of each group, and for comparison also gives data for the phytophages (where no division into specialists and generalists is possible). All the saprophages tend to be larger than the average for all syrphids, and hence their scores are all positive, denoting relatively large sizes. The more specialized groups living in tree-holes or in nests of social Hymenoptera are probably larger than those of generalized detritivores

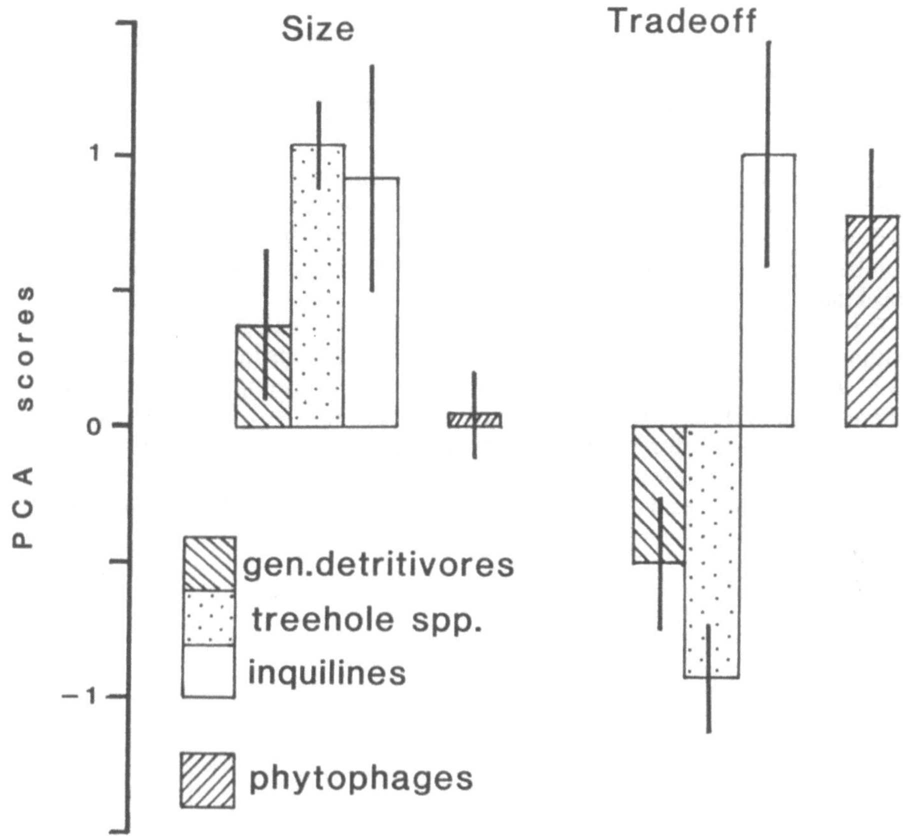


**Fig. 6.6.** Mean values (± SE) for "size" and "reproductive strategy" (see the text) for adult syrphids with phytophagous and saprophagous larvae. The saprophagous group has been split into three types according to the degree of specialization of larval habitat. gen., generalized.

($F_{2,39}$ = 2.58, $P$ = 0.09), conforming to the predicted pattern. Also the most specialized group, the inquilines, have very large eggs, while the other two groups are not different from one another ($F_{2,39} = 7.04$, $P < 0.003$). Again this tends to support the prediction that specialized species should lay fewer, larger eggs.

*Predators*

Predators are divided into five categories according to their larval habitat, and this is also a prey-range classification:

1. Leaf litter species probably feed on many different kinds of arthropod in the leaf litter, but appear to prefer aphids (e.g. *Melanostoma*)
2. Some species are generalized aphid predators, to be found feeding on many types of aphids on trees, shrubs and herbs (e.g. *Syrphus ribesii*)
3. Some occur only on herbs, again feeding only on aphids but still of many types (e.g. *Sphaerophoria*)

4. Other species are more restricted, being found only in tree-aphid colonies (e.g. *Dasysyrphus*)
5. Some species feed only on aphids in specialized enclosed habitats such as galls or underground (e.g. *Heringia*)

This order is assumed to be an order of increasing specialization in habitat, on the basis of field experience. It is an a priori order, but one that I am aware is open to dispute. Data are needed to quantify the relationship between habitat type and numbers and types of aphid colonies: I am unaware of the existence of any data of this kind.

Fig. 6.7 shows the average sizes and reproductive strategies of these groups. Although there are strong differences between the groups in size (Welch $F_{4,20} = 29.6$, $P << 0.001$), there is no obvious pattern conforming to the prediction. There are no differences in reproductive strategy ($F_{4,61} = 1.25$, n.s.). Therefore there is no support for the prediction that more specialized predators will be larger, and will lay fewer, larger eggs.

Among the species where we are certain that aphids are the only acceptable

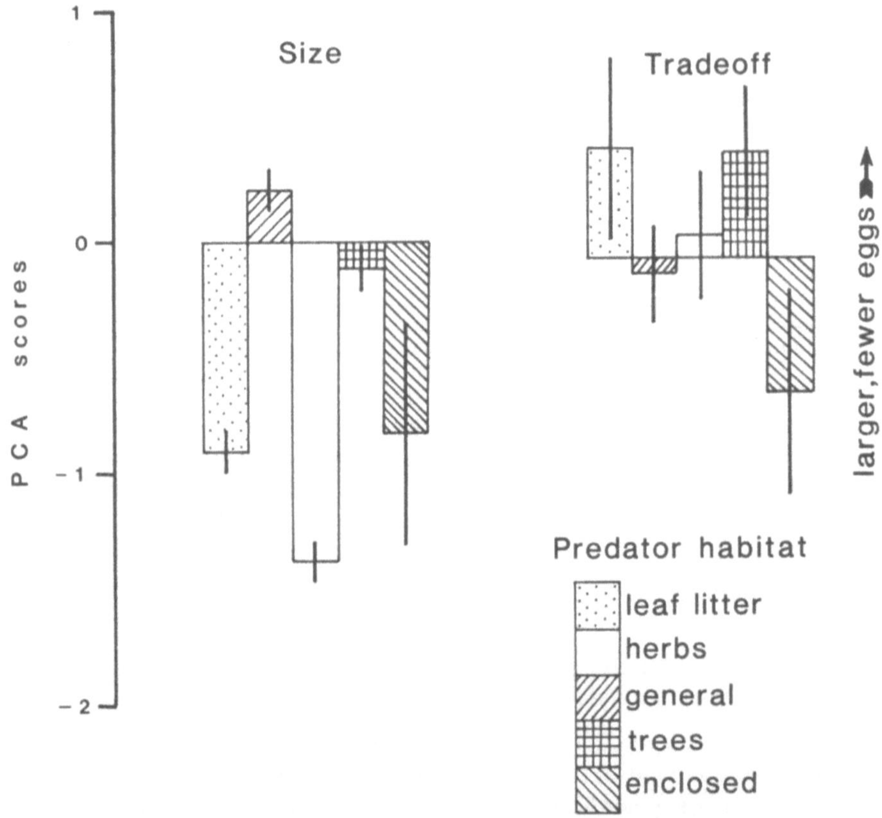


**Fig. 6.7.** Mean values (± SE) for "size" and "reproductive strategy" (see the text) for adult syrphids with predatory larvae. The data are split into groups according to larval habitat, and are displayed in order of increasing specialization of larval habitat.

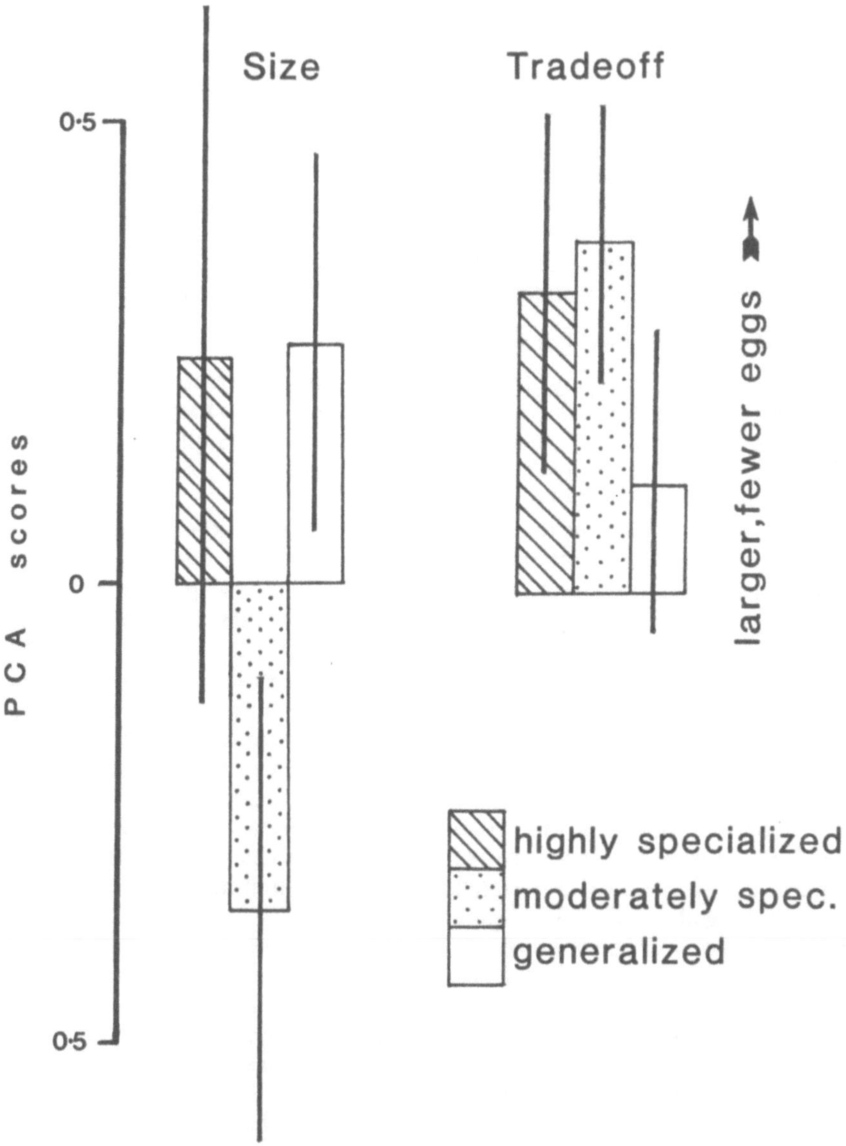


**Fig. 6.8.** Mean values (± SE) for "size" and "reproductive strategy" (see the text) for adult syrphids with predatory larvae whose range of prey within habitats can be assessed (see the text). The data are split into three groups differing in their degree of specialization, and are displayed in order of decreasing specialization. spec., specialist.

food, systematic sampling allows an accurate assessment of the prey range, from extreme specialists (monophages), to moderate specialists (oligophages), to generalists (polyphages). Details of the way in which these assessments were done, and the species in each group, are given by Owen and Gilbert (1989). Fig. 6.8 gives the mean sizes and reproductive strategies for the species of each category: there are no significant differences between groups in size (Welch $F_{2,6}$ = 1.81, n.s.) or reproductive strategy (Welch $F_{2,6}$ = 0.64, n.s.). Trends in the data do support the main prediction that specialists should lay fewer, larger eggs, but the variance is high, probably because of the low number of species that can definitely be assigned to the categories.

In conclusion, the prediction that more specialized species will lay fewer, larger eggs is supported by data from non-predators, but not from predators, although some of the data suggest that, given larger sample sizes, support might be forthcoming. The subsidiary prediction that specialists will also be large is also supported by evidence from non-predators, but not that from predators.

**Prediction 2a:** *Specialization is a derived trait.*

I have only a generic phylogeny of predators. One cannot characterize the degree of specialization of a genus when knowledge is patchy, and so I am unable to test this prediction quantitatively. Work is currently in progress on a species-level phylogeny of the genus *Platycheirus*, and, when complete, a more rigorous quantitative test can be carried out.

In spite of this, it is obvious that the whole trend of evolution has been one of increasing specialization (Rotheray and Gilbert 1989). All the plesiomorphic basal species are generalists (e.g. *Melanostoma*), probably living in the leaf litter as general zoophages, although taking aphids by choice. As one comes up the phylogenetic tree on the main "branch", so the species become grassland generalists (e.g. *Sphaerophoria*, *Platycheirus*), then generalist aphidophages on herbs, shrubs and trees (e.g. *Episyrphus*), and finally increasingly restricted to arboreal aphids (e.g. *Dasysyrphus*). As always, the pattern is not completely regular (e.g. the apparently highly specialized *Xanthandrus* (feeding on gregarious Lepidoptera) and *Xanthogramma* (feeding on aphids associated with ants?)). On the secondary branch of the evolutionary tree a parallel specialization has occurred: *Paragus* species appear to be herb-layer generalists, whilst members of the Pipizini are arboreal or subterranean specialists, often highly specialized to feed within the galls produced by aphids.

Thus, the history of the evolution of the predatory syrphids appears to be one of increasing specialization, from feeding generally on invertebrates to aphidophagy, from ground-layer aphids to tree aphids, from generalists to specialists.

I am able to use generic-level data to assess the role of phylogeny in influencing size and reproductive strategy, since 23 generic mean values for the Syrphinae can be calculated and used in Cheverud et al.'s (1985) model. The calculated value of the phylogenetic autocorrelation coefficient measures the degree to which the phylogenetic relatedness matrix accounts for variation in the trait being considered.

For size (scores on PCA1), the phylogenetic autocorrelation is 0.62, a very high value. This implies that more closely related species are more alike in size. From the pattern of mean sizes, phylogeny is clearly correlated with a general increase in body size, as predicted. The genera with the largest mean sizes are all towards the top of the tree (e.g. *Scaeva*, *Didea*, *Eriozona*, *Chrysotoxum*), whilst the smallest ones are basal (*Melanostoma*, *Platycheirus*, *Paragus* and all the Pipizini).

For scores on PCA2, i.e. position on the trade-off between egg size and number, there is an equally high phylogenetic autocorrelation coefficient, but it is negative (−0.62). Cheverud and Dow (1985) measured similar correlations in morphology for social groups of rhesus monkeys produced by fission, and also discovered negative values. They suggest that negative values should be rare, because this means that phylogenetic connections between genera produce dissimilarity, i.e. more closely related genera are more dissimilar. This has obvious implications for the process of fission itself, which must be one that divides a genus according to the most dissimilar elements. Ecological factors could be invoked here, e.g. character displacement, but this is mere speculation. Until more such analyses are carried out, we cannot know the full implications of negative phylogenetic autocorrelations.

**Prediction 3a:** *Generalists should compete, but specialists should not.*

**Prediction 3b:** *Generalists should have more stable populations.*

A test of the second part of this prediction is relatively straightforward. Fig. 6.9 shows the average stabilities of generalists, moderate specialists, and extreme specialists. Generalists do not have more stable populations, but are just as variable as extreme specialists. Populations of moderate specialists are more stable than either of the other two categories (Owen and Gilbert 1989).

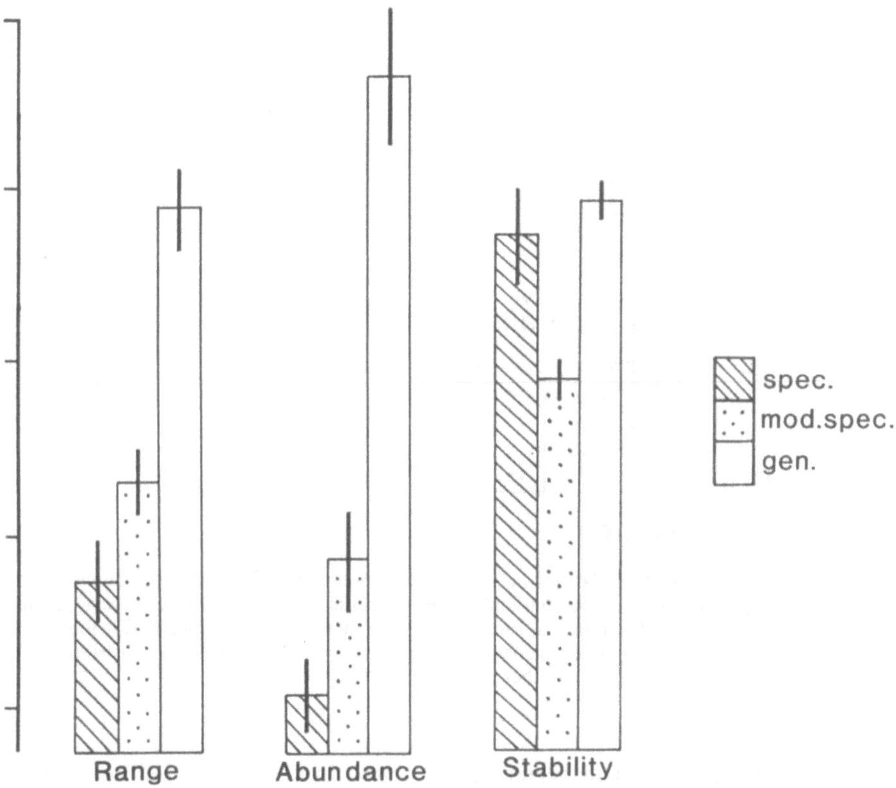


**Fig. 6.9.** Mean values (± SE) of the range (distribution of 10 km × 10 km squares: see Owen and Gilbert 1989), abundance (log annual numbers), and stability (S.D. of log census numbers) for populations of adult syrphids with predatory larvae whose range of prey can be assessed. Three groups are shown in order of decreasing degree of specialization. spec., specialist; mod., moderately; gen., generalist.

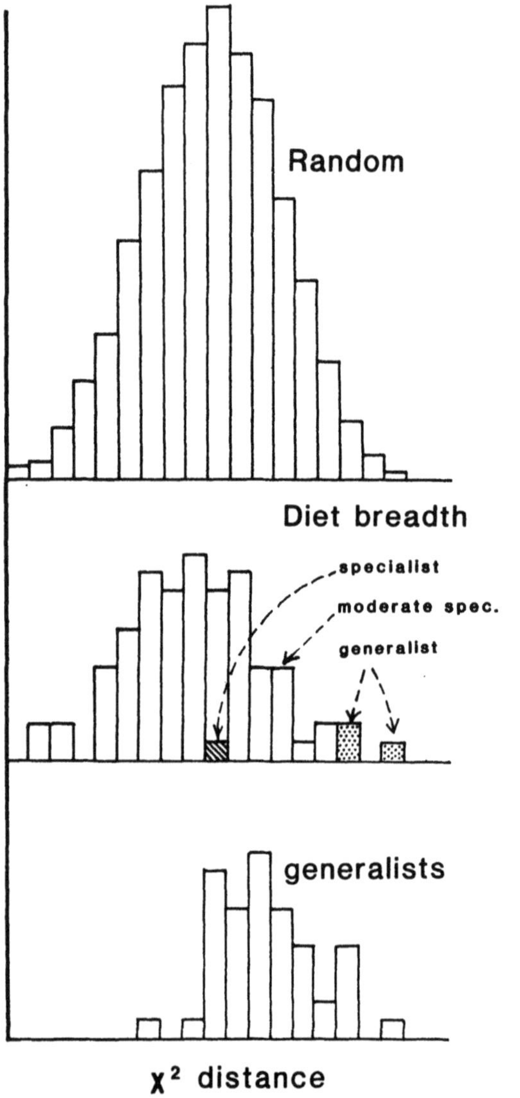


**Fig. 6.10.** Distribution of $\chi^2$ similarities in abundance between species in different categories of predatory larval specialization, and for randomly reshuffled residual abundances (see the text). *Upper distribution*, randomly reshuffled residuals for all species together; *middle distribution*, distribution of similarities *within* groups (e.g. between generalist and generalist); *lower distribution*, distribution of similarities between generalists and all other predators. spec., specialist.

Why should this be? The argument advanced in Owen and Gilbert (1989) mirrors other similar explanations (see Owen and Gilbert 1989) in suggesting that it is possible that the populations of extreme specialists are tied to fluctuations in their food supply, whilst populations of generalists take advantage of outbreaks of the commonest aphid species, which may fluctuate greatly from year to year (Taylor 1984). It is the moderately specialized predators that regularly switch to the common prey type of the range that they accept, and thus even out yearly fluctuations in food availability.

Testing for competition involves comparing the observed mean $\chi^2$ similarity of the group of species (e.g. generalists) against 100 random mean values generated as described above (see also Gilbert and Owen 1990). We expect the mean $\chi^2$ similarity to be unusually high between generalists, but not between either moderate or extreme specialists. This is exactly the pattern found. Fig. 6.10 shows the distribution of similarities. Only two of the 100 random mean similarities were larger than the observed mean similarity between the generalists. Mean similarities were not unusually large between moderate specialists, or between extreme specialists, or even between generalists and either type of specialist. This pattern is interpreted as implying that generalists compete with one another, but not with other predators. This is consistent with the prediction that part of the impetus towards specialization is avoidance of competition with generalists (although this implies no causality). Because generalists are "primitive" in a phylogenetic sense, this may be part of the reason why the history of the evolution of the predatory Syrphinae seems to have been one of increasing specialization.

## 6.5 Discussion

Futuyma and Moreno's (1988) recent review called for more phylogenetic information in the search for a general explanation for specialization. I concur strongly: a phylogeny imposes a pattern and an order to ecological features, narrowing the focus of hypothesis testing to particular sister groups.

The results of this comparative study of species of the Syrphidae show that there are clear patterns of specialization within predators that require explanation. Most of the predictions about size and reproductive strategy hold for non-predators, but fewer are supported by data from predators. In more detail:

1. Specialization of habitat in saprophagous species involves an increase in size, and a reproductive strategy of fewer, larger eggs, just as predicted. In predatory species, however, these predictions are not supported when one compares across different larval habitats (and assuming that these represent different degrees of specialization). For those predators where prey range is known with some confidence, there is also no support for the predictions, although specialized species do tend to have fewer, larger eggs.

2. I predicted that specialization should be an apomorphic character: the phylogenetic history of predatory syrphids is indeed one of increasing specialization. Plesiomorphic genera are leaf-litter or herb-layer generalists, and the more apomorphic the genus, the more its constituent species are confined in their prey range to a much narrower range of arboreal aphids.

Generic average size is strongly associated with this pattern of specialization as measured by Cheverud et al.'s (1985) method. Size increases in more apomorphic genera.

The average reproductive strategy of genera is also strongly associated with the pattern of phylogenetic relationships in that more closely related genera have more divergent reproductive strategies. This implies that the founding of new genera involves a splitting of the old genus into components with highly divergent reproductive patterns.

3. The prediction was made that generalists should compete with one another, but specialists should not. The suggestion from this, if supported, is that at least part of the selective pressure for greater specialization (cf. point 2, above) involves an escape from competition. This asymmetry was evident in long-term population data from an urban site in the UK: there was evidence that generalists do compete, resulting in reciprocal fluctuations in density, but there was no evidence that specialists compete either between themselves or with generalists.

Generalists are predicted to have more stable population densities because they can switch to the commonest prey types of any one year. This hypothesis is rejected: generalists and extreme specialists have equally unstable populations, while moderate specialists have relatively stable population densities.

This study demonstrates the utility of using a taxon such as the Syrphidae that contains examples of all types of life-cycle strategy. Detailed work on specialization of predatory syrphids is under way and, in the light of further refinement of the phylogenetic relationships between species, should bear fruit in the understanding of the basis of their specialization.

*Acknowledgements*. I dedicate this paper to my parents to acknowledge in a small way their immense influence and support during my education, and to my wife, Hilary, for her expert assistance, support and encouragement. I am indebted to my colleagues John Birks, Graham Rotheray and Jennifer Owen, without whose dedication and expertise this work would be impossible. Peter Stiling and two referees were very helpful in improving the text.

I thank Sally Corbet for her inspiring leadership in research. This work was greatly facilitated by the Harkness Foundation of New York, who by awarding me a Fellowship during 1982–1983 provided a superb opportunity to do field research where I wanted in the United States, and the marvellous gift of travel to me and my family. Special thanks go to the ecologists who made our visit possible by accepting me in their laboratories, and who provided a highly rewarding and stimulating environment: Bernd Heinrich (Vermont and Maine), Dan Simberloff and Don Strong (Florida), Stan Faeth (Arizona) and Bill Bradshaw (Oregon). I thank Bill Bradshaw also for stimulating my interest in life-histories; and Bill and Val Brown for asking me to speak at the Vancouver conference and to edit this volume. I am pleased to thank Chris Thompson for a preprint of his magnificent *World Catalogue*.

## References

Ashburner, M (1981) Entomophagous and other bizarre Drosophilidae. In: Ashburner M, Carson HL, Thompson JN (eds) The genetics and biology of *Drosophila*, vol. 3a. Academic Press, New York, pp 375–429

Atkinson W (1979) A field investigation of larval competition in domestic *Drosophila*. J Anim Ecol 48: 91–102

Bell G (1989) A comparative method. Am Nat 133: 553–571

Benson RB (1950) An introduction to the natural history of British sawflies (Hymenoptera, Symphyta). Trans Soc Br Entomol 10: 46–142

Berenbaum MR, Zangerl AR, Nitao JK (1986) Constraints on chemical coevolution: wild parsnips and the parsnip webworm. Evolution 40: 1215–1228

Bristowe CM (1988) What makes a predator specialise? Trends Ecol Evol 3: 1–2

Cheverud JM, Dow MM (1985) An autocorrelation analysis of genetic variation due to lineal fission in social groups of Rhesus macaques. Am J Phys Anthropol 67: 113–129
Cheverud JM, Dow MM, Leutenegger W (1985) The quantitative assessment of phylogenetic constraints in comparative analyses: sexual dimorphism in body weight among primates. Evolution 39: 1335–1351
Courtney SP (1981) Coevolution of pierid butterflies and their cruciferous foodplants. III. *Anthocharis cardamines* (L.). Survival, development and oviposition on different hostplants. Oecologia 51: 91–96
Courtney, SP (1982) Coevolution of pierid butterflies and their cruciferous foodplants. V. Habitat selection, community structure and speciation. Oecologia 54: 101–107
Courtney SP (1986) The ecology of pierid butterflies: dynamics and interactions. Adv Ecol Res 15: 51–151
Creager DB, Spruijt FJ (1935) The relation of certain fungi to larval development of *Eumerus tuberculatus* Rond. Ann Entomol Soc Am 28: 425–437
Derrickson EM, Ricklefs RE (1988) Taxon-dependent diversification of life-history traits and the perception of phylogenetic constraints. Funct Ecol 2: 417–423
Dixon WJ (1983) BMDP statistical software: manual. University of California Press, Berkeley, CA
Donisthorpe H (1927) The guests of British ants. Routledge & Sons, London
Felsenstein J (1985) Phylogenies and the comparative method. Am Nat 125: 1–15
Fitt GP (1986) The roles of adult and larval specialization in limiting the occurrence of five species of *Dacus* (Diptera: Tephritidae) in cultivated fruits. Oecologia 69: 101–109
Fox LR, Morrow PA (1981) Specialization: species property or local phenomenon? Science 211: 887–893
Futuyma DJ (1983) Selective factors in the evolution of host choice by phytophagous insects. In: Ahmad S (ed) Herbivorous insects: host seeking behavior and mechanisms. Academic Press, New York, pp 227–244
Futuyma DJ, Moreno G (1988) The evolution of ecological specialization. Annu Rev Ecol Syst 19: 207–233
Futuyma DJ, Peterson SC (1985) Genetic variation in the use of resources by insects. Annu Rev Entomol 30: 217–238
Garnett WB, Akre RD, Sehlke G (1985) Cocoon mimicry and predation by myrmecophilous Diptera (Diptera, Syrphidae). Fl Entomol 68: 615–621
Giesel JT (1986) Genetic correlation structure of life-history variables in outbred, wild *Drosophila melanogaster*: effects of photoperiod regimen. Am Nat 128: 593–603
Gilbert FS (1981) Foraging ecology of hoverflies (Diptera, Syrphidae): morphology of the mouthparts in relation to feeding on nectar and pollen in some common urban species. Ecol Entomol 6: 249–262
Gilbert FS (1985a) Ecomorphology of hoverflies (Diptera, Syrphidae). Proc R Soc Lond [Biol] 224: 91–105
Gilbert FS (1985b) Morphological relationships in hoverflies (Diptera, Syrphidae). Proc R Soc Lond [Biol] 224: 79–90
Gilbert FS (1986) Hoverflies (Cambridge Naturalists Handbooks 5). Cambridge University Press, Cambridge
Gilbert FS, Owen J (1990) Morphology and competition in hoverfly communities. J Anim Ecol 59: 21–40
Harvey PH, Clutton-Brock T (1985) Life history variation in primates. Evolution 39: 559–581
Hayami I (1978) Notes on the rates and patterns of size change in evolution. Paleobiology 4: 252–260
Hengeveld R (1989) Caught in an ecological web. Oikos 54: 15–22
Holloway JD, Hebert PDN (1979) Ecological and taxonomic trends in macrolepidopteran host plant selection. Biol J Linn Soc 11: 229–251
Horn H (1978) Optimal tactics of reproduction and life-history. In: Krebs JR, Davies NB (eds) Behavioural ecology. Blackwell Scientific Publishers, Oxford, pp 411–429
Jeffries MJ, Lawton JH (1984) Enemy free space and the structure of ecological communities. Biol J Linn Soc 23: 269–286
Kambysellis MP, Heed WB (1971) Studies of oogenesis in natural populations of Drosphilidae. I. Relation of ovarian development and ecological habits of the Hawaiian species. Am Nat 105: 31–49
Lawton JH, Strong DR (1981) Community patterns and competition in folivorous insects. Am Nat 118: 317–338
Montague JR, Mangan RL, Starmer WT (1981) Reproductive allocation in the Hawaiian Drosophilidae: egg size and number. Am Nat 118: 865–871
Moran NA (1988) The evolution of host-plant alternation in aphids: evidence for specialization as a dead end. Am Nat 132: 681–706

Ng D (1988) A novel level of interactions in plant–insect systems. Nature 334: 611–613
Owen J, Gilbert FS (1989) On the abundance of hoverflies. Oikos 55: 183–193
Price PW (1975) Reproductive strategies of parasitoids. In: Price PW (ed) Evolutionary strategies of parasitic insects and mites. Plenum Press, New York, pp 87–111
Rausher MD (1983) Ecology of host-selection behavior in phytophagous insects. In: Denno RF, McClure MS (eds) Variable plants and herbivores in natural and managed systems. Academic Press, New York, pp 223–257
Reznick D (1985) Costs of reproduction: an evaluation of the empirical evidence. Oikos 44: 257–267
Rollo D (1986) A test of the principle of allocation using two sympatric species of cockroaches. Ecology 67: 616–628
Rossiter MC (1987) Genetic and phenotypic variation in diet breadth in a generalist herbivore. Evol Ecol 1: 272–282
Rotheray GE (1988) Larval morphology and feeding patterns of four *Cheilosia* species (Diptera: Syrphidae) associated with *Cirsium palustre* L. Scopoli (Compositae) in Scotland. J Nat Hist 22: 17–25
Rotheray GE, Gilbert FS (1989) Phylogeny and systematics of European predacious Syrphidae from larval and puparial stages. Zool J Linn Soc 95: 27–70
Scriber JM (1983) Evolution of feeding specialisation, physiological efficiency, and host races in selected Papilionidae and Saturniidae. In: Denno RF, McClure MS (eds) Variable plants and herbivores in natural and managed systems. Academic Press, New York, pp 373–412
Shorrocks B, Rosewall J, Edwards K, Atkinson W (1984) Interspecific competition is not a major organizing force in many insect communities. Nature 310: 310–312
Simpson GG (1944) The major features of evolution. Columbia University Press, New York
Speight MCD (1987) External morphology of adult Syrphidae (Diptera). Tijdschr Entomol 130: 141–175
Stearns SC (1983) The influence of size and phylogeny on patterns of covariation among life history traits in the mammals. Oikos 41: 173–187
Strong DR, Southwood TRE, Lawton JH (1984) Insect and plants. Blackwell Scientific Publishers, Oxford
Swofford DL (1985) PAUP: phylogenetic analysis using parsimony version 2.4. Illinois Natural History Survey, Champaign, IL
Tauber CA, Tauber MJ (1987) Food specificity in predacious insects: a comparative ecophysiological and genetic study. Evol Ecol 1: 175–186
Taylor LR (1984) Assessing and interpreting the spatial distribution of insect populations. Annu Rev Ecol Syst 29: 321–357
Thompson FC (1989) World catalogue of the Syrphidae (Diptera). US Dept Agric [Held in the Smithsonian Institution]
Thompson JN (1988a) Evolutionary ecology of the relationship between oviposition preference and performance of offspring in phytophagous insects. Ent Exp Appl 47: 3–14
Thompson JN (1988b) Variation in preference and specificity in monophagous and oligophagous swallowtail butterflies. Evolution 42: 118–218
van Noordwijk A, de Jong G (1986) Acquisition and allocation of resources: their influence on variation in life history tactics. Am Nat 128: 137–142
Via S (1986) Genetic covariance between oviposition preference and larval performance in an insect herbivore. Evolution 40: 778–785
Wiklund C (1975) The evolutionary relationship between adult oviposition preferences and larval host plant range in *Papilio machaon* L. Oecologia 18: 185–197
Wiklund C, Karlsson B, Forsberg J (1987) Adaptive versus constraint explanations for egg-to-body size relationships in two butterfly families. Am Nat 130: 828–838
Zwölfer H (1975) Speciation and niche diversification in phytophagous insects. Verh Dtsch Zool Ges 67: 394–401

*Chapter 7*

# Opportunistic Diapause in the Subtropical Ground Cricket, *Dianemobius fascipes*

Sinzo Masaki

## 7.1 Introduction

Cricket life-cycles can be classified into two major types, homodynamic and heterodynamic (Masaki and Walker 1987). In the homodynamic type, growth and reproduction continue at similar rates or under similar thermal control ($Q_{10}$ = 2 to 3) throughout the year, while in the heterodynamic type they are retarded by the intervention of diapause and/or the response to seasonal cues such as photoperiod (Masaki 1978; Tanaka 1983, 1984). Since the life-cycles of extinct forms cannot be reconstructed from fossils, there is no direct means of knowing which one, homodynamic or heterodynamic, is the ancestral type. Only circumstantial evidence suggests that the homodynamic is ancestral.

Crickets are believed to have originated in the tropics (Alexander 1968), where temperature is higher than the development threshold all the year round. Although diapause might have evolved in tropical regions (Denlinger 1986; Tanaka et al. 1987), it seems, at least in some species of cricket, to be a derived trait, having evolved as a means of adaptation to temperate climates (Masaki 1984; Masaki et al. 1987). In our laboratory, we have found only homodynamic life-cycles in more than 15 tropical strains of several species of cricket collected from Southeast Asia. Although the information is still scanty, diapause has not yet been reported for any tropical cricket.

In contrast, most species of cricket from the temperate parts of the Japanese islands as well as North America are heterodynamic. Their development rates and temperature responses change as the active phase alternates with the diapause phase in their life-cycles (Masaki and Walker 1987). The diapause phase commonly occurs in the egg stage. Egg diapause is usually manifested as a remarkable delay of hatching when eggs are incubated at constant temperatures above the development threshold, making a sharp contrast to non-diapause development in the homodynamic tropical species (Fig. 7.1).

This broad latitudinal divergence in the life-cycle pattern is undoubtedly one of the crucial events in the evolutionary history of various cricket lineages. Many

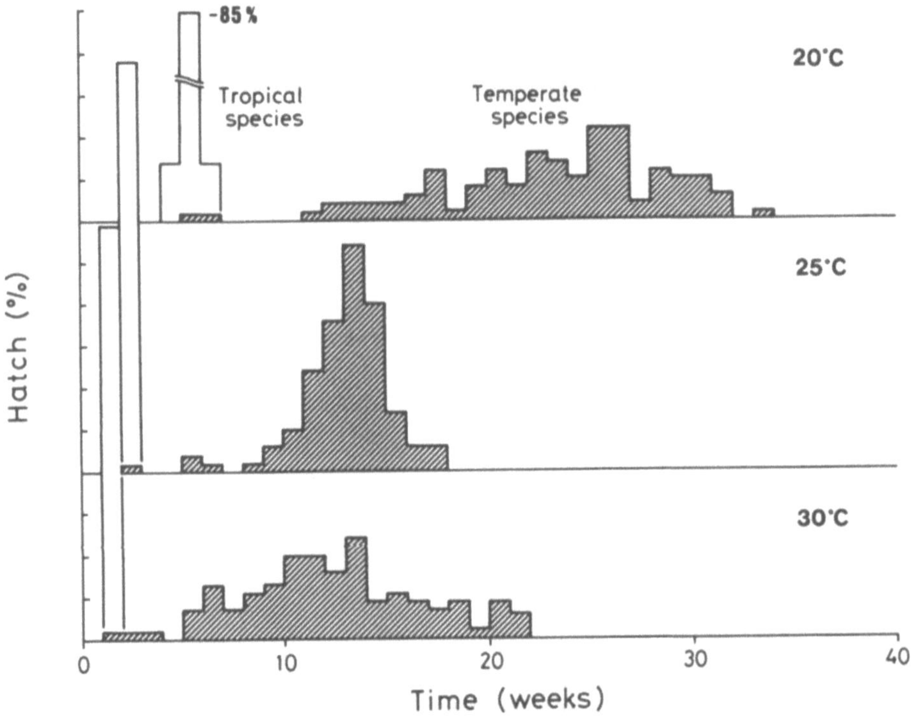


**Fig. 7.1.** Comparison of hatching distributions of a tropical strain of *Dianemobius fascipes* (*open bars*) from Khaoyai (18° N), Thailand, and a temperate strain of *D. nigrofasciatus* (*hatched bars*) from Hirosaki (40° N), Japan. In both strains, the eggs were laid by crickets reared under LD 12 : 12 h at 25–27 °C. One scale unit of the vertical axis is 10% for the tropical species and 5% for the temperate species.

species of cricket in the temperate parts of the Japanese islands are very closely related (or believed to be conspecific) to tropical populations. Had they not evolved diapause and associated mechanisms for seasonal adaptation, they would not have established their present distribution. Almost nothing is known about the origin of diapause, despite the fact that it is undoubtedly an evolutionary phenomenon of common occurrence (Danks 1987; Tauber et al. 1986). The crickets' system of seasonal adaptation perhaps undergoes a drastic reorganization from the homodynamic to the heterodynamic type as these insects disperse through the border between the tropical and the temperate regions. A possible way to approach the origin of diapause is therefore to explore cricket life-cycles in the subtropical and adjacent regions.

## 7.2 Ambiguity of the Subtropical Environment

The heterodynamic and homodynamic life-cycles of crickets are probably maintained under selection pressures exerted in temperate and tropical environments, respectively. For highly omnivorous insects such as crickets, one of the most important factors limiting growth and reproduction is temperature. Although there should be some interspecific and intraspecific variation, most cricket eggs of both tropical and temperate species stop embryogenesis at temperatures below 14 °C (Masaki and Walker 1987). If such low temperatures prevail during diapause, then diapause itself does not specifically increase the cost of delayed reproduction, for non-diapause eggs are also retarded in development. Diapause decreases winter mortality by enhancing cold hardiness and reducing metabolism, so that the more severe and longer the winter the greater would be its selective advantage. The relative contribution of diapause to fitness depends partly therefore on the environmental conditions during diapause. So far as the benefit of diapause exceeds the cost of delayed reproduction, diapause is favoured by natural selection, and the cost may be minimized by optimizing selection for timing the entrance into diapause (Taylor 1986). If the winter is mild and frequently allows growth and reproduction, diapause results in loss of time and heat available for development so that it may decrease fitness. Since the climatic conditions show a continuous north–south gradient, there should be a situation in which the benefit and cost derived from diapause are in a dynamic balance.

Background climatic conditions for such a situation are exemplified in Fig. 7.2,

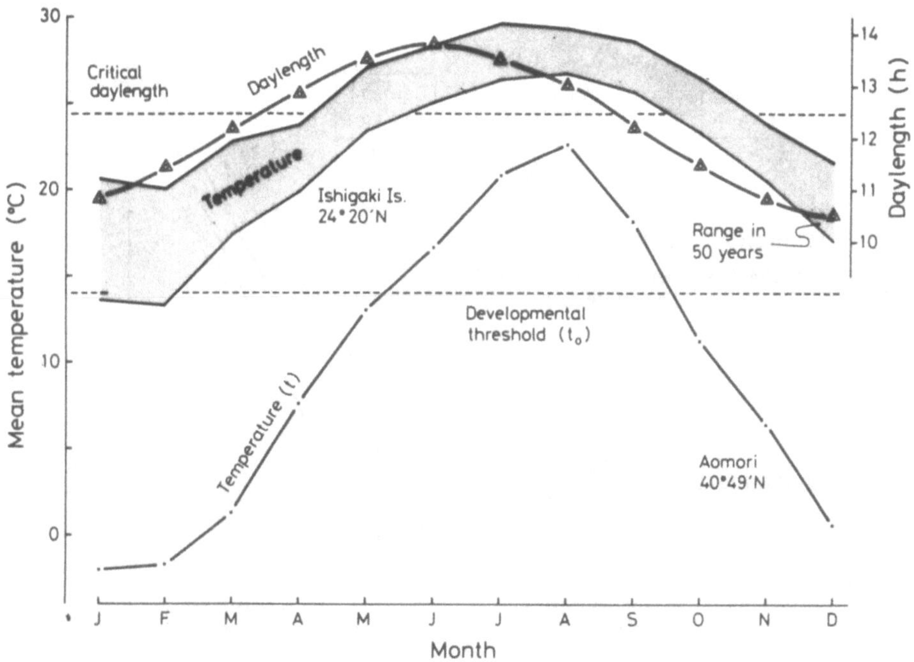


**Fig. 7.2.** Seasonality as represented by the monthly mean temperatures at two widely-separated localities in the Japanese islands. For Ishigaki Island, the highest and lowest monthly means of temperature during a 50-year period and the daylength (time from sunrise to sunset) are shown. Developmental threshold [$t_0$] is an estimation from the general regression of the development rate in the egg stage on temperature for various species of cricket (Masaki and Walker 1987). For critical daylength see Fig. 7.10.

which plots the monthly mean temperatures in the subtropical island of Ishigaki (24° N) together with those at Aomori (41° N) for comparison. In the northern locality, only a few months are available for growth and reproduction, and the crickets lie dormant as eggs during the rest of the year. At Ishigaki, on the contrary, the monthly mean temperature is normally higher than the development threshold throughout the year. However, the mean winter temperature may occasionally fall very close to the threshold. The daily minima could be still lower and suppress development. The probability of occurrence of such cases decreases farther south and virtually reaches zero in the tropical regions, as illustrated by the north–south gradient of the growing season (Fig. 7.3).

From the viewpoint of a cricket's seasonal adaptation the three climatic regions, the temperate, tropical and intermediate subtropical regions, are defined in Fig 7.3. The temperate region includes areas where the winter is consistently cold and diapause obviously makes a positive contribution to fitness every year. The tropical region covers areas where diapause may have a negative effect on fitness. In the intermediate subtropical region, the benefit of diapause varies between positive and negative values as the climatic conditions fluctuate from year to year.

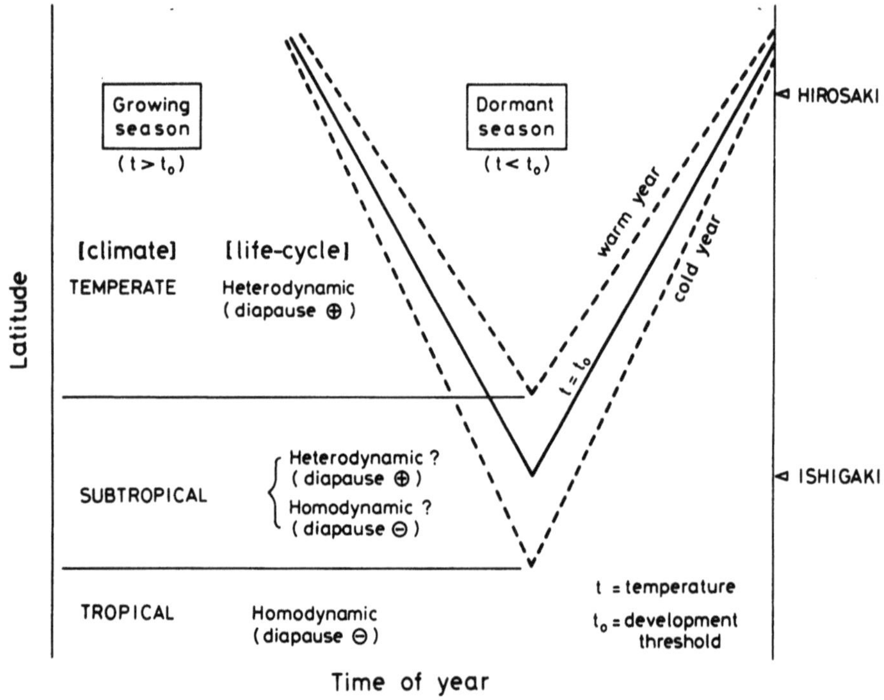


**Fig. 7.3.** Diagram showing the latitudinal gradient of the growing season and definitions of the three climatic regions. *Plus and minus signs*, in circles, following "*diapause*" indicate the selective advantage and disadvantage, respectively. The wedge of dormant season indicated by the *continuous line* is the mean, and *broken lines* show the range of fluctuation. The larger fluctuation toward the south is based on the correlation between the length of the growing season and its standard deviation presented by Roff (1983).

Since this fluctuation is unpredictable, evolution of a polymorphic life-cycle or stochastic polyphenism, which allows coexistence of diapause and non-diapause morphs in a population, is the most probable consequence of natural selection under such circumstances (Walker 1986). This is indeed the case in *Gryllus firmus* in Florida, as analysed in detail by Walker (1980) and Ibrahim and Walker (1980). Except for this example, we are quite ignorant of the seasonal adaptation of crickets in the subtropical region, which should be studied more intensively in order to get insight into the origin of heterodynamic life-cycles now prevalent in the temperate regions.

## 7.3 The Subtropical Ground Cricket

The ground cricket *Dianemobius fascipes* is particularly suitable for studying the evolution of seasonal adaptation, because it has a wide latitudinal range extending from Bali (8° S, Indonesia) to Ishigaki (24° N, Japan), and also because it is closely related to *Dianemobius nigrofasciatus*, which occurs in the temperate parts of Japan up to about 43° N. The two species are so similar in appearance that they had been considered conspecific for a long time (Masaki 1983). At first sight, this ground cricket in the subtropical island of Ishigaki seemed to have a polymorphic life-cycle similar to that of *Gryllus firmus* in Florida (Walker 1986). Previously, only a minor long-day type response of nymphal development and a very low incidence of egg diapause have been known as means of seasonal adaptation (Masaki 1983; Masaki et al. 1987). When the eggs were incubated at 25–27 °C, a very small proportion of them showed delayed hatching, and the possibility of overwintering in both the nymphal and egg stages were suggested. Moreover, the mild winter weather permits some nymphs to grow and some adults to reproduce. I was able to collect both nymphs (all stages except very young ones) and adults in Feburary 1984, at Ishigaki (my unpublished observations).

Recently, however, I have found that diapause eggs in this ground cricket may reach a substantial proportion under certain conditions. This diapause is extremely unstable compared with that widely occurring in the eggs of various temperate species of crickets including *D. nigrofasciatus* (Masaki 1960, 1983). I have suspected that this diapause constitutes a special adaptation to the ambiguous subtropical environment, and analysed its characteristics, particularly its termination in response to high temperature.

For this study I used a laboratory stock originally collected as adults in February 1984, at Ishigaki, and kept at 25–27° C. In maintaining stock cultures, I exposed successive generations alternately to long day (LD) 12 : 12 h and LD 13 : 11 h according to a special scheme of wing form selection for other experiments (Masaki and Watari 1989). I did not adopt any precautions to avoid unintentional selection on life-history traits. However, the characteristic diapause behaviour was maintained during the present experiments using eggs laid by adults of the fifth to tenth laboratory generations exposed to LD 12 : 12 h at 25–27 °C.

Adults were reared in groups of about 50 and allowed to lay eggs into the moist cotton plug of a water bottle put in each rearing jar. Cotton plugs were removed daily, kept in small plastic bottles and four to six replicates were allotted to each treatment without the eggs being counted. The bottles were opened daily and hatching nymphs were counted in order to assess the response. When no nymphs had hatched for more than 30 days I assumed that all viable eggs had hatched and stopped observation. In such a case, only some dead, discoloured eggs remained in the cotton plug, and usually few or no viable eggs were found.

## 7.4 Response to Temperature Rise

I noticed the peculiar response of the ground cricket eggs for the first time when I tried to hatch eggs at desired times by accelerating their development at a high temperature of 25 ± 1 °C after various periods of incubation at 20 ± 1 °C. These treatments gave the hatching curves presented in Fig. 7.4. In the top panel, there

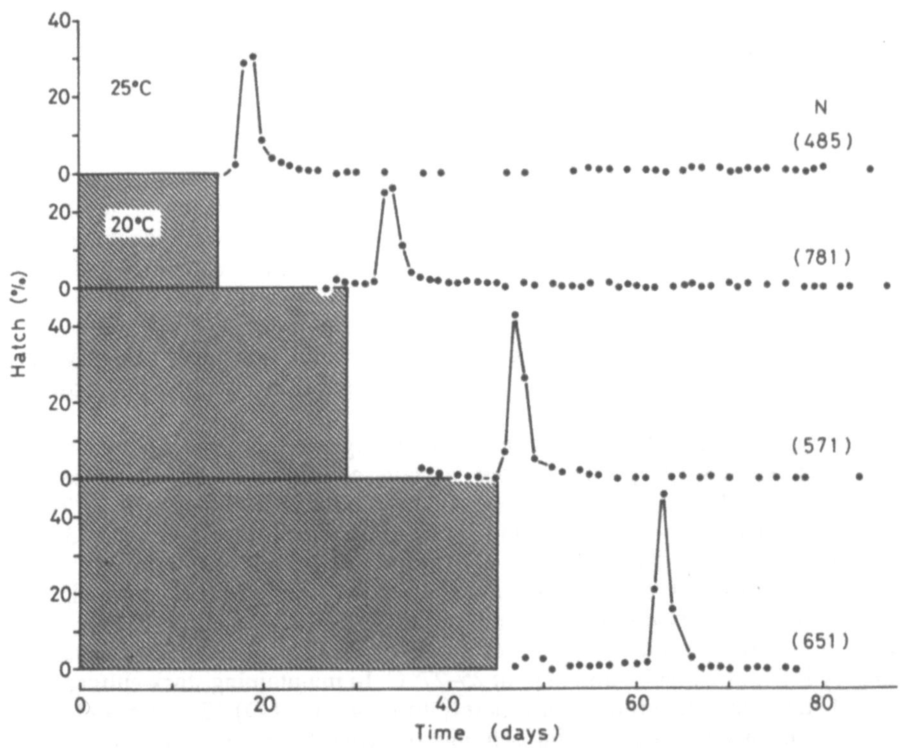


**Fig. 7.4.** Distribution of hatching eggs of *D. fascipes* first kept at 20 °C (hatched area) and then transferred to 25 °C (*open area*). *The top panel* represents control without preliminary exposure to 20 °C. *Numbers in parentheses* are sample sizes (*N* = numbers of hatching nymphs). The daily percentages of hatch in the total hatch are plotted. The eggs for this and all the subsequent experiments (Figs. 7.5–7.11) were laid by a parent generation reared in LD 12 : 12 h at 25 °C, except where otherwise indicated.

is a clear peak of hatching on days 18 to 19 of incubation at 25 °C. This is about equal to the incubation time of non-diapause eggs of conspecific tropical strains. A few eggs lagged behind the peak. In this subtropical strain, therefore, a small proportion of eggs may enter diapause while the majority do not – a situation just the reverse of what we observed in the northern ground cricket *D. nigrofasciatus* (Fig. 7.1; Masaki 1983).

There was, however, an unexpected response to temperature change. If eggs were first exposed to 20 °C before incubation at 25 °C, the hatching peak was delayed exactly for the duration of exposure to 20 °C, as demonstrated by the lower panels (Fig. 7.4). It would seem that the thermal threshold for embryogenesis is very close to 20°C. However, the graphs show a few eggs hatching before the peaks. Such precocious hatchings did not occur in the direct incubation at 25 °C (the top panel). These facts suggest the following. (a) The hatching peak at constant 25 °C represents the time required for non-diapause embryogenesis. (b) If (a) is true, those eggs forming the hatching peak after exposure to 20 °C started to develop at the time of transfer to 25 °C. (c) The eggs hatching before the peak time had therefore already started embryogenesis at 20 °C before transfer to 25 °C. This inference is supported by the fact that the hatching time (at 25 °C) of such precocious eggs became shorter with increasing length of the previous exposure to 20° C.

If the above inference is correct, there are two different types of egg. Eggs of one type do not develop at 20 °C at least for the first 45 days and start to do so as soon as they are transferred to 25 °C. However, the eggs of the other type can develop at 20 °C, which must therefore be above the developmental threshold. Then, why is the development of so many eggs retarded at 20 °C? A possible answer to this question is the occurrence of diapause at 20 °C and very quick termination of this egg diapause at a high temperature of 25 °C.

## 7.5 Response to Temperature Pulse

In order to confirm this hypothesis of an instant termination of diapause in response to a temperature rise, I kept groups of eggs first at 20 °C, and 30 days later exposed them to a short (1–4 days) pulse of high temperature, 25 °C (Fig. 7.5). The distribution of hatching at a constant 20 °C indicated two different groups of eggs. Eggs of one group formed a peak hatch at days 40–45, and eggs of the other group lagged behind and hatched over a long period of time extending from day 50 to 200 after oviposition. The first group represented non-diapause eggs and the second group diapause eggs. The fact that the expression of the non-diapause peak at 20 °C in this and subsequent experiments was clearer than in the previous one (Fig. 7.4) might be due to unavoidable selection against a few diapause eggs at 25 °C in successive generations of the stock culture. This did not seriously obstruct the present study, and a non-diapause peak could be used as a convenient reference point in assessing hatching responses of diapause eggs to various temperature conditions.

In all the other treatments with high-temperature pulses of 1, 2, 3 and 4 days, respectively, two clear peaks of hatching appeared. The first peak exactly corresponded to the non-diapause hatch around days 40–45 at a constant 20 °C.

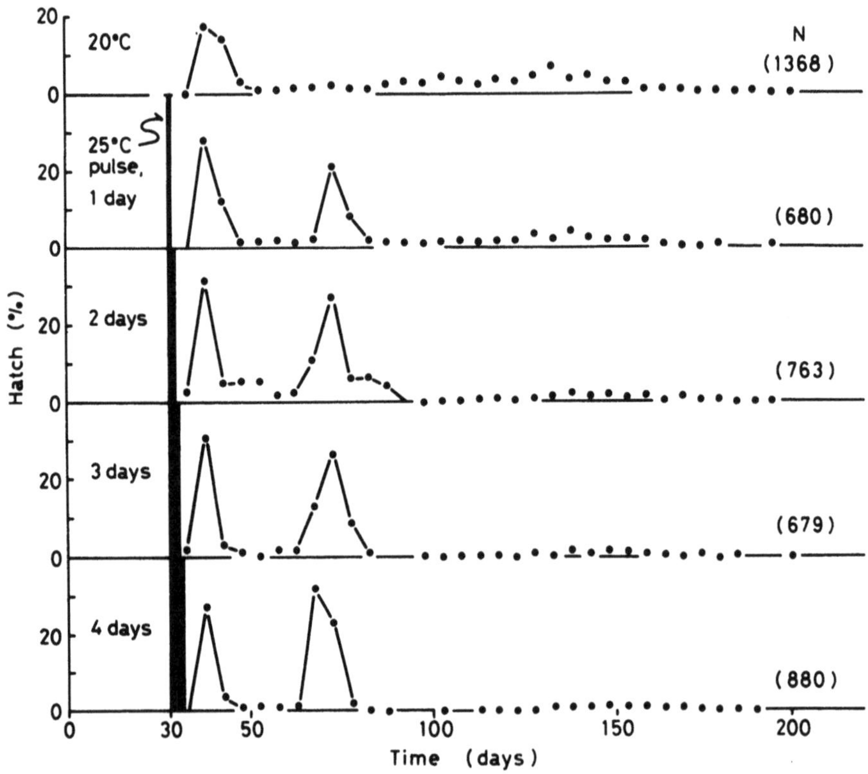


**Fig. 7.5.** Effects of high-temperature pulses (*vertical black bars*) of various durations given at day 30 of the egg stage on the distribution of hatching nymphs over time at *20 °C*. The total numbers hatching (*N*) are shown on the right.

The second peak appeared on days 70–75. The time lapse between this peak and the start of the high temperature pulse was 40–45 days, about equal to the duration of non-diapause development at 20 °C. The second peak of hatching was therefore due to very quick termination of diapause in response to the short high-temperature pulse. Even the shortest pulse of only 1 day was clearly effective.

This rapid termination is in sharp contrast to the maintenance of diapause for weeks or months even at 25 or 30 °C in the northern ground cricket *D. nigrofasciatus* (Fig. 7.1). Owing to this quick response, the eggs of the subtropical cricket are much more opportunistic than those of the northern cricket and would make better use of a temporary rise in temperature for development. Before concluding, however, two or three further questions should be answered. One question is whether the observed response is invoked by a high temperature itself or by an abrupt temperature change.

The answer is found in Fig. 7.6, which shows responses to 1-day pulses of various temperatures ranging from 10 to 35 °C by eggs otherwise kept at 20 °C. The first non-diapause peak occurred in all treatments, but the second peak induced by the pulse occurred only when the pulse temperature was 25, 30 or 35 °C. The low temperature pulses of 15 and 10 °C were not effective. I conclude therefore that the eggs terminate diapause as soon as temperature rises to 25 °C or higher.

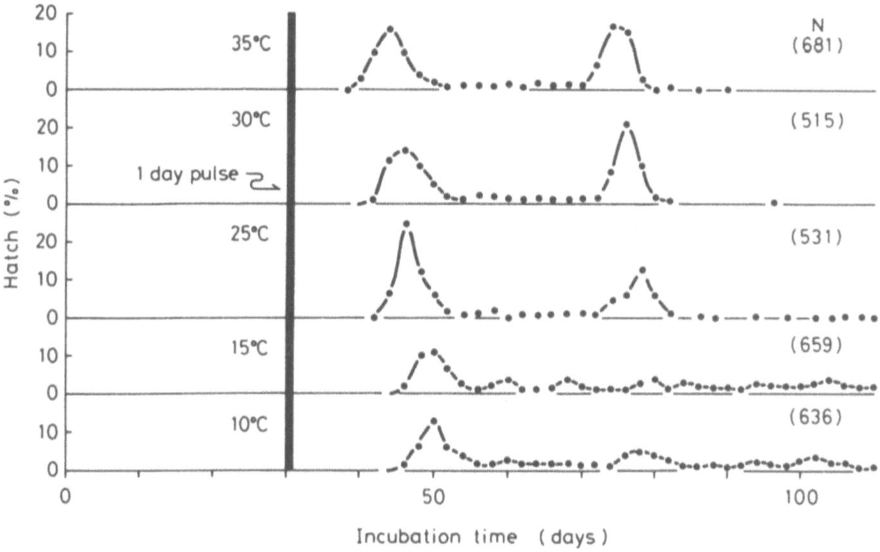


**Fig. 7.6.** Effects of pulses of different temperatures (*vertical black bars*) given at day 30 on the distribution of hatching nymphs over time at 20 °C. The total numbers hatching (*N*) are shown on the right.

## 7.6 Stage Sensitive to High Temperature

There remains one other question that should be answered before it can be concluded that this diapause is opportunistic: is the sensitivity to high temperature limited to a particular stage of embryogenesis or diapause development? If it is, the eggs cannot always be opportunistic in terminating diapause. To answer this question I gave a 2-day pulse of 25 °C after incubation at 20 °C for 1, 2, 3, 4 or 5 weeks. All treatments resulted in two clear hatching peaks (Fig. 7.7), and the second peak occurred later as a high temperature pulse was given later. This is strong evidence for the termination of diapause by a high temperature at any time during the first 5 weeks of the egg stage. It is then possible that, even when some eggs fail to respond to a temperature rise, they will have a further opportunity of responding to a later temperature rise.

To test this possibility, I repeated three 1-day pulses of 25 °C at 2-week intervals with eggs otherwise kept at 20 °C. The results are very clear, showing four peaks of hatch (Fig. 7.8). The first one occurred about 45 days after deposition and of course represented non-diapause eggs. This was followed by the second, third and fourth peaks, exactly at 2-week intervals, corresponding to the three high-temperature pulses given at 2-week intervals. Probably, these different peaks represent groups of eggs with different intensities of diapause. If so, heat pulses can be used to separate a sample of eggs into fractions with different intensities of diapause.

The polymodal distribution of hatching eggs thus obtained suggests a possible ecological consequence of this opportunistic termination of diapause. So far as the hatching distribution at a constant temperature of 20 °C suggests, the variation in the intensity of diapause is continuous and unimodal. This is

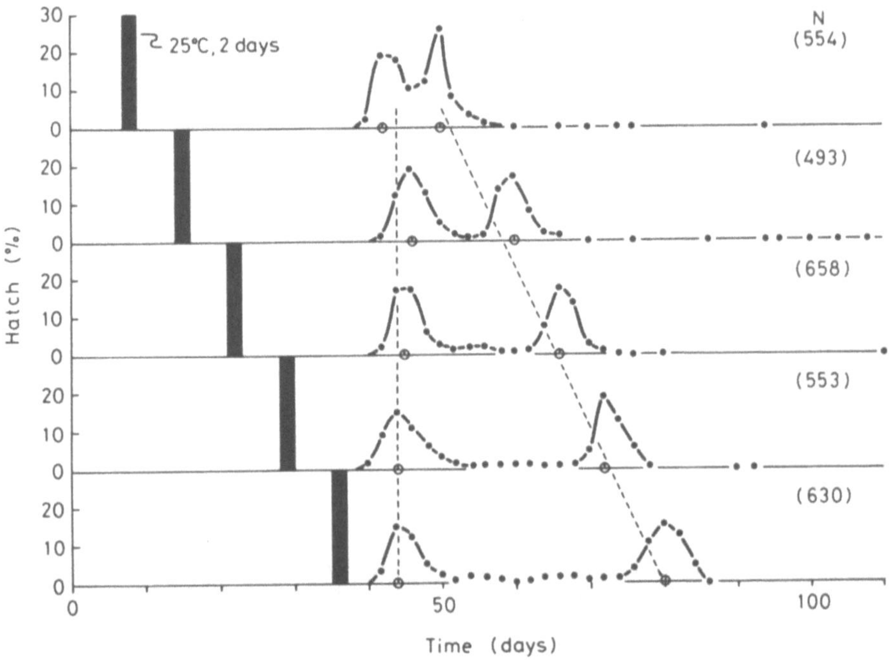


**Fig. 7.7.** Effects of 2-day pulses of a high temperature (*vertical black bars*) given at weeks 1, 2, 3, 4 and 5 of the egg stage on the distribution of hatching nymphs over time at 20 °C. The total numbers hatching (*N*) are shown on the right. *Broken lines* are drawn through the times of the first and second peaks, respectively.

converted into a polymodal distribution when diapause eggs are exposed repeatedly to suboptimal heat pulses at certain intervals. As a result, the eggs laid at the same time may be divided into several groups hatching at different times. This makes a bet-hedging tactic. When unpredictable weather conditions – cold, drought, etc. – kill vulnerable hatchlings, other groups remaining unhatched in the soil or having hatched earlier and grown to more tolerant stages will survive. Indeed, the benefit of the opportunistic tactic depends on unpredictable fluctuations in weather conditions and bet hedging is the other side of the same coin.

## 7.7 Interacting Effect of Photoperiod and Temperature

A further opportunistic character of egg diapause in this ground cricket can be seen in the mode of its induction. The egg diapause is primarily programmed by a short-day photoperiod affecting the female parents, and in fact all eggs used for the experiments described above were obtained under LD 12 : 12 h at 25–27 °C.

However, the occurrence of diapause largely depends on the incubation temperature (Fig. 7.9). Even when eggs were produced by adults reared at LD 12 : 12 h, diapausing eggs were very few at temperatures of 25 °C or higher. Only

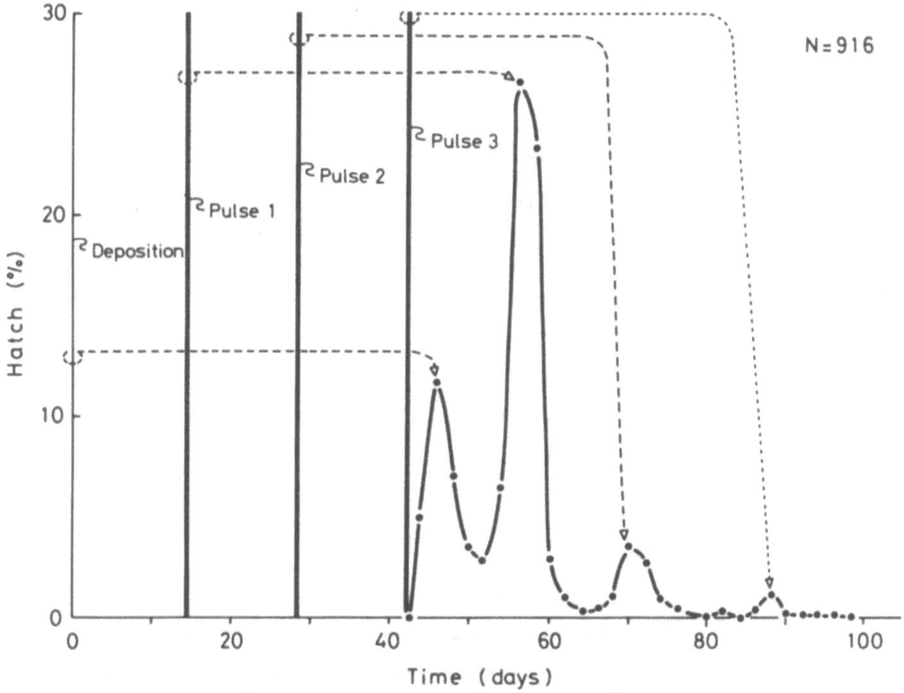


**Fig. 7.8.** Response of eggs of *D. fascipes* to repeated pulses of 25 °C (*vertical bars*). The eggs were otherwise kept at 20 °C. Each pulse caused a corresponding peak of hatch as indicated by a *broken line*. The first peak was due to non-diapause eggs. *N*, total number hatching.

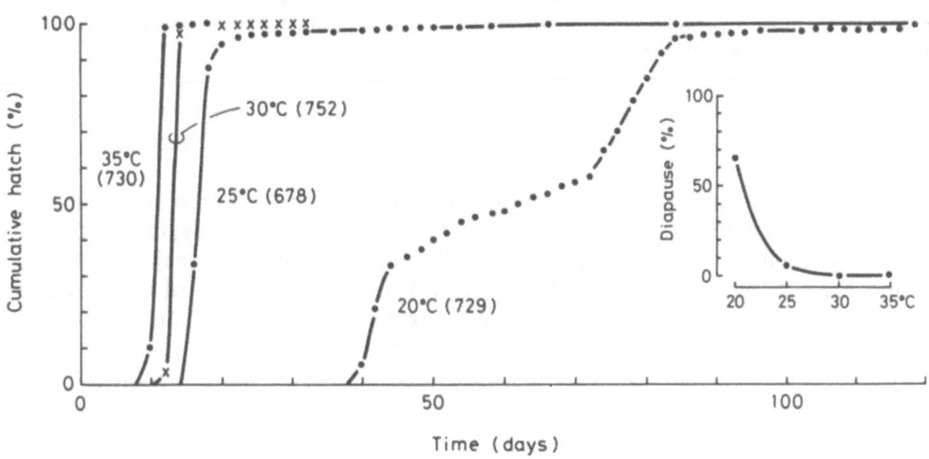


**Fig. 7.9.** The effect of temperature on hatching of eggs laid by short-day parents (LD 12 : 12 h, 26 °C). Eggs were exposed to the respective temperatures within 24 h of being laid. The early hatchings forming the steep sigmoid part of each curve were assumed to represent non-diapause development and the rest were regarded as diapausing. The percentage diapause thus estimated is plotted as a function of temperature in the inset. The total number hatching at each temperature is given in parentheses.

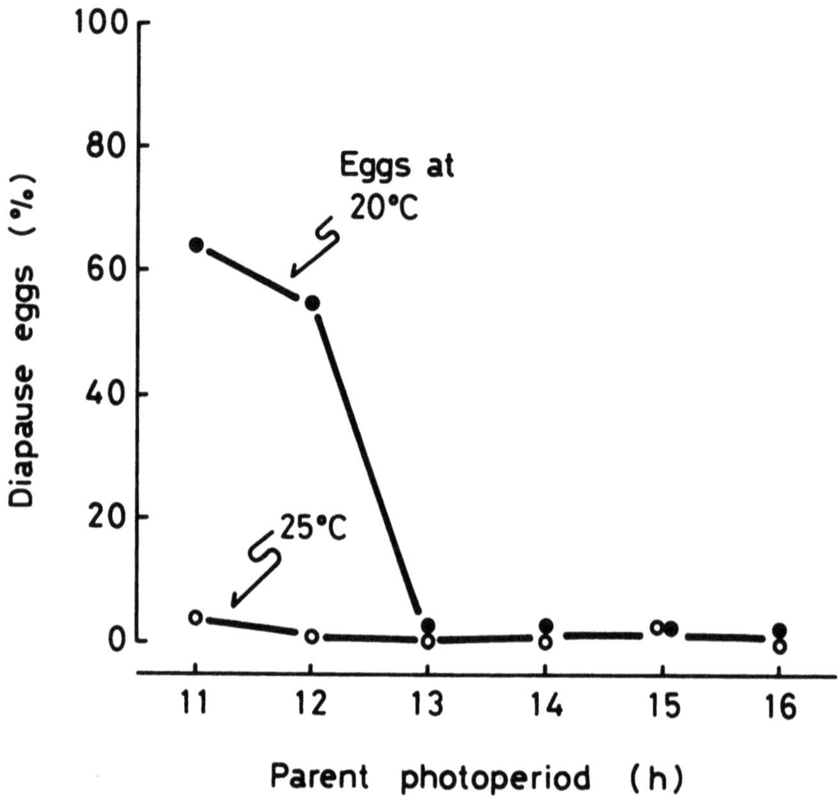


**Fig. 7.10.** The interacting effect of photoperiod and temperature on the induction of diapause in *D. fascipes*. The parent generation was reared in groups in various photoperiods at 26 °C and eggs were transferred to the respective temperatures within a day of deposition. Sample size for each datum point is 600–1032 eggs at 20 °C and 1124–3112 eggs at 25 °C.

at an incubation temperature of 20 °C, was a clear diapause response to short-day photoperiods observed (Fig. 7.10). The critical daylength was between 12 and 13 h.

In order to examine the effect of high incubation temperature in more detail, I exposed eggs within a day after being laid to 25 °C for brief periods ranging from 1 to 5 days before incubation at 20 °C. In all treatments only one peak of hatching appeared, between days 40 and 45 (Fig. 7.11). The proportion of eggs lagging behind the peak decreased with increasing exposure to 25 °C, and an exposure of only 4 days to 25 °C almost completely prevented diapause.

## 7.8 Discussion

Egg diapause in *D. fascipes* is thus primarily induced by short days in the parent generation. However, even after diapause has been programmed by the female parent before egg-laying, the eggs readily avert diapause in response to a high

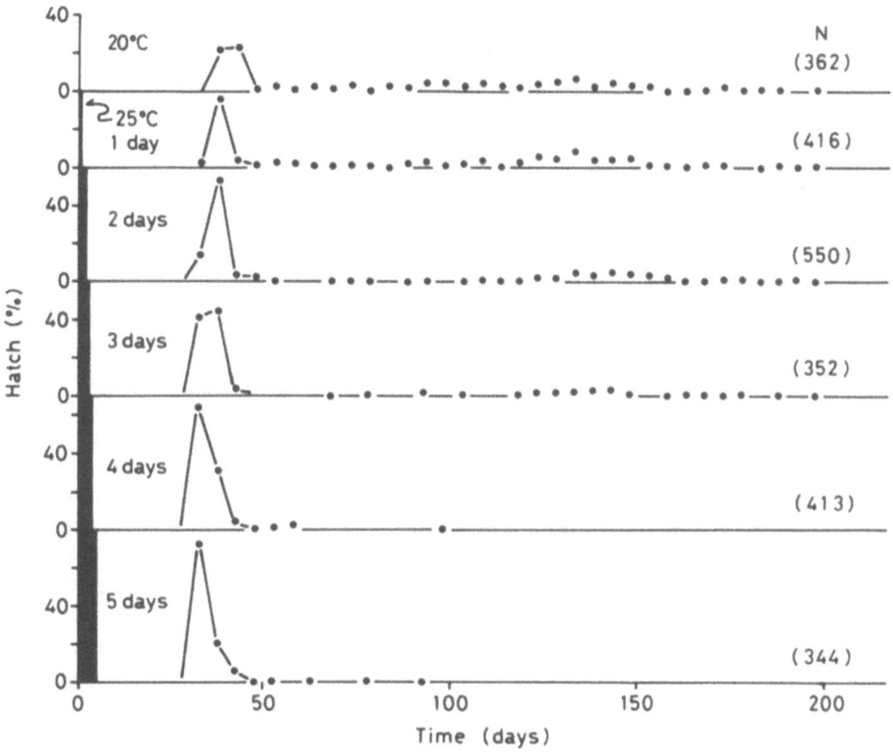


**Fig. 7.11.** Effect of high temperature (25 °C) in the early egg stage on the subsequent distribution of hatching nymphs over time at 20 °C. Total numbers hatching (*N*) are on the right.

temperature at the early stage of development. Even after the eggs have already entered diapause, they resume development at once whenever they are subject to high temperature. Therefore, the definitive cue (Roff 1983), daylength, opens the seasonal gate for diapause, but the actual manifestation of diapause inside this gate depends on the ambiguous cue (Roff 1983), temperature.

In order to infer the probability of opportunistic hatching of diapause eggs, the range of temperature and the time required for diapause elimination should be determined in more detail, and the temperature conditions near the soil surface should be measured in the natural habitat. Although such data are not yet available, it seems reasonable to assume that the ambiguous cue would exert the effect on diapause elimination most frequently at the beginning of the short-day season when temperature often exceeds 25 °C (Fig. 7.2). If the weather is mild, hatchlings can grow and survive together with nymphs that have hatched earlier. It may be that advanced nymphs emerge as adults even in winter and lay diapause eggs, because there is no evidence for reproductive diapause in adults collected in February in the field. Those eggs would hatch in response to the temperature rise in spring.

Despite the presence of egg diapause and photoperiodism, the population of this ground cricket partly continues to grow and breed all the year round in the subtropical environment with no fixed stage for overwintering. Therefore, its voltinism cannot be defined, and the growing season does not seem to be restricted. In the field, its life-cycle may appear to be a mixture of homodynamic and heterodynamic types: in the winter months there are active adults and nymphs of various stages, as shown by the field collection, and some diapausing eggs in the soil, as suggested by the present experiments. A question to be asked is: why has this egg diapause been evolved and maintained? At first glance, it seems that the selective advantages of continued activity and egg diapause in winter may be in a dynamic balance. Diapause in the egg stage is the commonest means of hibernation among the temperate species of crickets in the Japanese islands as well as in North America (Masaki and Walker 1987). This prevalent convergent evolution of life-cycles is probably due to the larger selective advantage of overwintering diapause in the egg than in the other life-stages. If, however, the existence of egg diapause in the subtropical island is ascribed to its high survival value through winter, it should not be opportunistically terminated.

Another hypothesis should therefore be considered. Since adults and nymphs of *D. fascipes* occur in winter, the egg diapause itself is not an exclusive means of adaptation to overwintering. It may, however, serve as a mechanism to protect hatchlings from being exposed to risky conditions. Owing to their small and soft body, the newly hatched nymphs are probably at the most critical stage in the life-cycle. If they are immobilized by a low temperature, for example, they would easily lose their water balance because of the large ratio of the surface area to the small mass of their body. They would become more tolerant as they grow. The opportunistic termination of diapause gives a selective advantage, if it decreases the mortality at hatching with a smaller cost of delayed hatching. Since individual life-cycles are not synchronized in continuously growing and breeding populations, there is little possibility of disruptive selection for non-diapausing (or nymph-overwintering) and egg-diapausing fractions "in situ". However, this situation may provide an important prerequisite for such life-cycle divergence to occur under appropriate conditions (Alexander and Bigelow 1960; Walker 1974).

The termination of egg diapause by high temperature is not a physiological property unique to this subtropical cricket. The diapause response of the tropical flesh fly *Poecilometopa spilogaster* to temperature is very similar to that described here, except that the flesh fly is not sensitive to photoperiod and enters diapause as a pupa (Denlinger 1974). This similarity may be a result of convergent evolution under similar selection pressure, i.e. unpredictable fluctuations of temperature up and down the development threshold. The absence of photoperiodism in the flesh fly is due to the virtual absence of seasonal change in daylength near the equator.

Further examples of diapause termination by high temperature are afforded by the temperate species of crickets *D. nigrofasciatus* (formerly included in *D. fascipes*; see Masaki 1983), *Loxoblemmus aomoriensis* (Masaki 1960) and *Teleogryllus commodus* (Browning 1952; Masaki et al. 1979) and the leaf beetle *Atrachya menetriesi* (Ando 1978), although their sensitivity to high temperature is much less than that of the subtropical ground cricket. Many other cases of the activation of diapausing insects or the acceleration of diapause termination by high temperature have been reviewed by Hodek and Hodkova (1988) (see also Danks 1987; Tauber et al. 1986). Therefore, the basic physiological mechanism for the diapause-terminating response of *D. fascipes* is probably shared with the temperate species. The subtropical diapause owes its highly opportunistic character mainly to its extremely low intensity. If the difference between the subtropical and more stable temperate diapause is not qualitative, the latter might have been derived from the former, i.e. they are homologous. To test this possibility, I have been selecting strains of *D. fascipes* with different intensities of egg diapause for several years, and could modify intensities of diapause in terms of its incidence and duration. In association with this modification, the responsiveness to a high temperature pulse has also changed (my unpublished observations). It seems highly possible therefore that the stabilized egg diapause in the temperate region has evolved directly from the opportunistic subtropical diapause. Although the subtropical diapause may be a means of preventing risky hatch, it can be an effective preadaptation to the severer winter of the temperate regions.

More intriguing is the origin of the opportunistic diapause in a homodynamic population. Repeated attempts to isolate diapause strains from several tropical populations of *D. fascipes* have not been successful as yet, mainly because of the difficulty in obtaining offspring from very rare individuals showing unusually long egg stages. There are two alternative possibilities: (a) the opportunistic diapause might have evolved as *D. fascipes* dispersed from its original locality in the tropics to the subtropical area; (b) it might have already existed in some tropical populations as a bet-hedging tactic under unpredictably changing conditions of environment. We have to explore more intensively the border between the tropical and subtropical regions in order to find a clue to the life-cycle evolution in this cricket lineage.

The opportunistic diapause as described here should not be considered to be a general means of adaptation to the ambiguous subtropical environment. At least two larger species of cricket, *Cardiodactylus novaeguineae* and *Xenogryllus marmoratus*, are univoltine and overwinter as eggs in a state of firm diapause in the subtropical island of Okinawa at 26–27° N (Oshiro and Tamashiro 1985; Oshiro et al. 1981). Several large-sized species of katydids in this island are also known to have univoltine life-cycles with very long periods of egg diapause (Oshiro et al. 1984, 1986, 1987; Uema and Oshiro 1980). Some of these univoltine orthopterans might have extended from the north, and their adaptation to the subtropical environment might be constrained by the pre-existing framework of univoltine life-cycle (Masaki 1978). As judged from the geographical pattern of distribution, this possibility can be excluded from *Cardiodactylus novaeguineae* which occurs widely in Southeast Asia but does not reach the temperate region. Although no information is available for the life-cycle of this species in tropical districts, the possibility is not ruled out that its firm, obligatory diapause has evolved as a means of adaptation to the subtropical climate. The large brown cricket, *Brachytrupes portentosus*, is also a typical tropical species and not known to occur in the temperate parts of East Asia. In Taiwan close to the northern limit of its distribution, it is strictly univoltine, overwintering as an active nymph (Sonan 1931). Although the physiological state of the overwintering nymph is yet to be examined closely, it seems probable that this cricket has some means of seasonal regulation of its nymphal growth by which its univoltine life-cycle is stabilized.

Since most of the subtropical univoltine orthopterans named here are large in adult size, they would probably take a longer time to mature than the ground cricket. This would constrain the evolutionary possibilities of their seasonal strategies. Theoretical works (Cole 1954; Roff 1983; Taylor 1986) suggest that the optimum voltinism or diapause character depends on interactions between different life-history parameters. Other features of lifestyle, for example the preference for food, the type of habitat, enemies and competitors, etc., may also be involved in shaping the seasonal life-cycle. It is not surprising therefore that different species of cricket exploit drastically different ways of life-cycle adaptation to the same subtropical environment.

## 7.9 Summary

Diapause would increase fitness where growth or reproduction is seasonally suppressed but decrease fitness where the environment is continuously favourable. In a subtropical climate, the selective advantage and disadvantage of diapause may be in a dynamic balance. The subtropical ground cricket *Dianemobius fascipes* is thus active both as adults and nymphs of various stages, even in winter, but partly dormant as eggs. The egg diapause is primarily programmed by the female parent in response to short days, but it is opportunistic in both its manifestation and termination. At a temperature higher than 25 °C it is both averted in the early egg stage and almost instantly terminated after induction. If several high-temperature pulses are given in succession at a certain interval of days, eggs with different intensities of diapause respond to different pulses in the sequence, showing a polymodal hatching curve. This makes for a bet-hedging tactic and, together with the opportunism, constitutes adaptation to the variable subtropical seasons. The selective advantage of this diapause seems to be to prevent hatching under risky conditions rather than to introduce a long period of dormancy, but it can be a preadaptation to hibernation in the temperate region and thus the origin of a more stable diapause.

*Acknowledgements*. I thank Y. Watari, M. Nakagawara, E. Seno and K. Tanaka for looking after the insects while I was absent from the laboratory, N. Sasaki for drawing figures, K. Ando and C. E. Gruenner for reading the manuscript and giving comments, F. Gilbert, J. C. Hartley and other anonymous reviewers for improvements. This study was supported in part by a grant (63480038) from the Ministry of Education, Science and Culture, Japan. The tropical strains of crickets were imported with permits (63–1255 and others) issued by the Ministry of Agriculture, Forestry and Fishery, Japan. Contribution no. 110 from Laboratory of Entomology, Hirosaki University.

### References

Alexander RD (1968) Life cycle origins, speciation, and related phenomena in crickets. Q Rev Biol 43: 1–41

Alexander RD, Bigelow RS (1960) Allochronic speciation in field crickets, and a new species, *Acheta veletis*. Evolution 14: 334–346

Ando Y (1978) Studies on egg diapause in the false melon beetle, *Atrachya menetriesi* Falderman (Coleoptera: Chrysomelidae). Bull Fac Agric Hirosaki Univ 30: 131–215

Browning TO (1952) On the rate of completion of diapause development at constant temperatures in the eggs of *Gryllulus commodus* Walker. Aust J Sci Res B5: 344–353

Cole LC (1954) The population consequences of life history phenomena. Q Rev Biol 29: 103–137

Danks HV (1987) Insect dormancy: an ecological perspective. Biological Survey of Canada Monograph Series no 1. Biological Survey of Canada, Ottawa
Denlinger DL (1974) Diapause potential in tropical flesh flies. Nature 252: 223–224
Denlinger DL (1986) Dormancy in tropical insects. Annu Rev Entomol 31: 239–264
Hodek I, Hodkova M (1988) Multiple role of temperature during insect diapause: a review. Entomol Exp Appl 49: 153–165
Ibrahim R, Walker TJ (1980) Diapause and nondiapause eggs laid daily by individual *Gryllus firmus* females (Orthoptera: Gryllidae). Fla Entomol 63: 510–512
Masaki S (1960) Thermal relations of diapause in the eggs of certain crickets (Orthoptera: Gryllidae). Bull Fac Agric Hirosaki Univ 6, 5–20
Masaki S (1978) Seasonal and latitudinal adaptations in the life cycles of crickets. In: Dingle H (ed) Evolution of insect migration and diapause. Springer, Berlin Heidelberg New York, pp 72–100
Masaki S (1983) Climatic speciation in Japanese ground crickets. GeoJournal 7.6: 483–490
Masaki S (1984) Unity and diversity in insect photoperiodism. Ciba Found Symp 104: 7–25
Masaki S, Walker T J (1987) Cricket life cycles. Evol Biol 21: 349–423
Masaki S, Watari Y (1989) Response to night interruption in photoperiodic determination of wing form of the ground cricket *Dianemobius fascipes*. Physiol Entomol 14: 179–186
Masaki S, Ando Y, Watanabe A (1979) High temperature and diapause in the eggs of *Teleogryllus commodus* (Orthoptera: Gryllidae). Kontyu 47: 493–504
Masaki S, Shirado I, Nagase A (1987) Tropical, subtropical and temperate life cycles in ground crickets. Insect Sci Applic 8: 475–481
Oshiro Y, Tamashiro I (1985) Studies on the singing insects in the Ryukyu Islands. Part 7. Life history of the Matsumushi cricket, *Xenogryllus marmoratus* Haan (Orthoptera, Gryllidae) in Okinawa Island. Biol Mag Okinawa 23: 17–24 (In Japanese with English summary)
Oshiro Y, Sakai T, Oshiro T (1981) Studies on the singing insects in the Ryukyu Islands. Part 5. Life history of the spotted tree cricket, *Cardiodactylus novaeguineae* Haan (Orthoptera: Gryllidae) in Okinawa Island. Biol Mag Okinawa 19: 25–32 (In Japanese with English summary)
Oshiro Y, Oshiro E, Maeshiro E (1984) Studies on the singing insects in the Ryukyu Islands. Part 6. Life history of the Okinawa katydid, *Gampsocleis ryukyuensis* (Orthoptera, Tettigoniidae) in Okinawa Island. Okinawa Nogyo 19: 39–47 (In Japanese with English summary)
Oshiro Y, Kiyokawa A, Oshiro M, Higashionna H (1986) Studies on the singing insects in the Ryukyu Islands. Part 9. Life history of the Taiwan-umaoi, *Hexacentrus* sp. (Orthoptera, Tettigoniidae) in Okinawa Island. Biol Mag Okinawa 24: 21–29 (In Japanese with English summary)
Oshiro Y, Oshiro M, Maeda Y, Itokazu E (1987) Studies on the singing insects in the Ryukyu Islands. Part 11. Life history of the Onaga-sasakiry, *Conocephalus gladiatus* (Orthoptera: Tettigoniidae) in Okinawa Island. Okinawa Nogyo 22: 23–30 (In Japanese with English summary)
Roff D (1983) Phenological adaptation in a seasonal environment: a theoretical perspective. In: Brown VK, Hodek I (eds) Diapause and life cycle strategies in insects. Dr W. Junk Publishers, The Hague, pp 253–270
Sonan J (1931) Studies on the big brown cricket and its control. Bull Formosan Cent Res Stat Dept Agric 86: 1–22 (In Japanese)
Tanaka S (1983) Seasonal control of nymphal diapause in the spring ground cricket, *Pteronemobius nitidus* (Orthoptera: Gryllidae). In: Brown VK, Hodek I (eds) Diapause and life cycle strategies in insects. Dr W. Junk Publishers, The Hague, pp 35–53
Tanaka S (1984) Seasonal variation in embryonic diapause of the striped ground cricket, *Allonemobius fasciatus*. Physiol Entomol 9: 97–105
Tanaka S, Denlinger DL, Wolda H (1987) Daylength and humidity as environmental cues for diapause termination in a tropical beetle. Physiol Entomol 12: 213–224
Tauber MJ, Tauber CA, Masaki S (1986) Seasonal adaptations of insects. Oxford University Press, New York Oxford
Taylor F (1986) Toward a theory for the evolution of the timing of hibernal diapause. In: Taylor F, Karban R (eds) The evolution of insect life cycles. Springer, Berlin Heidelberg New York, pp 236–257
Uema R, Oshiro Y (1980) Studies on the singing insects in the Ryukyu Islands. Part 4. Life history of the Taiwan noisy katydid, *Mecopoda elongata* Linne (Orthoptera: Tettigoniidae) in Okinawa Island. Biol Mag Okinawa 18: 25–30 (In Japanese with English summary)
Walker TJ (1974) *Gryllus ovisopis* n.sp.: a taciturn cricket with a life cycle suggesting allochronic speciation. Fla Entomol 57: 13–22
Walker TJ (1980) Mixed oviposition in individual females of *Gryllus firmus*: graded proportions of fast-developing and diapause eggs. Oecologia 47: 291–298
Walker TJ (1986) Stochastic polyphenism: coping with uncertainty. Fla Entomol 69: 46–62

*Chapter 8*

# Evolution of Mutualistic Life-Cycles: Yucca Moths and Fig Wasps

John F. Addicott, Judith Bronstein and Finn Kjellberg

## 8.1 Introduction

The study of mutualism is shifting from emphasizing the identification of systems that are mutualistic to attempting to understand the origin, maintenance and breakdown of mutualistic systems. Many different approaches are being used in this effort, including cost–benefit models (Roughgarden 1975), phylogenetic analyses (Thompson 1982), models of population dynamics (Addicott 1981; Wolin 1985), and the approach that we adopt in this chapter, the analysis of life-histories. We focus on two aspects of life-histories important for understanding the origin, maintenance and breakdown of mutualistic interactions: the probability of successful association between potential symbiont and host, and the comparative fitness of mutualistic and non-mutualistic symbionts when successfully associated with their host (Roughgarden 1975). More specifically, we ask two questions. (a) How does synchronization of life-cycles within and between symbiont and host populations influence the probability of successful association between mutualists and ultimately the maintenance of mutualism? (b) Under what conditions would "cheating" by potentially mutualistic individuals be favoured over strictly mutualistic behaviour? To address these questions we analyse the life-histories of species in two obligate pollination–seed predation mutualisms: the yucca–yucca moth interaction and the fig–fig wasp interaction.

Two factors affect the probability of association between symbionts (e.g. yucca moths or fig wasps) and their hosts (e.g. yuccas or figs). First, symbionts must be able to recognize their hosts. Recognition is usually based on various combinations of visual (e.g. Brown and Kodric-Brown 1979) or chemical (e.g. Dodson et al. 1969) cues. Second, there must be synchronization between the appropriate life-history stages of host and symbiont. For example, pollinators must be active when female flowers are receptive to pollen. Asynchrony of appropriate life-history stages between individual hosts and symbionts should lead to a lower probability of association, and decreased fitness for individual hosts and symbionts. Therefore, the first focus for our paper is how synchronization of life-cycles of mutualists is achieved. We consider synchronization of life-cycles between individual hosts and symbionts as well as among individuals within host populations and among individuals within symbiont populations.

"Cheaters" or "parasites" of mutualistic systems utilize services or resources provided by the host (or symbiont) without providing services or resources in return (Soberon Mainero and Martinez del Rio 1985). An individual of a normally mutualistic species that behaves in such a fashion is known as a "cheater", whereas a non-mutualistic species that utilizes mutualistic resources or services is known as an "aprovechado" (one who takes advantage) (Soberon Mainero and Martinez del Rio 1985). Individual plants that produce significantly less nectar than other individuals in the population but which receive equivalent attention from pollinators are examples of cheaters (Heinrich 1979); fish species that establish "cleaning" stations but remove flesh rather than parasites from fish that visit the stations are examples of "aprovechados" (Wickler 1968). The second focus of our paper is the conditions under which a mutualistic life-cycle could break down into a non-mutualistic life-cycle, producing either cheater morphs within a mutualistic species or "aprovechados".

There is not an extensive literature on the appropriate life-history traits of yuccas, yucca moths, figs and fig wasps. Therefore, our presentation is more a framework for how to think about evolution of life-histories in mutualistic systems rather than a summary of comprehensive data sets. We attempt to generate hypotheses about the fig and yucca systems, and by analogy mutualistic systems in general.

## 8.2 Natural History

The obligate mutualisms between figs and fig wasps and between yuccas and yucca moths share a basic aspect of their natural history. In both systems the benefits that the plants derive are in the form of pollination services (fertilization of ovules, transfer of pollen), whereas the cost of the interaction is consumption of some developing seeds by the offspring of the pollinators. With few exceptions (e.g. Brantjes 1976a, b; Pellmyr 1989), this combination of pollination and seed predation is unique to figs and yuccas. We briefly review the natural history of both systems. Further details are provided by Powell and Mackie (1966), Davis (1967), Janzen (1979a) and Bronstein and McKey (1989).

### *Figs and Fig Wasps*

Figs (*Ficus* spp., Moraceae) are trees, shrubs or vines distributed throughout the frost-free portions of the tropics and subtropics. They occur in relatively aseasonal environments, in the sense that plant growth and reproduction is possible year-round. However, many species exist where wet and dry seasons are pronounced, and a few species exist in climates with pronounced warm/cold seasonality (Bronstein 1989). There are approximately 700 species of figs divided into two major groups, one of which is gynodioecious and the other monoecious (Berg 1989). With the exception of the cultivated fig (*Ficus carica* L.), the biology of the gynodioecious figs is poorly known. In the remainder of this chapter, we consider only monoecious figs.

The inflorescence of monoecious figs (the syconium) has both male and female florets distributed on the inner surface of the invaginated receptacle. Access to the florets is limited by a narrow, bract-filled opening (the ostiole) at the distal tip of the syconium. Floret development within the syconium is protogynous, with female florets receptive several weeks prior to the maturation of male florets (Bronstein et al. 1990).

Figs are pollinated exclusively and obligately by fig wasps (Agaonidae, Chalcidoidea), which are also the primary seed predators of figs. There is generally a one-to-one relationship between species of figs and fig wasps (but see Michaloud et al. 1985). Agaonids are sexually dimorphic; males are wingless with greatly enlarged mouth parts, while females are winged and possess a more normal chalcidoid body form.

Inseminated female fig wasps leave the fig in which they matured carrying a load of pollen either in invaginations of intersegmental membranes (Galil and Neeman 1977) or in specialized pockets on their thorax and/or coxae (Galil and Eisikowitch 1974; Ramirez 1969). They then fly in search of a fig syconium that has receptive female florets. Such syconia are not usually present on the same tree on which the fig wasp matured. Female fig wasps live for only a few days and fig trees with receptive syconia are apparently attractive to fig wasps for only a few days (Bronstein 1989). Therefore, synchrony of wasp departure from one tree with receptivity of female florets in syconia on another tree is critical. Attraction of fig wasps to trees with receptive syconia is assumed to be olfactory (Barker 1985; Bronstein 1987; Condit 1947; Janzen 1979a).

After locating a syconium with receptive female florets, the female fig wasp enters the syconium through the ostiole. Once inside the syconium, she accomplishes both pollination and oviposition. She removes pollen from her thoracic or coxal pockets and applies it to the inner surface of the syconium, which is a continuous mat of stigmas. She oviposits by inserting her ovipositor through the length of the style, but style lengths are not uniform among florets. If her ovipositor is long enough to reach the ovary, she lays an egg and removes her ovipositor from the style. Otherwise, she removes her ovipositor from the style without having laid an egg. A female repeats the processes of pollination, insertion, oviposition and removal many times, but only within a single syconium. In this way ovaries are fertilized, but only a limited proportion receive an agaonid egg (Janzen 1979c).

Agaonid larvae develop within the ovaries, feeding on plant tissue. The sex ratio of agaonids within individual syconia is highly female-biased, in qualitative concurrence with local mate competition theory: the fewer females that enter a syconium, the greater the proportion of female offspring (Frank 1985; Herre 1985). Males emerge first, cut open the ovaries in which females are almost mature, mate with them, and then cut one or more exit holes through the wall of the ripening syconium. Females emerge, gather a load of pollen from the mature male flowers, and then exit the syconium to continue the cycle. The syconium ripens around the time that fig wasps begin to leave, and it becomes attractive to birds and mammals, which are the agents of seed dispersal for most figs (Janzen 1979a).

Throughout the remainder of the chapter, we use "figs" to refer to individual trees or species of monoecious members of the genus *Ficus*, "syconium" to refer to a fig inflorescence or infructescence, and "fig wasps" to refer to the agaonid pollinators of figs.

### *Yuccas and Yucca Moths*

Yuccas (*Yucca* spp., Agavaceae) are native to much of North and Central America. There are approximately 40 species of yucca, with the greatest diversity of species occurring in the southwestern United States and northern Mexico (Davis 1967; Webber 1953). Yuccas occur primarily in relatively arid, highly seasonal environments. They do not occur in the lowest, hottest deserts of North America, but they do occur at elevations of up to 2500 m in the Rocky Mountains of Colorado, and as far north as southern Alberta in Canada.

Yuccas are pollinated exclusively and obligately by yucca moths in two closely related genera, *Tegeticula* and *Parategeticula* (Incurvariidae, Prodoxinae) (Davis 1967). Female yucca moths possess maxillary tentacles, appendages unique to yucca moths, which function in collecting, transporting and transferring yucca

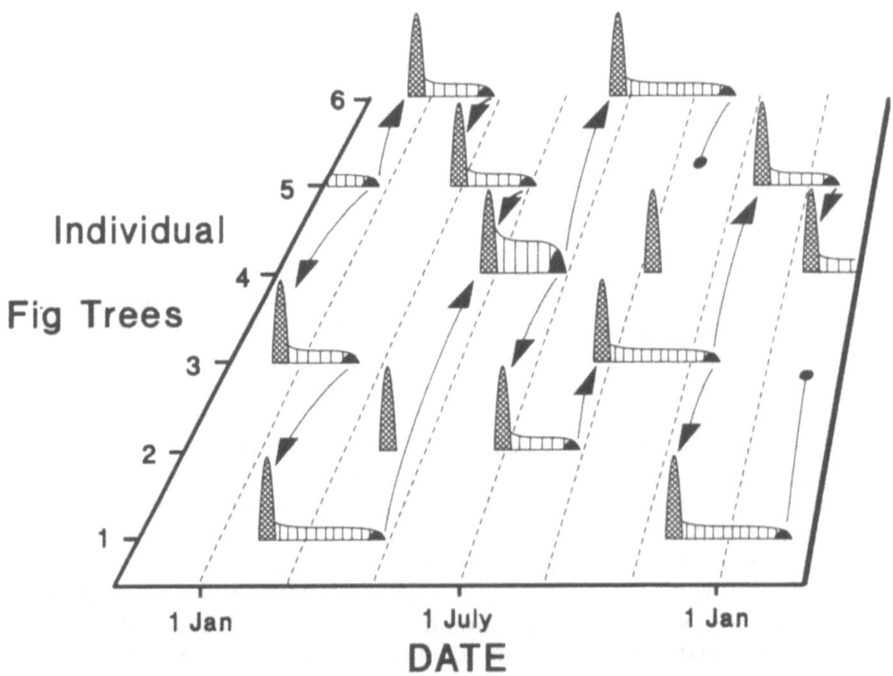


**Fig. 8.1.** Schematic representation of the interaction between fig wasps and monoecious figs. The implied third dimension represents the number of receptive syconia (*cross-hatching*), maturing syconia (*vertical stripes*), and ripe syconia (*black*) on individual fig trees. *Arrows* from one tree to another represent the movement of adult female fig wasps carrying pollen from ripe syconia to receptive syconia. *Circles* represent fig wasps unable to reproduce because no receptive syconia were available. Fig wasp eggs, larvae and pupae are located within the maturing syconia, and are not represented in the figure.

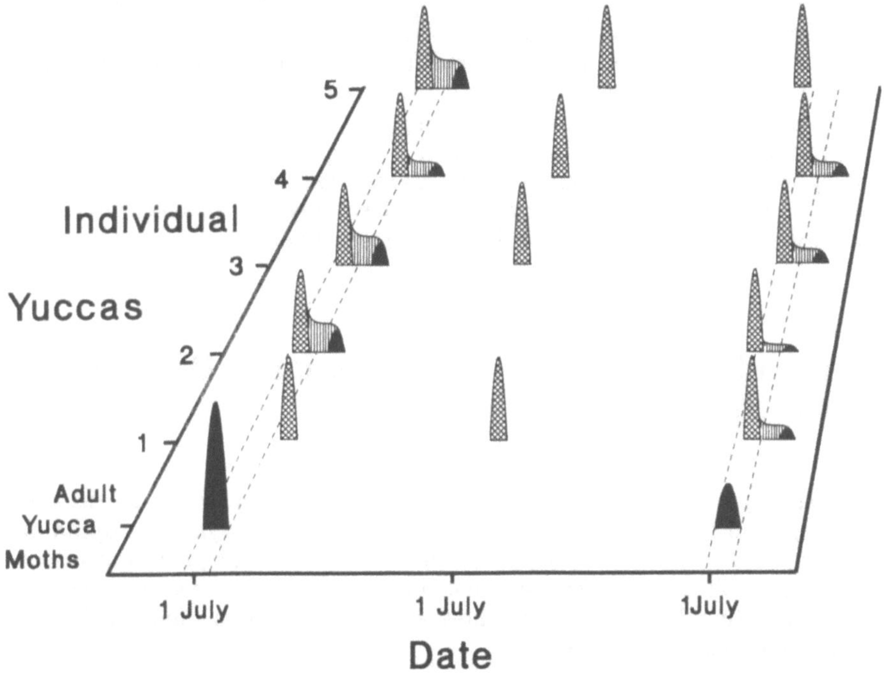


**Fig. 8.2.** Schematic representation of the interaction between yucca moths and yuccas. The implied third dimension represents the number of fresh flowers (*cross-hatching*), developing fruit (*vertical stripes*), and ripe fruit (*black*) for individual yuccas, and the number of adult yucca moths (*black*). *Broken lines* delineate the flight period of yucca moths. Yuccas that flower wholely or partially outside the flight season of yucca moths experience complete or partial reproductive failure. Yucca moth eggs and larvae in developing fruit and larvae and pupae in the soil are not explicitly represented in the figure. In the second flowering season, no yucca moths emerge, and yuccas fail to produce fruit. Prolonged diapause of yucca moth larvae in the soil allows the yucca moth population to persist. Fewer adult yucca moths in the third flowering season leads to lower fruit production per yucca.

pollen (Riley 1892). There are just three species of *Tegeticula* and one species of *Parategeticula* recognized in the literature (Davis 1967). Two of these four species are host specific, another pollinates two species of *Yucca*, while the fourth, *Tegeticula yuccasella*, was thought to pollinate all but two species of *Yucca*. However, *T. yuccasella* is actually a complex of species whose members also exhibit high, but not complete, host specificity (Miles 1983; J. F. Addicott, unpublished results). Throughout the remainder of this chapter, we use "yucca moths" to refer to moths in the genera *Tegeticula* and *Parategeticula*.

The interaction between yuccas and yucca moth begins in late spring, with newly eclosed yucca moths flying to yucca inflorescences where copulation occurs. Female yucca moths then gather a ball of pollen in their maxillary tentacles, fly to another yucca inflorescence, locate fresh flowers, oviposit into (or near) the yucca pistil, and actively transfer pollen to the receptive stigma (Riley 1892). These steps may be repeated in various sequences. Yucca moth larvae hatch in (or near) the yucca pistil and proceed to feed on the developing seeds, consuming 6–25 seeds (Addicott 1986; Keeley et al. 1984, 1986) before dropping to the soil in late summer and entering a pre-pupal diapause. Fruit maturation occurs shortly thereafter, and seeds are dispersed by either wind (Powell and Mackie 1966) or rodents (Wallen and Ludwig 1978).

The interactions between fig wasps and figs and between yucca moths and yuccas are portrayed in Figs. 8.1 and 8.2, respectively. Note the following: (a) adult fig wasps can be active at any time during the year, while adult yucca moths are active for only a short period each year; (b) reproduction is asynchronous among fig trees but synchronous among yucca infloresences; (c) fruit production in yuccas or figs may be complete or partial if they flower out of synchrony with the activity of adult yucca moths or fig wasps, respectively.

## 8.3 Increasing the Probability of Association Between Host and Symbiont

The probability of successful association between yuccas and yucca moths and between figs and fig wasps depends largely on the degree of synchronization of their life-cycles. In both systems, successful reproduction of plant and insect requires that a pollen-carrying adult female insect be active at the same time that pistils are receptive. We now consider how selection is likely to affect synchronization of life-cycles in these systems.

### 8.3.1 Yuccas and Yucca Moths

Synchrony of life-cycles within and between yuccas and yucca moths should be enhanced by (a) highly seasonal environments, (b) the obligate nature of the association, (c) problems of finding mates in low-density populations, and (d) resource limitation of fruit production leading to fruit abortion. However, asynchrony of life-cycles within and between yuccas and yucca moths would be enhanced by (a) lack of continuity of symbiosis between host and symbiont, and (b) high variation of environments between years.

*Seasonality*

Seasonality constrains the synchronization between the life-cycles of yuccas and yucca moths. In the highly seasonal environments in which most yucca species exist, conditions exceed the limits of tolerance of particular life-history stages of yuccas or yucca moths during some seasons. Yuccas experience two climatic constraints on flowering time: cold temperatures and drought. At higher elevations and latitudes, individual yuccas that flower too early or too late in the season are likely to suffer catastrophic reproductive failure because of frost. At lower and hotter locations, later flowering individuals may not have sufficient water for successful flowering or fruiting. Tolerances of yucca moths are probably even more narrow. Since most yucca moths are nocturnal ectotherms (Powell and Mackie 1966), their activity is limited to those seasons and elevations during which night-time temperatures are sufficient for flight activity (Cruden et al.

1976; Powell 1984). However, because they do not feed as adults, and because desiccation could seriously affect the longevity of adults (Kingsolver 1984), they may be absent from localities or seasons during which temperatures are high and humidity low.

### *Obligate Mutualism*

Because the interaction between yuccas and yucca moths is obligate for both, there is strong stabilizing selection for synchrony of flight and flowering periods (Rau 1945). Yucca moths that are active in the absence of yucca flowers experience complete lifetime reproductive failure. Any yucca that flowers before or after the flight period of moths will not be pollinated (see Fig. 8.2). For monocarpic yuccas (e.g. *Y. whipplei*) this results in complete lifetime reproductive failure (Powell and Mackie 1966). However, if sexual reproduction fails for polycarpic yuccas, an individual may flower in a subsequent year, and it has a higher probability of reproducing vegetatively (Kingsolver 1986). Partial reproductive failure of yuccas can occur when individuals flower earlier or later than the main flight season of yucca moths (Aker 1982a; Powell and Mackie 1966) (see Fig. 8.2). Therefore, the limit on flight and flowering imposed by seasonality should be further tightened to increase synchrony between flowering and flight.

### *Mate-Finding in Low-Density Populations*

Even shorter flight seasons would be favoured if female fecundity is lower early and late in the flight season as a result of delays in finding mates. Yucca moth densities can be very low, with less than one individual per inflorescence (Kingsolver 1984; Powell 1984). In addition, moth densities are particularly low during the early and late parts of the flight season (Aker 1982a; Rau 1945). Since male yucca moths appear to be relatively sedentary (Powell 1984), mating by female yucca moths could be delayed in low-density populations. The lifespan of yucca moths is generally just 2–5 days (Kingsolver 1984; Powell 1984; Rau 1945), and therefore a delay in mating would decrease the number of eggs that could be laid during the lifespan of a given female. This should favour even more synchronous emergence of yucca moths.

### *Fruit Abortion and Resource Limitation*

Another factor favouring shortening of the flight season is larval mortality associated with resource-limited fruit production and the resulting abortion of excess fruit (Aker 1982b; Aker and Udovic 1981; Udovic and Aker 1981). When yucca moth densities are high, yucca fruit production becomes limited not by availability of pollen but by other resources required for maturation of fruit (Udovic and Aker 1981). As a consequence considerable numbers of yucca fruit abort (Udovic and Aker 1981; Addicott 1985), causing the mortality of yucca moth eggs and larvae contained therein. Yucca moths that fly later are more likely to encounter flowers which, even if pollinated, are unlikely to be retained as fruit, since aborted yucca fruit are usually those initiated later on a given inflorescence (Aker and Udovic 1981). This should select for skewed flight seasons, with most individuals flying early. The emergence pattern of *T. yuccasella* on *Y. filamentosa* is skewed (Rau 1945), as predicted, but flight activity of *T. maculata* on *Y. whipplei* is symmetrical (Aker 1982a). Note that increased synchrony of yucca moth flight activity should further enhance the density of yucca moths for short periods of time, continuing to ensure that yucca reproduction remains resource-limited.

Resource limitation of fruit production and abortion of pollinated fruit does not occur in all species and populations of yuccas (Addicott 1985; Aker 1982a) or all individuals within a population (Aker 1982a). Therefore, we predict that yucca moths associated with populations of yuccas that consistently are pollen-limited should have longer and less skewed flight seasons.

### *Continuity of Symbiosis and Environmental Variation*

Two factors counter the selection for short, synchronous flight and flowering seasons of yucca moths and yuccas: (a) symbiosis between yuccas and yucca moths is not continuous, and (b) the environment is highly variable from year to year. Yucca moths are symbiotic with yuccas for only about 2 months per year, during their adult, egg and larval stages. The remainder of the year is spent in the soil in a pre-pupal diapause which can even last more than 1 year (Riley 1892). Since yuccas and diapausing yucca moths are not in close association, the environments experienced by one may not provide an error-free prediction of the environments experienced by the other, and therefore, perfect synchrony of flight and flowering would be hard to obtain.

The more variable the environment from year to year, the more critical would be the lack of continuity of symbiosis between yucca moths and yuccas. The proportion of individuals flowering can be highly variable from year to year (McKelvey 1938; Kingsolver 1984; Powell 1984; Riley 1892) and is apparently associated with variation in the availability of moisture from fall, winter and spring rains (Campbell and Keller 1932; Smith and Ludwig 1976). In environments in which yucca flowering is highly variable from year to year, such as the Mojave Desert (for *Y. schidgera*), synchrony between flight and flowering should be poorer than in less variable environments, such as on the Colorado Plateau (for *Y. kanabensis*).

Rau (1945) observed the arrival of *T. yuccasella* in perfect synchrony with the initiation of flowering, and interpreted this to indicate tight synchrony of emergence of yucca moths with flowering in yuccas. However, synchrony of arrival of moths with initiation of flowering is not necessarily indicative of synchrony with emergence. Moreover, synchrony of flight and flowering is very poor in other yucca–yucca moth systems (Kingsolver 1984; Powell 1984; Powell and Mackie 1966). Testing the hypothesis of decreased synchrony between yuccas and yucca moths in more variable environments will require more extensive observations, and censusing yucca moths away from flowering yuccas by using emergence traps or light traps.

Given that discontinuous symbiosis and environmental variation make unlikely the perfect synchronization of initiation of flowering and flight seasons, yuccas that produce flowers more asynchronously on an inflorescence should have a higher probability of receiving at least some pollination/oviposition visits from yucca moths (Aker 1982b). Individual yucca inflorescences produce flowers for as few as 3–4 days in *Y. schottii* (Powell 1984) and as long as 3–6 weeks in *Y. whipplei* (Powell 1984). The potential importance of a long flowering period is shown by individual plants that produce all their fruit from late flowers, as a result of just catching the beginning of the yucca moth flight season (Kingsolver 1984). Therefore, yuccas in more variable environments should produce flowers over longer periods of time. Variation in the timing and availability of yucca flowers from year to year, including the complete absence of flowers, should select for yucca moths that produce young that emerge over longer periods of time within a season or over more seasons.

In summary, duration and synchrony of flight and flowering should be constrained by the highly seasonal environments of yuccas, strongly reinforced by the obligate nature of the interaction between yucca and yucca moth, and further reinforced by problems of mate location in low-density populations and by fruit abortion. Conversely, the independence of yucca moths and yuccas during considerable periods of time and high variation in environmental conditions among years, provides selection for longer flowering and emergence periods, and for prolonged diapause.

## 8.3.2 Figs and Fig Wasps

Synchronization of life-cycles within and between figs and fig wasps can occur at three different levels: (a) within individual syconia, (b) among trees within populations, and (c) among syconia within trees. Synchrony of wasp emergence with maturity of male florets within individual syconia is high and should be maintained by the obligate interaction between figs and fig wasps. Synchrony of flowering and wasp production among trees within populations will be affected by competition for pollinators, competition among pollinators, and problems associated with flowering in low-density populations. Synchrony of flowering and wasp production within trees (crown synchrony) should be a function of the need to attract fig wasps, with synchronous crops providing a more powerful attractant if wasps are available, but asynchronous crops having a higher probability of flowering when at least one other tree is producing fig wasps.

### *Synchrony Within Syconia*

Synchrony of development of male florets and female fig wasps is under very strong selection due to the obligate nature of the association. Wasps that mature and leave the syconium prior to maturation of the stamens will experience lower reproductive success, because their offspring (which are seed predators) would starve. Wasps that mature well after anthers mature may not have access to pollen and are also likely to perish when ripe syconia are consumed by vertebrate frugivores (Janzen 1979a).

*Synchrony of Flowering and Fruiting Among Trees*

In order for figs and fig wasps to reproduce, some fig trees must bear receptive syconia at the same time that other fig trees are producing pollen and fig wasps (see Fig. 8.1). Since most fig species occur in relatively aseasonal environments, in the sense that growth and reproduction of figs and fig wasps is possible year-round, synchronization of life-cycles could occur in two ways. Reproduction of figs could be seasonal and synchronous among trees within a population, with maturation of one crop of syconia requiring 1 year and occurring at the time of receptivity of the next crop. Although such a life-history should be possible, it is not known to occur in figs. Instead, flowering, wasp production, and fruiting are asynchronous among trees within a population, and flowering and fruiting trees are present continuously throughout the year (see Fig. 8.1). We now examine the selection pressures that would tend to increase or decrease the degree of asynchrony of flowering and fig wasp production among trees.

Increased levels of asynchrony of flowering among trees would be favoured, because syconium maturation is frequently pollen-limited (Bronstein 1988a): the more wasps that arrive at a tree to pollinate it, the more mature syconia it is likely to produce (see Fig. 8.1). Therefore, for any fixed number of pollen- and wasp-producing figs, trees that flower when fewer neighbouring trees flower should attract more pollinators and have higher seed production. This supposition is supported by the observation that fig wasps nearly always enter the nearest available receptive syconia whether these are on the same tree on which they matured (Kjellberg et al. 1987) or a different tree (Kjellberg et al. 1988). Conversely, for a fixed number of receptive trees, trees that mature fruits and release pollen-carrying wasps when fewer other trees are doing so should have higher male fitnesses.

Asynchrony of maturation of fig wasps emerging from different trees should also be advantageous for fig wasps. For a fixed number of receptive trees, wasps emerging from fruiting trees when fewer other figs are producing fig wasps are more likely to be the first foundresses into receptive syconia. Although some workers have assumed that all foundresses would leave similar numbers of offspring (Frank 1985; Herre 1985), the first foundress into a syconium should have access to a greater number of ovaries for oviposition and should achieve higher fecundity (Janzen 1979a). However, there are not yet any data on the relative success of foundresses as a function of their order of arrival in receptive syconia, and therefore this hypothesis remains to be tested.

Asynchrony of flowering and wasp production among trees may become disadvantageous, if it results in flowering when few or no pollen-producing trees are available or releasing pollen-carrying wasps when few or no receptive trees are available (Bronstein et al. 1990) (see Fig. 8.1). Field evidence to support this hypothesis is not available, but the hypothesis is supported by the following predictions from conceptual and simulation models. First, the number of wasps arriving at a receptive tree should be a function of the number and size of crops producing fig wasps at that time (Bronstein 1989), and therefore flowering when there are few trees producing wasps is disadvantageous. Second, when the number of trees in a population is low, a significant proportion of them should fail to attract pollinators (resulting in failure in both female and male functions) or to donate pollen (resulting in failure in male function) (Bronstein et al. 1990). Third, if the number of trees in a population is very low, all trees with pollinated syconia could mature those syconia when there are no receptive trees nearby, leading to the local extinction of the fig wasp population (Bronstein et al. 1990).

The density of fig trees may be low continuously because of a lack of their relatively specific establishment sites (e.g. Michaloud and Michaloud-Pelletier 1987). Alternatively, the density of fig trees may be effectively low on a seasonal basis because of adverse climatic conditions. Both long duration and long intervals between flowering episodes decrease the number of trees in a population likely to be receptive or producing fig wasps at any given time, and therefore effectively lowers the density of the population. Within fig species, the duration of reproductive episodes is longer in cooler evironments (Wharton et al. 1980; Galil and Eisikowitch 1968) and during cooler seasons (Baijnath and Ramcharun 1988). Intervals between crops should also be longer in cool sites or seasons, because of longer times required for the accumulation of resources required for reproduction (Milton et al. 1982).

The proximate causes of asynchrony among trees are the somewhat variable duration and the extremely variable intervals between reproductive episodes (Bronstein et al. 1990). Variable durations of flowering and variable intervals between flowering among individual trees could be a function of genetic differences among individuals, but it is more likely that they arise because of age-dependent processes or chance events associated with asynchrony among trees (Michaloud 1988). For example, shorter durations and shorter intervals between flowering episodes occur when crop size is small (Michaloud 1988), and shorter intervals between flowering episodes occur for larger trees (Milton et al. 1982).

### *Synchrony Among Syconia Within Trees*

Synchronous production of syconia within individual fig trees would be favoured if it increases the probability of attracting fig wasps (Janzen 1979a; Bronstein 1989). Since flowering and fruiting are asynchronous among fig trees (see above), the average nearest-neighbour distance between pollen-producing trees and receptive trees is greater than the average nearest-neighbour distance for all fig trees in the population (Janzen 1979a). In order effectively to attract pollen-carrying fig wasps from greater distances, thus attracting more fig wasps, individual fig trees should produce synchronous crops of receptive syconia, maximizing the olfactory cues provided to fig wasps (Bronstein 1989; Janzen 1979a). For a given number of wasp-attracting and wasp-producing trees, the number of fig wasps arriving at a tree should be a function of the total size of its crop of syconia and their synchrony of development. There is great variation in the number of fig wasps arriving at trees (Bronstein 1987, 1988a; Michaloud et al. 1985), but the contribution that numbers of receptive syconia per tree makes to this is unknown.

However, less synchronous flowering and wasp production among syconia within trees should be favoured in populations of figs in which the density of trees is low, or in seasonal environments where the number of flowering/fruiting trees may be seasonally low. In low-density populations/seasons, reproductive failure of trees and fig wasps is likely (see above). Asynchrony of syconium receptivity and maturation within crops or overlap between successive crops should be favoured. For example, trees that produce relatively asynchronous crops of receptive syconia are more likely to experience at least partial reproductive success by being attractive to fig wasps for longer periods of time. Trees that produce relatively asynchronous crops of mature syconia are more likely to experience at least partial reproductive success by producing pollen-carrying fig wasps for longer periods of time. With extreme asynchrony among syconia within trees, some syconia could be receptive at the same time pollen-carrying wasps were emerging from other syconia on the same tree. This could result in a low level of self-pollination and the maintenance of the fig wasp population even when fig density is low (Ramirez 1970; Janzen 1979a). Alternatively, production of syconia on individual trees could remain synchronous, with discrete but overlapping crops of syconia. Figs that occur in sparse populations, perhaps because of a lack of suitable host trees for establishment, and figs in marginal habitats should be more likely to have some kind of crown asynchrony (Ramirez 1970; Janzen 1979a). However, this prediction has not been confirmed (Bronstein 1989).

## 8.4 Disruption of Mutualistic Life-Cycles

We now consider the circumstances under which a mutualistic life-history could break down to a non-mutualistic life-history, leading either to a polymorphism within a single species for mutualistic and cheating behaviour or to speciation leaving separate mutualist and "aprovechado" species. We examine two aspects of the yucca–yucca moth and fig–fig wasp mutualisms that would affect the probability of the development of cheaters or "aprovechados" from mutualists. First, the probability that cheating behaviour would be a viable alternative to mutualistic behaviour should be directly related to the frequency of larval mortality caused by fruit/syconium abortion. Cheaters/"aprovechados" would experience higher survivorship of their young, if they oviposited in fruit/syconia at a time when most fruit/syconium abortion had already occurred. We examine whether fruit/syconium abortion is likely to occur, given (a) the degree of pollinator or resource limitation experienced by figs or yuccas, (b) the probability of significant overexploitation of ovules within fruit/syconia by fig wasps or yucca moths, and (c) the existence of protogyny in figs and its absence in yuccas. Second, if fruit/syconium abortion does occur, the development of cheating from mutualism should be more likely when flowering is seasonal rather than continuous. With seasonal flowering, there should be selection for mutualistic behaviour when flowers predominate but non-mutualistic behaviour when fruit/syconia predominate.

### 8.4.1 Figs and Fig Wasps

There should be very limited selection for cheaters to arise from pollinating wasps associated with monoecious fig species, because there are no strong selection pressures for syconium abortion to occur, and flowering and any syconium abortion that does occur should be relatively continuous rather than seasonal in the population as a whole. Female fig wasps that enter and oviposit in syconia that subsequently abort experience complete reproductive failure. Therefore, if abortion of syconia were common, there could be strong selection for female fig wasps to oviposit in maturing, pollinated syconia that would be less likely to abort. Such females would be cheaters on the system, since they would not be providing a pollination service, but instead would be ovipositing into ovaries already pollinated by another fig wasp. However, three lines of evidence suggest that syconium abortion should be rare in figs.

First, mature syconium production in figs is frequently pollen-limited. Young syconia abort but in very few species do these syconia contain pollinating fig wasp foundresses (Bronstein 1988a). As we discussed above, pollen limitation is probably a function of the small effective size of populations of figs and the high degree of dispersion of figs resulting from asynchronous flowering and fruiting among trees. The chances of a receptive tree being close to a wasp-producing tree are small. Under conditions of pollen limitation, abortion of pollinated syconia should be maladaptive, because by doing so the fig tree would decrease both its male and female components of fitness (Bronstein 1988a).

Second, selective abortion of syconia is unlikely to occur, because exploitation of an excessively high proportion of ovaries by fig wasps is unlikely. The average proportion of ovaries attacked is high at 41%–77%, but the variance is low: few syconia have fewer than 25% or more than 85% of their ovaries attacked by agaonids or their parasites (Janzen 1979c). Although the precise mechanisms remain to be understood, the ratio of style length to ovipositor length seems to place an upper limit on the number of ovaries into which fig wasps can place eggs (Bronstein 1988b). And, since each fig wasp larva feeds in only a single ovary, high-intensity oviposition would eventually result not in an increasing loss of seeds within the syconium but instead in competition for oviposition sites. This prediction conforms well to the observation in *F. pertusa* that syconia entered by more foundresses actually produce fewer fig wasps (Bronstein 1988b).

Third, the existence of protogyny in figs limits the benefits that a fig would acquire through the abortion of these syconia in which a high proportion of ovaries contain fig wasp eggs. Although exploitation of ovaries by fig wasps decreases the female component of fitness of a fig, this cost is nearly compensated by an increase in the male fitness of figs, since female fig wasps carry pollen from the syconium. However, there is not a one-to-one trade-off between female and male fitness of figs, because only female fig wasps carry pollen. Male fig wasps contribute only indirectly to increasing the male fitness of figs by facilitating the exit of females from the ripening syconium. But the number of males required for this function is probably very small. Because of haplodiploid sex determination, the small number of foundresses entering syconia (Janzen 1979b), and mating by fig wasps within syconia, the proportion of male fig wasps is generally only 10%–15% (Frank 1985; Herre 1985). However, in fig species with larger syconia, there are many foundresses per syconium (Janzen 1979a), and the proportion of male offspring approaches 50% (J. F. Addicott and J. L. Bronstein, unpublished results). This is one of the few examples of the interests of fig wasps (adjustment of sex ratio based upon local mate competition) not corresponding with the interest of figs (loss of seeds but a corresponding gain in pollen transfer) (Herre 1989). Therefore, if selective abortion of heavily exploited syconia occurs, it should be more common in fig species with large syconia.

In the absence of syconium abortion, there should be little or no advantage for fig wasps to cheat by ovipositing in maturing pollinated syconia. If fig trees flower asynchronously year-round and if the density of trees is high, then a female agaonid would usually have access to trees with receptive syconia as well as trees with maturing pollinated syconia. Since female fig wasps ovipositing in receptive figs should still have access to more ovaries, pollinating behaviour should be selected over cheating behaviour. However, cheating could be advantageous any time the number of trees with receptive syconia is low. Cheaters would have a greater chance of finding a tree with syconia suitable for oviposition, since oviposition could occur in maturing as well as receptive syconia. Therefore, we predict that cheating fig wasps are more likely to exist on species of figs where there is a seasonal period of low flowering or on species with very low-density populations.

Another way for female fig wasps to cheat would be for them to enter receptive syconia to oviposit without pollinating. However, this pattern of cheating is unlikely for two reasons. First, the bract-filled ostiole seals shut several hours after the first foundresses enter a syconium (Janzen 1979a), probably eliminating the possibility that such a cheating strategy could evolve. Second, a female fig wasp that does not pollinate would suffer complete reproductive failure unless a pollinating fig wasp also enters the same syconium. Therefore, there would be strong selection pressure against cheaters being the first to enter a syconium, since in fig species with small syconia a high proportion of syconia would not be entered by another fig wasp (Herre 1989; Janzen 1979b), and the syconium would abort unpollinated. Therefore, we predict that if cheaters do exist among fig wasps, they should be more common in fig species with larger syconia that are more likely to be entered by many foundresses.

Non-pollinating fig wasps could avoid the risk of complete reproductive failure if they entered only syconia which had already been entered by other (presumably pollinating) fig wasps. However, even here there should be no advantage to cheating. Although there are no data on the relative success of foundresses as a function of their order of entry into a syconium (see above), the second or subsequent female agaonid to enter a syconium should have access to fewer ovaries and should experience decreased fecundity in comparison to the first foundress (Janzen 1979a), because there is a limited number of ovaries accessible to fig wasps within a syconium. This effect would be more likely to occur in species of figs with small syconia. Therefore we predict that cheaters derived from mutualists should be more common on figs with large syconia.

There is very limited evidence for the existence of cheaters or "aprovechados" derived from pollinating fig wasps. This could reflect the difficulty of observing the behaviour of pollinating/ovipositing wasps within syconia, or a lack of looking for cheaters/"aprovechados" among fig wasps. However, there is no evidence for the existence within species of individuals that facultatively cheat or morphs that obligately cheat. There is only one example of an agaonid "aprovechado" that utilizes developing ovaries without pollinating (Galil and Eisikowitch 1968). This species is apparently derived from a pollinating agaonid (Janzen 1979a).

A diverse group of non-pollinating sycophagine chalcidoid wasps (primarily Torymidae) interact in the fig system. Most do not enter the syconium, but instead oviposit through the receptacle wall some time after pollination has occurred. They are either cleptoparasites of the agaonids, seed predators, or gall formers (for reviews, see Janzen 1979a; Bronstein 1986). To the extent that torymids of figs are seed predators, they could be considered "aprovechados" of the mutualism. However, their existence is not relevant to the problem of the breakdown of mutualism, because they are presumably not derived from agaonids and are therefore equivalent to other seed predators or parasitoids in this system.

### 8.4.2 Yuccas and Yucca Moths

In yuccas there should be seasonally varying selection for cheating, because fruit abortion is common and flowering is seasonal. Consider first the same three factors that would affect the probability of mortality of yucca moth larvae caused by fruit abortion.

First, fruit set is frequently resource-limited in yuccas and consequently fruit abortion is common (Aker 1982b; Addicott 1985). In *Y. whipplei* approximately 51% of all initiated fruit eventually abort (Udovic and Aker 1981). Resource limitation and fruit abortion are likely to exist in a positive feedback loop with the oviposition behaviour of yucca moths. If abortion occurs it will select against moths pollinating and ovipositing in flowers that will abort, resulting in increased synchrony of emergence of yucca moths (see above). However, pollination synchrony would reduce pollen limitation on individual inflorescences, further increasing the likelihood of fruit abortion on them. Given that fruit abortion occurs, female yucca moths should pollinate many flowers and place only a few eggs in each pistil, resulting in further reductions of pollen limitation and increased probabilities of fruit abortion. For example, female *T. maculata* place only two to three eggs in each pistil that they pollinate before moving to another pistil (Aker 1982a).

Second, because of the mating system of yuccas, there is no selection against yuccas selectively aborting fruit in which yucca moth larvae exploit a high proportion of ovules. In yuccas there is no direct link between loss in female fitness, as the result of seed predation, and gain in male fitness, as the result of pollen transfer: pollen transfer is performed by the individual that oviposits, not (as in figs) by her offspring that destroy seeds. Therefore, an individual yucca plant is not constrained against "defecting" against the yucca moth larvae in its developing fruit. The yucca has already achieved all the gain in its male fitness that it can for that flowering cycle, and abortion of overexploited fruit would serve to decrease its loss of female fitness through the allocation of resources to ovules that will produce seeds rather than yucca moth larvae.

Third, yucca moth larvae can exploit the entire crop of seeds within a given fruit (Addicott 1986; Rau 1945). Although there is an apparent aversion to oviposition in locules that have already received an oviposition (Kingsolver 1984; Aker and Udovic 1981), there is no direct limit imposed on the number of eggs that can be placed in a pistil. For example, more than 60 eggs of some species of the *T. yuccasella* complex are placed in a single pistil (J. F. Addicott, unpublished data). Any more than 12–15 larvae per fruit results in the complete destruction of the seeds in a given pistil (Rau 1945; Addicott 1986), because yucca moth larve are not restricted to feeding in a single seed or even in a single locule.

The combination of high levels of fruit abortion and the existence of seasonal flowering in yuccas should provide strong selection for the development of cheaters from mutualistic yucca moths. First, seasonal flowering of yuccas results in yucca moths experiencing different floral/fruit environments depending upon when they fly (Powell and Mackie 1966). Moths flying early will encounter only fresh flowers. Moths flying progressively later will encounter fewer plants with fresh flowers and more plants with developing fruit. If fruit abortion occurs, there would be differential selection for mutualists or cheaters as a function of when moths fly. Mutualistic behaviour would be favoured early, since a female can ensure the survival of many of her offspring by pollinating flowers and ovipositing in them. Individuals ovipositing into, but not pollinating, fresh flowers would not be assured that the particular flower would ever be pollinated and retained by the plant. Conversely, late in the season individuals that oviposit into fruit would be favoured, since developing fruit are more likely to be retained by the plant than would be newly pollinated flowers (Aker and Udovic 1981). Therefore, cheating by yucca moths should occur and should be temporally dependent, with moths flying early being predominantly mutualistic and moths flying later being predominantly cheaters.

In comparison with fig wasps, there is extensive evidence for the existence of cheaters/"aprovechados" derived from mutualistic yucca moths. There is facultative cheating by individual *T. maculata* interacting with *Y. whipplei* (Aker and Udovic 1981). Moths with normally developed maxillary tentacles and full pollen loads oviposit in well-developed fruit rather than in fresh flowers, but only late in the flowering season and only on inflorescences that still have at least a few fresh flowers (Aker and Udovic 1981).

There are also at least four species of "aprovechados" within the *T. yuccasella* complex. Females of these "aprovechados" lack maxillary tentacles for carrying pollen (Davis 1967), oviposit into developing fruit rather than into fresh flowers, and fly later than sympatric members of the *T. yuccasella* complex that pollinate (J. F. Addicott, unpublished results). These "aprovechados" are more closely related to other members of the *T. yuccasella* complex than to other species of *Tegeticula* or to *Parategeticula pollinifera*, and cheating has arisen independently at least twice within the *T. yuccasella* complex (J. F. Addicott, unpublished results). These "aprovechados" are also widespread geographically, frequent (but not ubiquitous) on many different *Yucca* hosts, and numerically abundant where they occur (J. F. Addicott, unpublished results).

## 8.5 Conclusions

In this paper we have addressed the maintenance and breakdown of mutualistic systems by examining two fundamental aspects of the life-histories of symbiotic mutualistic species: how mutualists increase the probability of successful association between them, and under what conditions the mutualists could become cheaters. The obligate pollination–seed predation mutualisms involving figs and fig wasps, and yuccas and yucca moths experience different constraints and selection pressures for increasing the probability of successful association between host and symbiont. In one case it is achieved through extreme synchrony of life-cycles within and between mutualists, and in the other case through strong asynchrony. In yucca moths, there should be strong stabilizing selection for short, synchronous flight and flowering periods, imposed by environmental constraints, obligate mutualism, difficulty of finding mates in low-density populations, and the necessity to pollinate and oviposit early so as to avoid fruit abortion by yuccas. These selection presssures are countered by the lack of year-round association between yuccas and moths and by highly variable environments. In the case of fig wasps there should be selection for synchrony of reproduction within trees and asynchrony of reproduction among trees imposed by protogyny in figs and competition for and among pollinating fig wasps. However, selection for synchrony within trees and asynchrony among trees will be countered by complete or partial reproductive failure of figs or wasps when the density of trees is low.

The fig and yucca systems differ markedly in the extent to which cheaters or "aprovechados" derived from mutualists are known to occur. They are very common in the yucca–yucca moth system, but almost unknown and presumably rare in the fig–fig wasp system. These observations are consistent with our hypotheses for what factors should favour the development of cheaters from mutualists. These arguments are based largely on selection pressures imposed by larval mortality caused by fruit or syconium abortion, which is, and is expected to be, common in yuccas but rare in figs. With fruit abortion and seasonal flowering in yuccas, individual yucca moths that fly at the end of the flight season could increase their fitness by being cheaters rather than mutualists. With syconium abortion being a rare phenomenon in figs, selection for cheating would have to be based upon other factors. However, neither cheating by entering syconia but failing to pollinate nor cheating by ovipositing in maturing syconia appears to be favoured over mutualistic behaviour. This does not imply that "aprovechados" could not exist in this system, as numerous species of torymids exist and are either seed predators or parasites of agaonids. However, it does imply that mutualists are unlikely to give rise to cheaters or "aprovechados".

Our analysis has suffered from a lack of adequate data sets against which to compare our predictions. Therefore, part of the contribution of this paper has been to suggest life-history variables that need to be observed in the fig and yucca systems. For example, understanding the probability of cheaters arising from mutualistic fig wasps depends upon knowing whether the order in which an agaonid foundress enters a syconium affects her fecundity. Our arguments have also led to numerous predictions for patterns that could be easily observed that would provide support or refutation of our arguments. For example, cheating fig wasps should be much more common on figs with large syconia than on figs with small syconia. Therefore, our analysis provides a basis for understanding synchrony and cheating in the fig and yucca systems based upon presently available knowledge, but it also identifies critical variables or observations that would further refine our understanding of the maintenance and breakdown of these two pollination–seed predation mutualisms.

# References

Addicott JF (1981) Stability properties of 2-species models of mutualism: simulation studies. Oecologia 49: 42–49

Addicott JF (1985) Competition in mutualistic systems. In: Boucher DH (ed) The biology of mutualism. Oxford University Press, Oxford, pp 217–247
Addicott JF (1986) Variation in the costs and benefits of mutualism: the interaction between yuccas and yucca moths. Oecologia 70: 486–494
Aker CL (1982a) Spatial and temporal dispersion patterns of pollinators and their relationship to the flowering strategy of *Yucca whipplei* (Agavaceae). Oecologia 54: 243–252
Aker CL (1982b) Regulation of flower, fruit and seed production by a monocarpic perennial, *Yucca whipplei*. J Ecol 70: 357–372
Aker CL, Udovic, D (1981) Oviposition and pollination behavior of the yucca moth, *Tegeticula maculata* (Lepidoptera: Prodoxidae), and its relation to the reproductive biology of *Yucca whipplei* (Agavacae). Oecologia 49: 96–101
Baijnath H, Ramcharun S (1988) Reproductive biology and chalcid symbiosis in *Ficus burtt-davyi* (Moraceae). Monogr Syst Bot MO Bot Gard 25: 227–235
Barker NP (1985) Evidence of a volatile attractant in *Ficus ingens* (Moraceae). Bothalia 15: 607–611
Berg CC (1989) Classification and distribution of *Ficus*. Experientia 45: 605–611
Brantjes NBM (1976a) Prevention of superparasitism of *Melandrium* flowers (Caryophyllaceae) by *Hadena* (Lepidoptera). Oecologia 24: 1–6
Brantjes NBM (1976b) Riddles around the pollination of *Melandrium album* (Mill.) Garcke (Caryophyllaceae) during the oviposition by *Hadena bicuris* Hufn. (Noctuidae, Lepidoptera), II. Proc K Ned Akad Wet Ser C 79: 127–141
Bronstein JL (1986) Coevolution and constraints in a neotropical fig–pollinator wasp mutualism. PhD thesis. University of Michigan, Ann Arbor, MI
Bronstein JL (1987) Maintenance of species-specificity in a neotropical fig–pollinator wasp mutualism. Oikos 48: 39–46
Bronstein JL (1988a) Fruit production in a monoecious fig: consequences of an obligate mutualism. Ecology 69: 207–214
Bronstein JL (1988b) Mutualism, antagonism, and the fig–pollinator interaction. Ecology 69: 1298–1302
Bronstein JL (1989) A mutualism at the edge of its range. Experientia 45: 622–636
Bronstein JL, McKey D (1989) The fig/pollination mutualism: a model system for comparative biology. Experientia 45: 601–604
Bronstein JL, Gouyon P-H, Gliddon C, Kjellberg F, Michaloud G (1990) Ecological consequences of flowering asynchrony in monoecious figs: a simulation study. Ecology (in press).
Brown JH, Kodric-Brown A (1979) Convergence, competition, and mimicry in a temperate community of hummingbird pollinated flowers. Ecology 60: 1022–1035
Campbell RS, Keller JG (1932) Growth and reproduction of *Yucca elata*. Ecology 13: 364–374
Condit IJ (ed) (1947) The fig. Chronica Botanica, Waltham, MA
Cruden RW, Kinsman S II, Stockhouse RE, Linhart YB (1976) Pollination, fecundity, and the distribution of moth-flowered plants. Biotropica 8: 204–210
Davis DR (1967) A revision of the moths of the subfamily Prodoxinae (Lepidoptera: Incurvariidae). US Natl Mus Bull 255: 1–170
Dodson CH, Dressler RL, Hills HG, Adams RM, Williams NH (1969) Biologically active compounds in orchid fragrances. Science 164: 1243–1249
Frank SA (1985) Hierarchical selection theory and sex ratios: II. On applying the theory, and a test with fig wasps. Evolution 39: 949–964
Galil J, Eisikowitch D (1968) On the pollination ecology of *Ficus sycomorus* in East Africa. Ecology 49: 259–269
Galil J, Eisikowitch D (1974) Further studies on pollination ecology in *Ficus sycomorus*. II. Pocket filling and emptying by *Ceratosolen arabicus* Mayr New Phytol 73: 515–528
Galil J, Neeman G (1977) Pollen transfer and pollination in the common fig (*Ficus carica* L.). New Phytol 79: 163–171
Heinrich B (1979) Bumblebee economics. Harvard University Press, Cambridge, MA
Herre EA (1985) Sex ratio adjustments in fig wasps. Science 228: 896–898
Herre EA (1989) Coevolution of reproductive characteristics in twelve species of New World figs and their pollinator wasps. Experientia 45: 637–646
Janzen DH (1979a) How to be a fig. Annu Rev Ecol Syst 10: 13–51
Janzen DH (1979b) How many parents do the wasps from a fig have? Biotropica 11: 127–129
Janzen DH (1979c) How many babies do figs pay for babies? Biotropica 11: 48–50
Keeley JE, Keeley SC, Swift CC, Lee J (1984) Seed predation due to the yucca–moth symbiosis. Am Midl Nat 112: 187–191

Keeley JE, Keeley SC Ikeda DA (1986) Seed predation by yucca moths on semelparous, iteroparous and vegetatively reproducing subspecies of *Yucca whipplei* (Agavaceae). Am Midl Nat 115: 1–9
Kingsolver RW (1984) Population biology of a mutualistic association: *Yucca glauca* and *Tegeticula yuccasella*. PhD dissertation, University of Kansas
Kingsolver RW (1986) Vegetative reproduction as a stabilizing feature of the population dynamics of *Yucca glauca*. Oecologia 69: 380–387
Kjellberg F, Gouyon P-H, Ibrahim M, Raymond R, Valdeyron G (1987) The stability of the symbiosis between dioecious figs and their pollinators: a study of *Ficus carica* L. and *Blastophaga psenes* L. Evolution 41: 693–704
Kjellberg F, Doumesche B, Bronstein JL (1988) Longevity of a fig wasp (*Blastophaga psenes*). Proc K Ned Akad Wet Ser C. 91: 117–122
McKelvey SD (1938) Yuccas of the Southwestern United States, Part 1. The Arnold Arboretum of Harvard University, Jamaica Plain, MA
Michaloud G (1988) Aspects de la réproduction des figuiers monoïques en forêt équatoriale africaine. Thesis, Université des Sciences et Techniques du Languedoc, Montpellier, France
Michaloud G, Michaloud-Pelletier S (1987) *Ficus* himi-épiphytes (Moraceae) et arbres supports. Biotropica 19: 125–136
Michaloud G, Michaloud-Pelletier S, Wiebes JT, Berg CC (1985) The co-occurrence of two pollinating species of fig wasp and one species of fig. Proc K Ned Akad Wet Ser C 88: 93–120
Miles NJ (1983) Variation and host specificity in the yucca moth, *Tegeticula yuccasella* (Incurvariidae): a morphometric approach. J Lepid Soc 37: 207–216
Milton K, Windsor DM, Morrison DW, Estribi MA (1982) Fruiting phenologies of two neotropical *Ficus* species. Ecology 63: 752–762
Pellmyr O (1989) The cost of mutualism: interactions between *Trollius europaeus* and its pollinating parasites. Oecologia 78: 53–59
Powell JA (1984) Biological interrelationships of moths and *Yucca schottii*. Univ Calif Berkeley, Publ Entomol 100: 1–93
Powell JA, Mackie RA (1966) Biological relationships of moths and *Yucca whipplei* (Lepidoptera: Gelechiidae, Blastobasidae, Prodoxidae). Univ Calif, Berkeley, Publ Entomol 42: 1–59
Ramirez BW (1969) Fig wasps; mechanisms of pollen transport. Science 163: 580–581
Ramirez BW (1970) Host specificity of fig wasps (Agaonidae). Evolution 24: 681–691
Rau P (1945) The yucca plant, *Yucca filametosa*, and the yucca moth, *Tegeticula (Pronuba) yuccasella* Riley: an ecologico-behavior study. An MO Bot Gard 32: 373–394
Riley CV (1892) The yucca moth and yucca pollination. Annu Rept MO Bot Gard 3: 99–159
Roughgarden J (1975) Evolution of marine symbiosis – a simple cost–benefit model. Ecology 56: 1201–1208
Smith SD, Ludwig JA (1976) Reproductive and vegetative growth patterns in *Yucca elata* Engelm. (Liliaceae). Southwest Nat 21: 177–184
Soberon Mainero J, Martinez del Rio C (1985) Cheating and taking advantage in mutualistic associations. In: Boucher DH (ed) The biology of mutualism. Oxford University Press, Oxford, pp 192–216
Thompson JN (1982) Interaction and coevolution. Wiley, New York
Udovic D (1981) Determinants of fruit set in *Yucca whipplei*: reproductive expenditure vs. pollinator availability. Oecologia 48: 389–399
Udovic D, Aker C (1981) Fruit abortion and the regulation of fruit number in *Yucca whipplei*. Oecologia 49: 245–248
Wallen DR, Ludwig JA (1978) Energy dynamics of vegetative and reproductive growth in Spanish bayonet (*Yucca baccata* Torr.). Southwest Nat 23: 409–422
Webber JM (1953) Yuccas of the southwest. US Dept Agric Agric Monogr 17
Wharton RA, Tilson JW, Tilson RL (1980) Asynchrony in a wild population of *Ficus sycomorous* L. S Afr J Sci 76: 378–480
Wickler W (1968) Mimicry in plants and animals. McGraw-Hill Company, New York
Wolin CL (1985) The population dynamics of mutualistic systems. In: Boucher DH (ed) The biology of mutualism. Oxford University Press, Oxford, pp 248–269

*Chapter 9*

# Testing Hypotheses About the Evolution of the Mean Phenotype in Temporally Variable Environments

Fritz Taylor

## 9.1 Introduction

It is almost axiomatic in the study of natural selection that fitness functions often exhibit intermediate optima (Endler 1986), i.e. selection is stabilizing or normalizing, or what I shall call here optimizing, and that the optimum varies between generations. This form of selection has been studied in many interesting systems, for example reproductive effort (Schaffer 1974; Hastings and Caswell 1979; Bulmer 1985), the timing of seed germination (Cohen 1966; Bulmer 1984; Venable and Brown 1988) and flowering in plants (Cohen 1971; King and Roughgarden 1982), dormancy in arthropods (Cohen 1970; Taylor 1986a, b), and clutch size in birds (Boyce and Perrins 1987).

A critical theoretical question pertaining to variable optimizing selection is prediction of the phenotypic mean. The generally accepted procedure for dealing with this problem involves solving for the phenotype maximizing the geometric mean fitness. This result provides the correct result for sexual populations when the fitness function and phenotypic distribution are symmetric and with autocorrelation, but no age structure (Orzack 1985). The arithmetic mean of the optimum, however, yields the same result (F. Taylor, unpublished results). When the fitness function or phenotypic distribution are asymmetric, the geometric mean criterion does not predict the long-term mean phenotype, which in these instances is predicted as the phenotype for which the mean fitness differential is zero (F. Taylor, unpublished results).

Independent of the method used to predict the asymptotic mean phenotype, the empirical problem remains – over how many generations must the mean be measured to estimate the true mean with a desired level of precision? In this chapter I investigate the consequences of an initial observation that temporally variable selection generates high levels of autocorrelation in the mean phenotype. This autocorrelation creates substantial problems in testing hypotheses about the long-term mean phenotype and raises questions about the meaning of adaptation. Earlier work has dealt with the bias caused by autocorrelation in estimating realized heritability from the results of directional selection (Hill 1972a, b), but its role in variable selection in nature has apparently not been considered explicitly.

## 9.2 The Model

The three elements in this discrete-generation model interact as follows: the *phenotypic distribution* in each generation is acted upon by the *fitness function*, which has an optimum. The optimum phenotype is chosen randomly each generation from a *distribution of optima*.

In more detail, the model works in the following manner. The phenotypic distribution is gaussian with a standard deviation, $\sigma_z$, that is a constant 10 phenotypic units and this scales all other elements of the model. The constancy of the phenotypic standard deviation assumes that selection is an ongoing process and that the phenotypic distribution has arrived at an equilibrium variance caused by a balance between mutation and selection (Lande 1975; Turelli 1988). The mean phenotype, $\bar{z}$, changes with each generation and is the variable of central interest. The fitness function has the shape of a normal distribution, which gives a process called nor-optimal selection, characterized by an optimum, $z^*$, and a constant "standard deviation", $\sigma_w$, which measures the strength of selection. A small value of $\sigma_w$ causes fitness to decline rapidly from the optimum and therefore implies strong selection. The mean phenotype is determined in successive generations by the following formula (Lande 1976; Emlen 1984):

$$\bar{z}_{t+1} = \bar{z}_t - (\bar{z}_t - z_t^*)h^2 \frac{\sigma_z^2}{\sigma_z^2 + \sigma_w^2} \qquad (9.1)$$

where $t$ refers to generation $t$ and $h^2$ specifies the heritability of the phenotype. In each generation, the optimum is chosen at random from a gaussian distribution characterized by a mean, $\mu^*$, and a standard deviation, $\sigma^*$. The algorithm for the generation of random normal deviates was given by Press et al. (1986). Where relevant, autocorrelation, $\rho^*$, was introduced into the time series of optima as described below.

The main parameters that will be varied in this model are the strength of selection, $\sigma_w$, variability of the optimum, $\sigma^*$, autocorrelation of the optimum, $\rho^*$, and heritability, $h^2$.

## 9.3 Results

### 9.3.1 Generation of Autocorrelation by Selection

A time series of optima exhibiting a temporal pattern of white noise (i.e. having no autocorrelation) generates a smoothed time series of mean phenotypes with much less fluctuation (i.e. having high levels of autocorrelation; cf. Fig. 9.1a and

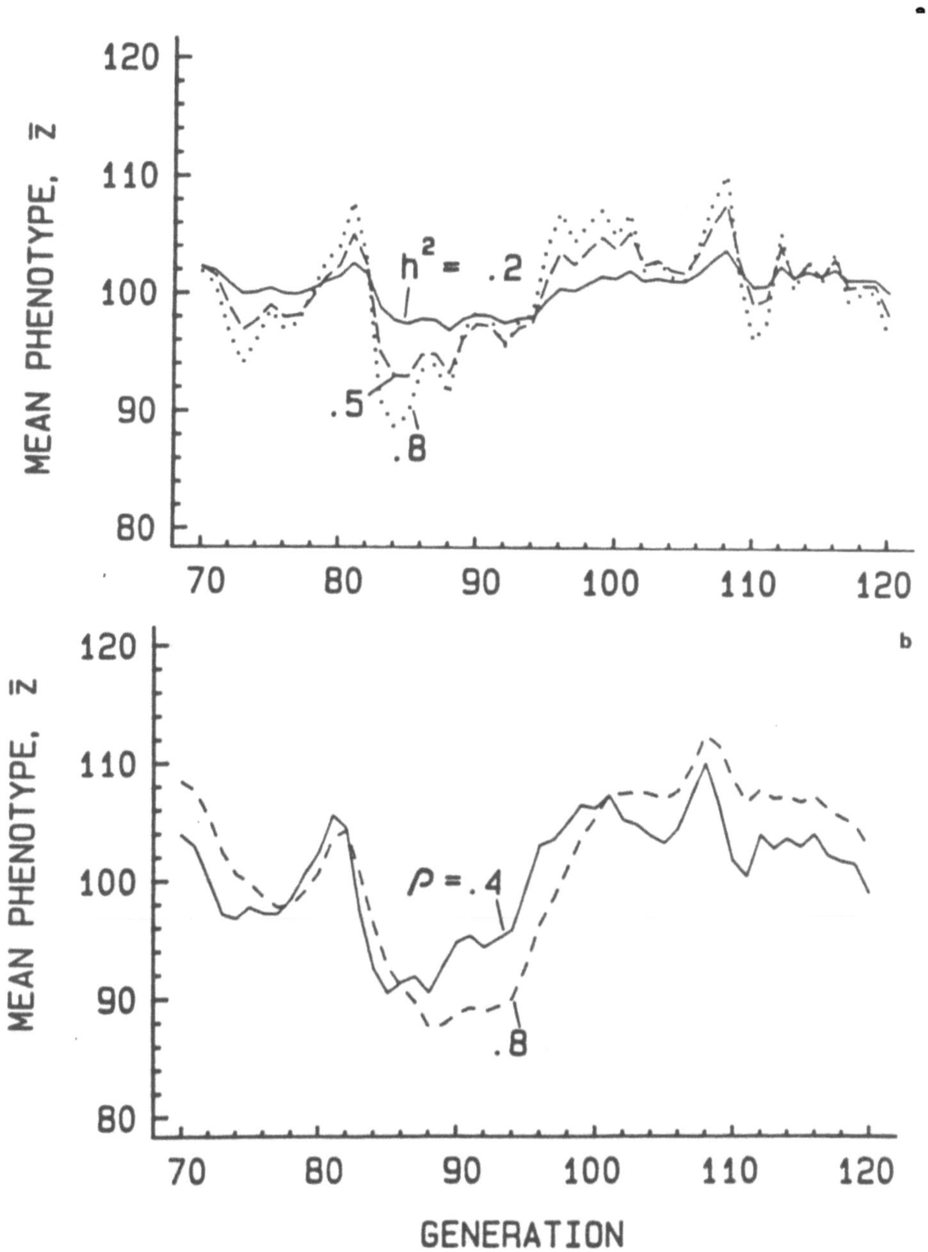


**Fig. 9.1.** Representative time series of the mean phenotype. Intervals of 50 generations were chosen for single realizations of the model. All were generated by the same sequence of random numbers used to choose the optimum and for all $\sigma^* = \sigma_w = 15$ phenotypic units. **a** Series of mean phenotypes generated by optima having zero autocorrelation. **b** Series of mean phenotypes generated by optima with variance computed before the autocorrelations indicated were added. Here $h^2 = 0.5$. Note that the mean phenotype traverses farther for longer periods from the long-term mean of 100 phenotypic units than in **a**.

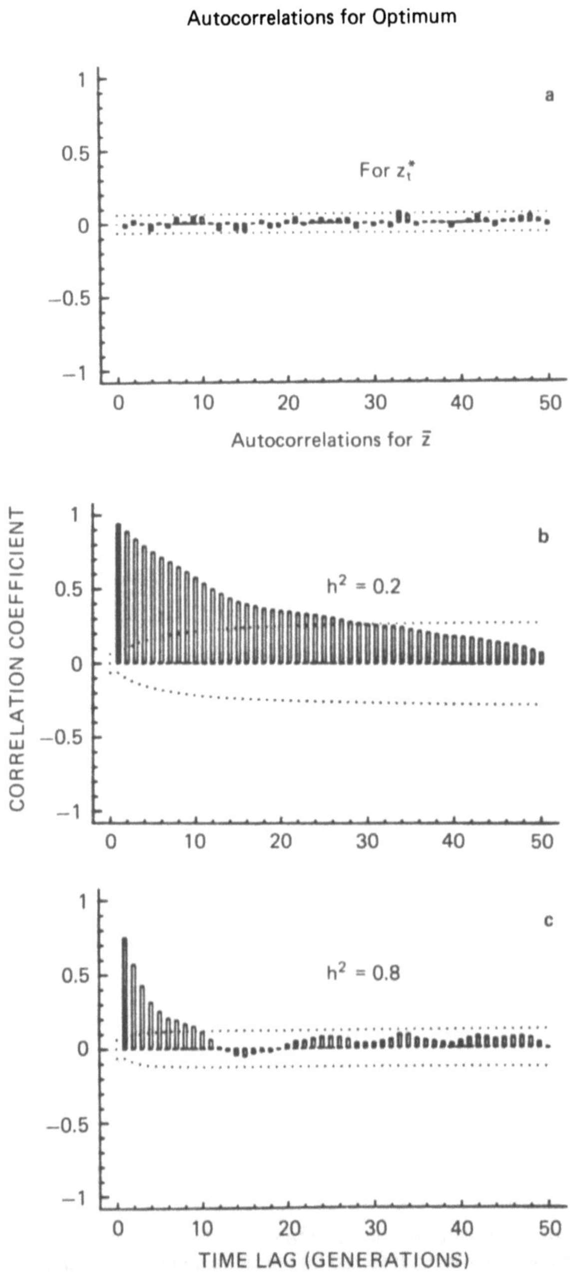


**Fig. 9.2.** Autocorrelation functions for various time series. *Bars* extending beyond the *broken curves* are significant at the $P = 0.05$ level. **a** For the optimum randomly chosen from a normal distribution with $\sigma^* = 15$. **b** For the mean phenotype when $\sigma_w = \sigma^* = 15$ and $h^2 = 0.2$. **c** For the mean phenotype as before but $h^2 = 0.8$. In **b** and **c** each function is computed for 1200 values of the mean.

b). The magnitude of the fluctuations is largely influenced by the value of heritability. In Fig. 9.1a, for example, the mean phenotype in a population with $h^2 = 0.2$ remains very close to the long-term mean, $\mu^* = 100$ phenotypic units, whereas a population with $h^2 = 0.8$ and experiencing the same series of $z^*$ values traverses far from the long-term mean. This pattern is revealed more dramatically by the autocorrelation functions associated with each time series (Fig. 9.2). As expected, the time series for $z^*$ lacks significant correlations at all time intervals (Fig. 9.2a). The series for $h^2 = 0.2$ begins with an extremely high autocorrelation between adjacent generations, $r = 0.94$, and the magnitude of the correlation declines slowly, remaining significant up to a time lag of 28 generations (Fig. 9.2b). The time series for $h^2 = 0.8$ also starts with a high correlation between adjacent generations, $r = 0.76$, but in this case the correlation declines rapidly and is significant only for nine generations (Fig. 9.2c).

These results can be derived analytically by first observing that Eq. (9.1) is a first-order autoregressive process of the form:

$$\bar{z}_{t+1} = (1-a)\bar{z}_t + az^*_t \tag{9.2}$$

where $a = h^2\sigma_z^2/(\sigma_z^2 + \sigma_w^2)$ so that $0 \leqslant a \leqslant 1$. The standard analysis of such processes (Chatfield 1975) gives the asymptotic variance of $\bar{z}$ as:

$$\sigma_{\bar{z}}^2 = \frac{a(\sigma^*)^2}{2-a} \tag{9.3}$$

and the autocorrelation coefficient as:

$$\rho_{\bar{z}} = 1-a \tag{9.4}$$

It is of interest that, whereas the variance of $\bar{z}$ is influenced by the level of environmental variability measured as $\sigma^*$, the autocorrelation is not (because $\sigma^*$ cancels out when the covariance is standardized by $\sigma_{\bar{z}}^2$, which includes $\sigma^*$.) Thus, given that $\sigma_z = 10$, $\rho_{\bar{z}}$ is only a function of $h^2$ and $\sigma_w$ (Fig. 9.3). As noted earlier in the simulation results, the levels of autocorrelation generated by variable optimizing selection are always high. Low values of $h^2$ combined with weak selection (high $\sigma_w$) produce the highest levels of autocorrelation. These combinations, however, also yield the lowest standard deviations for $\bar{z}$ (Fig. 9.3). This trade-off will be important below in considerations of testing hypotheses about $\bar{z}$ because the autocorrelation is only of consequence when the variance is large.

## 9.3.2 Convergence of the Long-Term Phenotypic Mean

For each set of parameters, the simulation produced time series of the mean phenotype for 300 generations, which were replicated 500 times. Thus, for each set of parameters the simulation can be thought of as generating an array with 300 columns, corresponding to generations, and 500 rows, corresponding to replicate runs. All simulations were produced using the same series of random numbers to generate values of $z^*$. From such an array, I performed two types of analysis. First, to study convergence of the standard deviation of the mean phenotype, I computed the standard deviation of $\bar{z}$ for $t$ generations, i.e. across a row in the

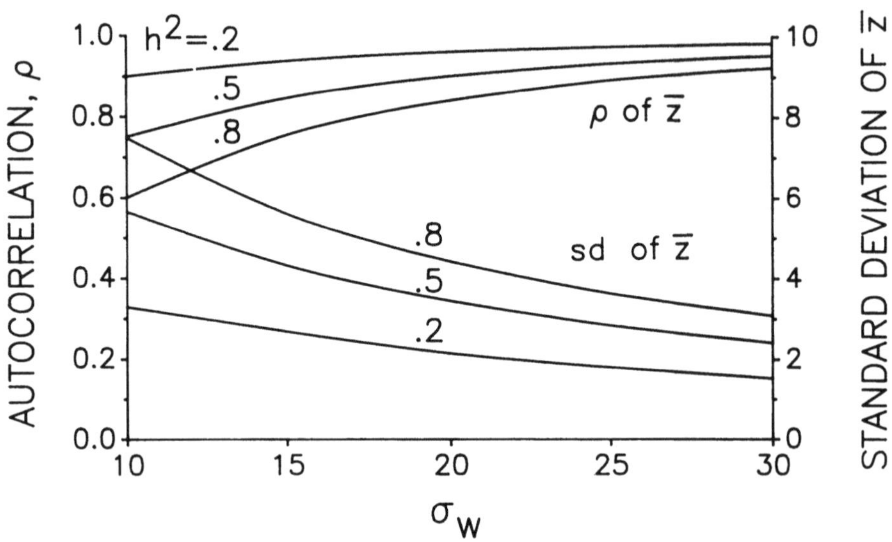


**Fig. 9.3.** Autocorrelation and standard deviation of $\bar{z}$ as a function of the strength of selection, $\sigma_w$. In computing the standard deviation, $\sigma^* = 15$ phenotypic units.

array just described, and determined the mean standard deviation across replicate runs. Second, to study convergence of the mean phenotype, I computed the mean of $\bar{z}$ for $t$ generations and then determined the standard deviation across replicate runs. I shall concentrate here on convergence of the long-term mean phenotype and briefly consider convergence of the standard deviation of the mean.

### 9.3.3 Convergence of the Standard Deviation of the Mean Phenotype

With autocorrelation the convergence of the standard deviation of the mean phenotype may be slow (Fig. 9.4). It is slowest for $h^2 = 0.2$ where the standard deviation is 0.8 phenotypic units after five generations, 1.4 after 20, 2.2 after 100 generations and is still far from converging. In this case, however, the asymptotic standard deviation is so low, because of the slight response to selection with $h^2 = 0.2$ (Fig. 9.1a), that error in estimating the standard deviation is not likely to be as important as for larger values of $h^2$. For a high value of $h^2 = 0.8$ the standard deviation rises rapidly from 3.3 phenotypic units at 5 generations, to 4.0 at 10, to 4.6 at 20 generations with a value of 5.4 very close to an asymptote at 100 generations. The most troublesome heritability is the intermediate of $h^2 = 0.5$, for which convergence is slow and the asymptotic standard deviation is relatively large. Here the standard deviation is only 2.0 at 5 generations, 3.2 at 20 generations, and reaches 4.1 phenotypic units after 100 generations. In this instance, an estimate after 5 generations, which is a relatively long time in studies of natural populations, estimates the standard deviation as less than half its final value.

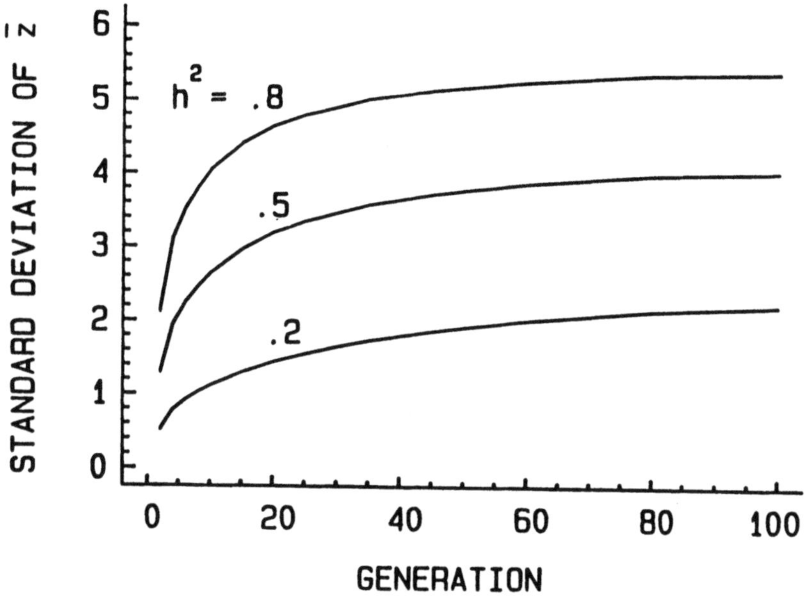


**Fig. 9.4.** Convergence of the standard deviation of the mean phenotype ($\bar{z}$) for three heritabilities. This is case 1 with $\sigma^* = \sigma_w = 15$ phenotypic units.

## 9.3.4 Convergence of the Long-Term Mean Phenotype

The standard deviation across replications of the long-term mean (the mean of the mean phenotype computed across generations) becomes smaller and the long-term mean converges on the mean of the distribution of optima, $\mu^*$, as more generations are included. After enough generations, this standard deviation becomes small enough that one can infer with a certain probability that one additional realization of the simulation for this number of generations will lie within a desired interval, called a prediction interval (Neter and Wasserman 1974). That additional realization can be interpreted as a single time series, as would be observed in nature. The autocorrelation that exists within the time series across generations does not affect the computation across replications, because the replications are independent.

The approach taken here is to determine the number of generations over which the mean must be measured (assuming no error in estimation) for the 95% prediction interval, which is taken to be 2×1.96 (or 3.92) times the standard deviation computed over the 500 replicate runs of the simulation, to become less than some specified width. These intervals were scaled by the phenotypic standard deviation, $\sigma_z = 10$ phenotypic units, and were selected as 1, 1.5 and 2.0 times $\sigma_z$. As an example, consider Fig. 9.5, from a simulation in which $\sigma^* = \sigma_w =$ 15 phenotypic units. In the figures and tables, I use the term confidence, instead of prediction, interval because of the more common usage of confidence intervals and their formal similarity. When $h^2 = 0.2$, any of the desired widths of the confidence interval is achieved in 2 generations, which is, of course, known in this simulation but would not be in nature. When $h^2 = 0.5$, a width of the confidence interval of 2 $\sigma_z$ phenotypic units is achieved after 2 generations, 1.5 $\sigma_z$

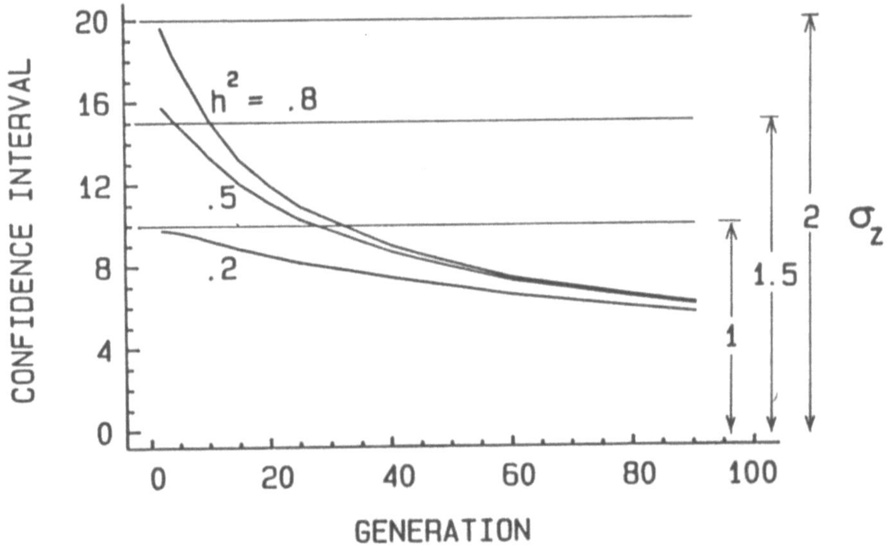


**Fig. 9.5.** Convergence of the confidence interval for the mean of the phenotypic mean for three heritabilities and three criteria of convergence. This is case 1 with $\sigma^* = \sigma_w = 15$ phenotypic units.

after 4, and 1.0 $\sigma_z$ after 28 generations (Table 9.1). The time for convergence to the desired widths is somewhat greater for $h^2 = 0.8$, being 2, 10 and 32 generations for 2, 1.5 and 1.0 $\sigma_z$, respectively. It is not clear at this time which criterion for convergence would be appropriate for any particular study, but for present purposes a width of 2 $\sigma_z$ will be considered "reasonable" and one of 1 $\sigma_z$ as stringent.

I shall compare three analyses in quantifying the effect of autocorrelation on

**Table 9.1.** Generations for confidence interval for the mean of the phenotypic mean to be a specified multiple, $c$, of $\sigma_z$. Case 1, no autocorrelation of the optimum

| | $h^2$: | 0.2 | | | 0.5 | | | 0.8 | | |
|---|---|---|---|---|---|---|---|---|---|---|
| | $c$: | 1 | 1.5 | 2 | 1 | 1.5 | 2 | 1 | 1.5 | 2 |
| $\sigma^*$ | $\sigma_w$[a] | | | | | | | | | |
| 5 | 25 | 2 | 2 | 2 | 2 | 2 | 2 | 2 | 2 | 2 |
| | 15 | 2 | 2 | 2 | 2 | 2 | 2 | 2 | 2 | 2 |
| | 10 | 2 | 2 | 2 | 2 | 2 | 2 | 2 | 2 | 2 |
| 10 | 25 | 2 | 2 | 2 | 2 | 2 | 2 | 2 | 2 | 2 |
| | 15 | 2 | 2 | 2 | 4 | 2 | 2 | 10 | 2 | 2 |
| | 10 | 2 | 2 | 2 | 11 | 2 | 2 | 13 | 4 | 2 |
| 15 | 25 | 2 | 2 | 2 | 8 | 2 | 2 | 22 | 2 | 2 |
| | 15 | 2 | 2 | 2 | 28 | 4 | 2 | 32 | 10 | 2 |
| | 10 | 20 | 2 | 2 | 32 | 11 | 2 | 34 | 13 | 6 |
| 20 | 25 | 2 | 2 | 2 | 43 | 2 | 2 | 53 | 14 | 2 |
| | 15 | 38 | 2 | 2 | 56 | 19 | 4 | 58 | 23 | 10 |
| | 10 | 52 | 11 | 2 | 59 | 23 | 10 | 60 | 26 | 13 |

[a] Values of $\sigma_w$ are ordered from large to small, i.e from weak to strong selection, to correspond to $\sigma^*$, which is ordered from low to high variability in the optimum, so that instances of high numbers of generations will tend to be at the lower end of the table.

the time to convergence of the confidence interval. In case 1, the phenotypic evolution model is used with no autocorrelation involved in generating the series of optima, $z^*$. In cases 2 and 3, autocorrelation is produced using a first-order autoregressive process (Chatfield 1975):

$$z_{t+1}^* - \mu^* = \rho^* (z_t^* - \mu^*) + e_t \tag{9.5}$$

where the $e_1, \ldots, e_n$ are independent, identically distributed normal variates with mean zero and variance $(\sigma^*)^2$, and $\rho^*$ is the autocorrelation coefficient for the time series of the optimum, $z^*$.
Setting:

$$z_1^* = e_1/(1-(\rho^*)^2)^{1/2}$$

then the variance of $z^*$ including autocorrelation is:

$$\text{Var}\,(z^*) = (\sigma^*)^2/[1-(\rho^*)^2] \tag{9.6}$$

In case 2, the variance of $z^*$ is computed after the introduction of autocorrelation. Thus, $\sigma^*$ is computed given the desired value of Var($z^*$) as follows:

$$\sigma^* = \{\text{Var}(z^*)\,[1-(\rho^*)^2]\}^{1/2} \tag{9.7}$$

The interpretation of this case is that the variance of the optimum is measured directly and includes the autocorrelation.

In case 3, $\sigma^*$ is specified, so Var ($z^*$) is determined by equation (9.3) before the introduction of autocorrelation. This is interpreted as simulating an interaction among trophic levels in which a lower trophic level creates the optimum for the level above. For example, the timing of germination or leaf production of a plant might respond to the time of the last hard frost in the spring, which may vary randomly between years with no autocorrelation. The mean time of germination or leaf production should show autocorrelation in response to selection by the hard frost and these times might act to produce optimal times for herbivorous insects to begin development. Thus, as in case 3, the time series of optima experienced by the herbivore contains a component of autocorrelation imposed on uncorrelated variability in the time of hard frost. As illustrated in Fig. 9.1b, the addition to the optimum of autocorrelation at the high levels produced by selection makes the time course of the mean phenotype very sluggish compared to that without autocorrelation (Fig. 9.1a).

### *Case 1: No Autocorrelation in $z^*$*

As shown in Table 9.1, the three parameters determining the time to convergence are: variability of the optimum, $\sigma^*$, the strength of selection, $\sigma_w$, and heritability, $h^2$. Generally, the time to convergence increases as any of these parameters increases. For $h^2 = 0.2$, convergence times are very low except for the most stringent criterion, $c = \sigma_z$, in highly variable environments with strong selection. For $h^2 = 0.8$, convergence times are still small for $c = 2\sigma_z$ except at extreme parameter combinations.

### *Case 2: $\sigma^*$, Computed after Introduction of Autocorrelation*

This analysis requires consideration of a fourth parameter, $\rho^*$, measuring autocorrelation of $z^*$ (Table 9.2). High levels of autocorrelation ($\rho^* = 0.6$ or $0.8$) combined with moderate to high heritabilities ($h^2 = 0.5$ or $0.8$) dramatically increase the time to convergence of the confidence intervals for reasonable combinations of parameters. For example, now with Var ($z^*$) = 10 phenotypic units and $\sigma_w = 10$, i.e. a fairly variable environment with weak selection, convergence to $c = 2\sigma_z$ takes 16 generations ($h^2 = 0.8$ and $\rho^* = 0.8$) instead of the 2 ($\rho^* = 0$) previously. For a variable environment with strong selection (Var ($z^*$) = $\sigma_w$ = 15), convergence can take a very long time, 29 generations for $\rho^* = 0.6$ and 70 generations for $\rho^* = 0.8$.

### *Case 3: $\sigma^*$, Computed before Introduction of Autocorrelation*

Here the effects of autocorrelation on time to convergence are yet more dramatic, especially for large values of $\rho^*$ as can be seen in Fig. 9.6. For moderate to high levels of autocorrelation ($\rho^* = 0.6$ or $0.8$) and heritability ($h^2 = 0.5$ or $0.8$), combined with moderate variability ($\sigma^* = 10$ or $15$) and strong selection ($\sigma_w = 15$), convergence times are all high (Table 9.3). If a more stringent criterion for convergence than $c = 2\sigma_z$ were desired, then times to convergence could be very large indeed. In both cases 2 and 3, convergence times are particularly sensitive to the increase in $\rho^*$ from 0.6 to 0.8.

**Table 9.2.** Generations for confidence interval for the mean of the phenotypic mean to be a specified multiple, $c$, of $\sigma_z$. Case 2, in which $\sigma^*$ is determined after introduction of autocorrelation

| | | $h^2$: | 0.2 | | | 0.5 | | | 0.8 | | |
|---|---|---|---|---|---|---|---|---|---|---|---|
| | | $c$: | 1 | 1.5 | 2 | 1 | 1.5 | 2 | 1 | 1.5 | 2 |
| Var $(z^*)^{\frac{1}{2}}$ | $\sigma_w$ | $\rho^*$ | | | | | | | | | |
| 5 | 15 | 0.0 | 2 | 2 | 2 | 2 | 2 | 2 | 2 | 2 | 2 |
| | | 0.4 | 2 | 2 | 2 | 2 | 2 | 2 | 2 | 2 | 2 |
| | | 0.6 | 2 | 2 | 2 | 7 | 2 | 2 | 8 | 4 | 2 |
| | | 0.8 | 2 | 2 | 2 | 22 | 2 | 2 | 25 | 2 | 2 |
| 10 | 10 | 0.0 | 2 | 2 | 2 | 2 | 2 | 2 | 2 | 2 | 2 |
| | | 0.4 | 2 | 2 | 2 | 7 | 2 | 2 | 22 | 2 | 2 |
| | | 0.6 | 2 | 2 | 2 | 40 | 2 | 2 | 50 | 11 | 2 |
| | | 0.8 | 57 | 2 | 2 | 122 | 36 | 2 | 129 | 48 | 16 |
| 10 | 15 | 0.0 | 2 | 2 | 2 | 4 | 2 | 2 | 10 | 2 | 2 |
| | | 0.4 | 2 | 2 | 2 | 27 | 5 | 2 | 32 | 10 | 2 |
| | | 0.6 | 35 | 2 | 2 | 53 | 17 | 2 | 55 | 21 | 8 |
| | | 0.8 | 119 | 31 | 2 | 131 | 52 | 22 | 133 | 54 | 25 |
| 15 | 15 | 0.0 | 2 | 2 | 2 | 28 | 4 | 2 | 32 | 10 | 2 |
| | | 0.2 | 22 | 2 | 2 | 43 | 13 | 2 | 48 | 18 | 7 |
| | | 0.4 | 57 | 2 | 2 | 73 | 27 | 10 | 75 | 32 | 15 |
| | | 0.6 | 120 | 35 | 2 | 132 | 53 | 24 | 134 | 55 | 29 |
| | | 0.8 | 284 | 119 | 52 | 294 | 131 | 68 | 295 | 133 | 70 |

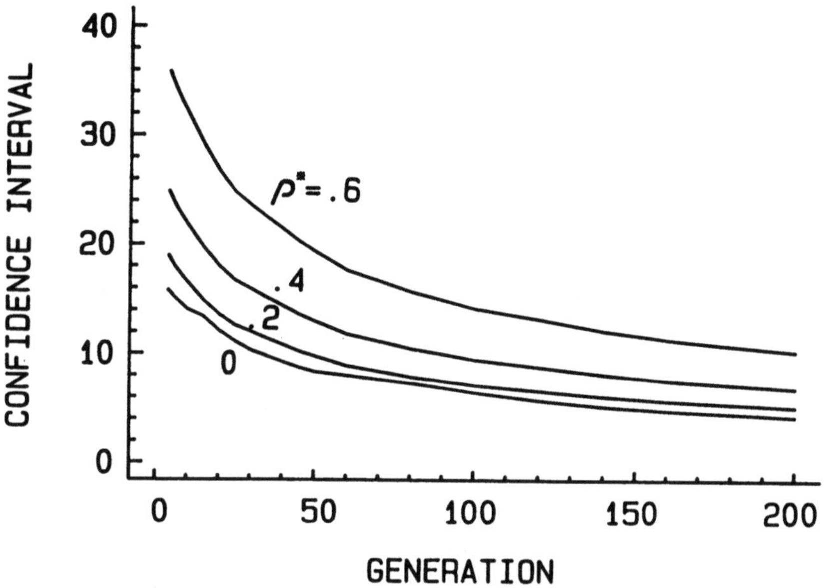


**Fig. 9.6.** Convergence of the confidence interval for the mean of the phenotypic mean for various levels of autocorrelation in case 3. Here $\sigma^* = \sigma_w = 15$ phenotypic units.

**Table 9.3.** Generations for confidence interval for the mean of the phenotypic mean to be a specified multiple, $c$, of $\sigma_z$. Case 3, in which $\sigma^*$ is determined before introduction of autocorrelation

| | | $h^2$: | 0.2 | | | 0.5 | | | 0.8 | | |
|---|---|---|---|---|---|---|---|---|---|---|---|
| | | $c$: | 1 | 1.5 | 2 | 1 | 1.5 | 2 | 1 | 1.5 | 2 |
| $\sigma^*$ | $\sigma_w$ | $\rho^*$ | | | | | | | | | |
| 5 | 15 | 0.0 | 2 | 2 | 2 | 2 | 2 | 2 | 2 | 2 | 2 |
| | | 0.4 | 2 | 2 | 2 | 2 | 2 | 2 | 4 | 2 | 2 |
| | | 0.6 | 2 | 2 | 2 | 13 | 2 | 2 | 18 | 4 | 2 |
| | | 0.8 | 73 | 2 | 2 | 87 | 31 | 8 | 89 | 35 | 14 |
| 10 | 25 | 0.0 | 2 | 2 | 2 | 2 | 2 | 2 | 2 | 2 | 2 |
| | | 0.4 | 2 | 2 | 2 | 16 | 2 | 2 | 30 | 2 | 2 |
| | | 0.6 | 2 | 2 | 2 | 78 | 15 | 2 | 86 | 29 | 7 |
| | | 0.8 | >300 | 105 | 2 | >300 | 153 | 76 | >300 | 159 | 84 |
| 10 | 15 | 0.0 | 2 | 2 | 2 | 4 | 2 | 2 | 10 | 2 | 2 |
| | | 0.2 | 2 | 2 | 2 | 14 | 2 | 2 | 19 | 4 | 2 |
| | | 0.4 | 10 | 2 | 2 | 36 | 8 | 2 | 39 | 13 | 4 |
| | | 0.6 | 74 | 8 | 2 | 88 | 35 | 13 | 90 | 38 | 18 |
| | | 0.8 | >300 | 148 | 76 | >300 | 159 | 90 | >300 | 162 | 92 |
| 15 | 15 | 0.0 | 2 | 2 | 2 | 28 | 4 | 2 | 32 | 10 | 2 |
| | | 0.2 | 25 | 2 | 2 | 47 | 14 | 2 | 50 | 19 | 8 |
| | | 0.4 | 74 | 10 | 2 | 88 | 36 | 14 | 91 | 39 | 18 |
| | | 0.6 | 198 | 74 | 24 | 208 | 88 | 46 | 210 | 90 | 49 |
| | | 0.8 | >300 | >300 | 197 | >300 | >300 | 208 | >300 | >300 | 209 |

## 9.4 Discussion

The results in this chapter follow from the initial observation that selection generates high levels of autocorrelation in the time series of the mean phenotype (Figs. 9.1, 9.2 and 9.3). As a consequence, it has been found that the standard deviation of the mean phenotype, when estimated over few generations (10 or fewer), especially for intermediate heritabilities, may substantially underestimate the true, asymptotic, standard deviation (Fig. 9.4). In the absence of autocorrelation, the time for convergence of the confidence interval for the mean phenotype is low for unexceptional combinations of parameters including highly variable environments (high $\sigma^*$) and strong selection (low $\sigma_w$) (Table 9.1). Addition of autocorrelation in the time series of the optimum, $z^*$, markedly increases the time to convergence (Table 9.2). This effect is especially striking when the analysis takes into account the potential for accumulation of autocorrelation between trophic levels (Table 9.3). Thus, under what appear to be reasonable assumptions, one can conclude that tests of the mean phenotype in temporally variable environments may require measurements over many generations, 10 or more generations being common and more than 100 being possible.

A question that immediately arises is: why isn't the situation worse? Specifically, why is convergence so fast in the absence of autocorrelation in the optimum (Table 9.1)? The answer is that nor-optimal selection, as used here, is surprisingly weak. To quantify the strength of selection, we can use selection intensity, which is the displacement of the mean of the selected parents from the mean of the unselected population (known as the selection differential) standardized by dividing by the phenotypic standard deviation. The selection intensity associated with nor-optimal selection is given by the following formula obtained by observing that $\Delta\bar{z} = h^2S$ in Eq. (9.1), where $S$ is the selection differential. Then the selection intensity is:

$$i = -(\bar{z}-z_t^*)\frac{\sigma_z^2}{\sigma_z^2 + \sigma_w^2} \tag{9.8}$$

Using this formula, Fig. 9.7 shows selection intensity as a function of the displacement of the optimum from the mean phenotype and the ratio of the strength of selection ($\sigma_w$) to the phenotypic standard deviation ($\sigma_z$). Even strong selection, in which the ratio $\sigma_w$ to $\sigma_z$ is 1.5 (i.e. $\sigma_w = 15$, see above), does not produce a selection intensity of 1 until the optimum differs from the phenotypic mean by more than three phenotypic standard deviations. Consequently, a given level of variability in the optimum translates into much smaller variability in the mean phenotype, even when the heritability is high (see Fig. 9.1).

Autocorrelation in the optimum significantly increases the convergence time by keeping the mean phenotype away from the long-term phenotypic mean for extended periods (Fig. 9.1b). This effect is most apparent when all of the four parameters in the model take on relatively large values, i.e. when variability and autocorrelation of the optimum, strength of selection and heritability are large.

The fact that the mean phenotype must be measured over many generations to obtain a satisfactory estimate of the true long-term mean will be important in testing evolutionary theory in two ways. Most obviously it may be difficult to test a theory that makes predictions about the mean phenotype in a particular environment. Several theories referred to in Section 9.1 have been developed

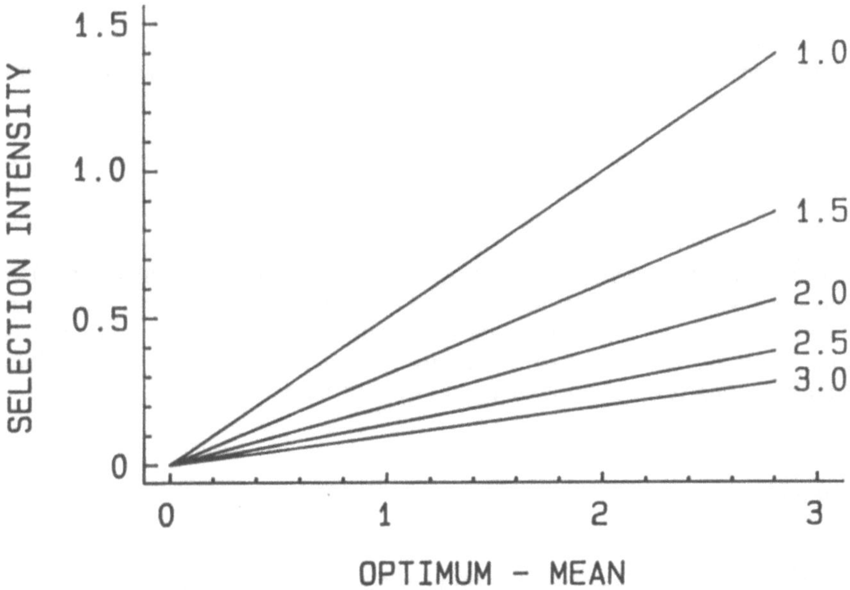


**Fig. 9.7.** Selection intensity as a function of the deviation of the optimum phenotype from the mean phenotype standardized by dividing by the phenotypic standard deviation, i.e. $(z^* - \bar{z})/\sigma_z$. Each line represents a different value of the ratio, $\sigma_w/\sigma_z$, as labelled.

specifically to generate predictions about long-term mean phenotypes in variable environments and they will all be subject to the "curse of autocorrelation" addressed in this chapter.

Second, this estimation problem might arise generally as well in comparisons of phenotypic means among populations. In variable environments, especially in the absence of positive spatial autocorrelation, it is quite possible for two populations to have the same long-term mean but be judged as significantly different when the short-term means happen to be traversing on opposite sides of the long-term mean. Acceptance of a significant difference is additionally likely because, with autocorrelation, the standard error of the mean may be grossly underestimated (Fig. 9.4).

Obstacles to testing theories caused by autocorrelation aside, the fact of autocorrelation generated by selection and potentially amplified through trophic level interactions forces one to question the notion that the property of being adapted implies a close fit to the environment. Over the long run the mean phenotype reflects its history of selection. If the fitness function and phenotypic distribution are symmetric, the asymptotic mean phenotype should equal the long-term mean of the optimum even with autocorrelation. If either is asymmetric, the asymptotic mean phenotype is determined by a balancing of selection intensities over time (F. Taylor, unpublished results). In both cases, however, when the parameters considered above are large enough, the observed mean phenotype will be far from the current optimum most of the time and assessment of agreement between observation and prediction, as argued in this chapter, will be difficult.

Thus, we are led to ask: where is the real world? More concretely, we need to specify the domain of application for the results presented in this chapter. Ideally we should desire long-term data on the following quantities: the optimum of the fitness function with some measure of the strength of selection, the mean and standard deviation of the phenotype, and heritability. We are ignoring, for the purposes of argument, the potential importance of genetic and phenotypic correlations. Such extensive information is not likely to be available at the present for any trait, therefore we must assess the utility of less complete data. Two types would seem to be of considerable interest. First, long-term data on the mean and standard deviation of any phenotypic trait in populations with non-overlapping generations will address questions about levels of variability scaled to the phenotypic standard deviation and about autocorrelation. Second, long-term data on events that are likely to relate to an optimum of selection will address questions about variability and autocorrelation in selection. Sources of these data would include long-term phenological and meteorological records.

## 9.5 Summary

Temporally variable stabilizing, or optimizing, selection creates high levels of autocorrelation in the mean phenotype. As a consequence, the standard deviation of the mean phenotype will tend to be underestimated and the long-term mean phenotype may, under reasonable assumptions, converge slowly, necessitating measurements over many generations to obtain an acceptable estimate of the true mean. These results imply that in sufficiently variable environments precise predictions of evolutionary theory will be more difficult to test than seems to have been appreciated previously. The difficulty of testing theories increases as the following parameters increase in ways that are quantified in this chapter: variability and strength of selection, heritability, and autocorrelation of the optimum. Aside from the problems of statistical inference, the effect of strong autocorrelation is such that, even though over the long term the phenotypic mean of a population may conform to a theoretical prediction, most of the time this mean is far from the predicted mean. On this view, the property of being adapted is a long-term phenomenon that can be very difficult to verify. It will be necessary to specify the domain of application of this view. Specifically, under what circumstances in nature do the parameters listed combine to make these conclusions relevant?

*Acknowledgements.* I want especially to thank Ron Schrader and an anonymous reviewer for their assistance with the statistical theory used in this chapter. Advice in setting up the model was also provided by Joe Felsenstein, Russ Lande, Monty Slatkin and Michael Turelli. The following people made helpful comments on an earlier draft: Stephen Kucera, Ted Nusbaum and Ron Schrader. This research was supported by NSF grant BSR–8506768.

### References

Boyce MS, Perrins CD (1987) Optimizing great tit clutch size in a fluctuating environment. Ecology 68: 142–153

Bulmer MG (1984) Delayed germination of seeds: Cohen's model revisited. Theor Popul Biol 26: 367–377
Bulmer MG (1985) Selection for iteroparity in a variable environment. Am Nat 126: 63–86
Chatfield C (1975) Analysis of time series: theory and practice. John Wiley and Son, New York
Cohen D (1966) Optimizing reproduction in a randomly varying environment. J Theor Biol 16: 1–14
Cohen D (1970) A theoretical model for the optimal timing of diapause. Am Nat 104: 389–399
Cohen D (1971) Maximizing final yield when growth is limited by time or by limiting resources. J Theor Biol 33: 299–307
Emlen JM (1984) Population biology: the coevolution of population dynamics and behavior. Macmillan Publishing Company, New York
Endler J (1986) Natural selection in the wild. Princeton University Press, Princeton, NJ (Mongr Popul Biol 21)
Hastings A, Caswell H (1979) Role of environmental variability in the evolution of life history strategies. Proc Natl Acad Sci USA 76: 4700–4703
Hill W (1972a) Estimation of realized heritabilities from selection experiments. I. Divergent selection. Biometrics 28: 747–765
Hill W (1972b) Estimation of realized heritabilities from selection experiments. II. Selection in one direction. Biometrics 28: 767–780
King D, Roughgarden J (1982) Graded allocation between vegetative and reproductive growth for annual plants in growing seasons of random length. Theor Popul Biol 22: 1–16
Lande R (1975) The maintenance of genetic variability by mutation in a polygenic character with linked loci. Genet Res 26: 221–235
Lande R (1976) Natural selection and random genetic drift in phenotypic evolution. Evolution 30: 314–334
Neter J, Wasserman W (1974) Applied linear statistical models. Richard D. Irwin, Inc., Homewood, IL
Orzack SH (1985) Population dynamics in variable environments. V. The genetics of homeostasis revisited. Am Nat 125: 550–572
Press WH, Flannery BP, Teudolsky SA, Vetterling WT (1986) Numerical recipes: the art of scientific computing. Cambridge University Press, Cambridge
Schaffer WM (1974) Optimal reproductive effort in fluctuating environments. Am Nat 108: 783–790
Taylor F (1986a) The fitness functions associated with diapause induction in arthropods. I. The effects of age structure. Theor Popul Biol 30: 76–92
Taylor F (1986b) Toward a theory for the evolution of the timing of hibernal diapause. In: Taylor F, Karban R (eds) Evolution of insect life cycles. Springer, Berlin Heidelberg New York
Turelli M (1988) Population genetic models for polygenic variation and evolution. In: Weir BS, Eisen EJ, Goodman MM, Namkooney G (eds) Quantitative genetics. Sinauer, Sunderland, MA, pp 601–618
Venable DL, Brown JS (1988) The selective interactions of dispersal, dormancy, and seed size as adaptations for reducing risk in variable environments. Am Nat 131: 360–384

*Section III*

# Coordination of Life-Cycle Traits

# Introduction

William E. Bradshaw

As amply demonstrated in Sect. II, insects do not live in uniform environments or experience uniform selection. The intensity and direction of selection may change in space and time. The chapters of this section amplify this theme and also consider insect life-cycles in the context of the broader community.

In Chap. 10, Valerie Brown examines insect herbivore life-histories during plant succession. During the process of succession, we are too ready to see associated herbivores as passive passengers on their supportive hosts through successional time. To some extent, Brown shows that this image is appropriate: guilds of insects, their seasonal life-cycles, and their life-history syndromes clearly change during succession. Even more intriguingly, however, Brown argues that herbivores may affect the progress of succession by altering the rate at which it proceeds and the specific directions it may take.

In the succeeding chapter, Christer Solbreck, David Anderson and Jonas Förare look at the coordination of migration into the life-cycle syndromes of lygaeid bugs. This coordination occurs within the context of both the geographical climate and local microhabitat in which they live. Migration by flight involves a behavioural tendency to flight accompanied by both the wings and the wing musculature to accomplish this flight. Development of the latter two structures necessarily reallocates resources that potentially contribute to growth, development, survivorship and reproduction in both space and time. These trade-offs result in strongly contrasting but none the less coherent adaptive syndromes of migratory and non-migratory morphs, races and species.

Brian Shorrocks (Chap. 12) next considers interaction within and between species in space and time. Habitats in nature are frequently distributed as patches in the broader environment and many of these patches persist for only one or a few generations of the insects that inhabit them. Shorrocks reviews his previous models and observations that lead to the conclusion that, as either individual species become more aggregated or the environment becomes more dissected into ephemeral patches, indefinite coexistence of the constituent species becomes progressively more likely. Shorrocks then extends his "probability refuge model" to selection acting differentially on separate genotypes. Aggregation of genotypes in a patchy environment can maintain even inferior alleles in the population. Patchy, ephemeral habitats and aggregation of genotypes can therefore serve to maintain a variety of alternative life-histories, even when there is uniformity of selection across patches.

The vast majority of insects considered in this symposium are uni- or multivoltine. Most of us are aware of the very long life-cycle exemplified by cicadas, but Wolfgang Wipking in the last paper provides us with an analysis of the intermediate life-cycles of semivoltine moths. In this paper, Wipking extends his early work to three species of moth whose seasonal life-cycles span the range from one to three years. Of particular interest in the present paper is the description of pseudobivoltinism, where separate morphs of the same overwintering cohort may fly in the spring, or aestivate and fly in the fall. "Spreading the risk" in this remarkable group of moths thus extends from between seasons within years to between years.

*Chapter 10*

# Insect Herbivores, Herbivory and Plant Succession

Valerie K. Brown

## 10.1 Introduction

Succession is a characteristic feature of most plant communities and is clearly seen when any habitat is disturbed. The temporal and directional changes in plant communities, which constitute succession, are well documented and have been described in a range of different habitats and geographical regions (e.g. Oosting 1942; Beckwith 1954; Olson 1958; Bazzaz 1968; Tramer 1975; Southwood et al. 1979; Brown and Southwood 1987). Accounts of the associated changes in animal communities are much rarer and there are relatively few studies spanning a period of several years or even comparing habitats of different successional age (but see Southwood et al. 1979; Fox and Fox 1986; Brown and Southwood 1983, 1987). Invertebrates, and especially insects, are more amenable to a quantified sampling programme at different stages of succession than are some organisms (e.g. birds) and can therefore be compared between habitats (Brown et al. 1988a) or even in geographical regions (see Brown et al. 1987b). Although patterns of insect abundance through time are in themselves interesting, it is an understanding of their interaction with the plant community which can contribute to further insight into the mechanisms underlying plant succession. It is well known that large vertebrate herbivores can have a substantial impact on vegetation dynamics (e.g. Watt 1960; Gibson et al. 1987). However, invertebrate herbivory has only recently been considered in this context. In order to consider the role of insect herbivores in plant succession, the composition, abundance and special attributes of the herbivores must be understood. The variable nature of the life-cycle strategies of insects is well known, and the change in life-cycles during plant succession was the subject of a paper at the preceding symposium in this series in Hamburg (see Brown 1986). On this basis, we are now in a position to pose the question "how do insect herbivores, with different life-cycles, influence the plant community in which they are found?". This chapter will outline trends in abundance and life-cycle strategies of insects and will then consider the impact of the herbivores on successional plant communities and the fitness of individual species. The chapter will focus on an experimental study of secondary succession on acidic, sandy soil in southern Britain which was specifically designed to study the various interactions between insects and their host plants.

## 10.2 Insect Communities

Insect species number and the abundance of individuals generally increase with successional age, although these trends are not always translated into changes in alpha diversity (Southwood et al. 1979; Brown and Southwood 1987). Of more relevance than absolute abundance in the structuring of successional communities is the relative contribution of the different trophic types or guilds (Root 1973). On the basis of an analysis of some 34 500 individuals, Brown and Southwood (1987) compared the relative contributions of the major guilds (see Moran and Southwood 1982) in sites of different successional age. Guild structure varied along the successional gradient. In the early stages, the phytophages and particularly sap-feeders were dominant, while in later succession a group including the tourists (insects with no permanent association with the host plant, but merely visiting it for the purpose of resting, basking or sexual display) became numerically dominant. At all stages in succession, chewing insects contributed little to the phytophagous guilds.

### 10.2.1 Life-Cycle Strategies

The interaction of the herbivores with the plant community is inextricably involved with their life-cycle strategies. Successional trends in a number of separate traits have been described by Brown (1986, 1990). An important feature of this work was that the analysis was based not only on number of species but also on number of individuals. From the standpoint of community dynamics, the latter is of overriding importance.

Successional trends in five life-cycle traits are summarized in Fig. 10.1. These are all of potential importance in a consideration of the role of herbivory in natural plant communities. The high occurrence of multivoltinism in early succession (Fig. 10.1a) can be related to the rapid turnover of plant structures, individuals and species which provides suitable nutrition throughout the growing season. This is in direct contrast to the mid- and late-successional species, where the host plant may provide food in a suitable form only for a limited period each year (e.g. Feeny 1970), a feature favouring selection for a univoltine life-cycle.

Dispersal ability, as indicated by the degree of macroptery, was high in early succession (Fig. 10.1b) and has been coupled with the ability to immigrate into and exploit ephemeral, highly heterogeneous resources and to emigrate once these become depleted. In more stable habitats, there is a tendency for selection to favour aptery or brachyptery (see Southwood 1962; Vepsäläinen 1978), a trend which was seen in mid succession. However, in late succession there was a return to macroptery. This has been considered by Waloff (1983) to be due to the high

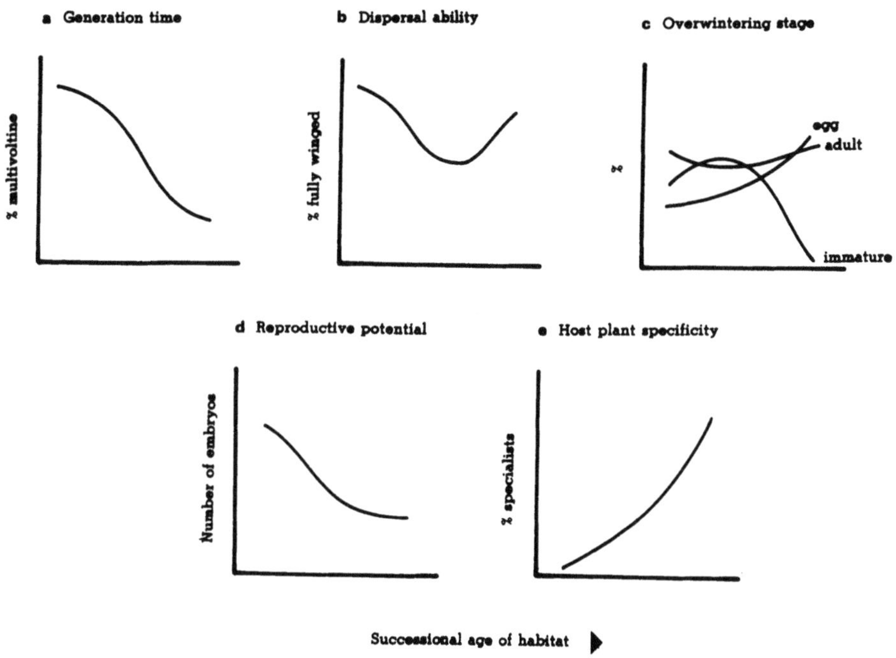


**Fig. 10.1.** Successional trends in five life-cycle traits (from Brown 1990). In **a**, **b**, **c** and **e** % is % of individuals.

spatial and architectural complexity of late-successional habitats, with resources for insects being distributed in three dimensions tending to favour macroptery. An alternative explanation, which warrants further consideration, is that macroptery in late-successional species results from increased predation pressure in these complex habitats and the need for trivial (cf. migratory) movement. The advantages of aptery or brachyptery, in terms of increased resources for reproduction, would appear to be associated only with mid-successional species which can locate their more uniformly distributed resources by movement other than flight.

The utilization of the resources and structure offered by the host plant has implications for the stage at which overwintering occurs. There were some distinct successional trends amongst species (Fig. 10.1c). In species associated with the very early stages of succession, overwintering as an adult is an advantageous strategy, since insects have to disperse rapidly in order to locate their food source and lay eggs. In early and mid succession, overwintering as an immature was the most common strategy and may be related to the fact that many perennial plant species die back in winter and suitable foliage for egg-laying and feeding is often not apparent until relatively late in the season, thereby reducing the time available for development. Thus, after a feeding period, overwintering as an immature utilizing the litter layer or grass tussocks, which were predominant at that stage in succession, is appropriate. In the case of insect species associated with late-successional shrubs and trees, the structure of the plant is already present in spring, and insects overwintering as eggs can exploit the early flush of spring foliage for larval feeding. Additionally, the structural complexity of the late-successional habitat provides many "safe sites" for adult overwintering, accompanied by early-season reproduction.

Although reproductive potential has featured prominently in life-cycle theory, there are no empirical data sets to demonstrate trends at the community level. Fig. 10.1d is based on a study by Brown and Llewellyn (1985) where aphids associated with a range of plant species, characteristic of different stages in succession, were dissected to determine their reproductive potential. In general, reproductive potential declined with the successional age of the habitat with aphids associated with the early-successional annual forbs and grasses having a higher reproductive potential than those associated with the mid-successional perennial species. Those associated with shrubs and trees consistently had a lower reproductive potential. Finally, the degree of host-plant specificity varies with successional age. Brown (1985, 1986) described a means of categorizing insect species according to the range of host plants exploited as food. "Generalists" referred to species associated with more than one family of plants, whereas three grades of "specialists" ($S_1$, $S_2$ and $S_3$) referred to species feeding on a single species, genus or family of plants respectively. Fig. 10.1e summarizes these data and shows that the degree of feeding specialization increased along the successional gradient, with specialist feeders generally dominating the late successional habitats.

## 10.3 Impact of Insect Herbivores

From the above life-cycle traits, with ruderal and early-successional insect species being able to invade a resource rapidly (by adult overwintering and high-dispersal ability), produce many generations a year each with high reproductive potential and feed on a wide range of host plants, one might predict that herbivores associated with such habitats may have more impact on their host plants than those associated with later-successional communities and that their effect may be manifest more quickly.

To test this prediction requires long-term field experiments. While the effects of above-ground chewing insects are often clearly manifest in the damage they cause to a plant community, the effects of sap-feeding species are more difficult to ascertain. Moreover, the latter were dominant, at least in this particular succession (see Brown and Southwood 1987). For this reason, manipulative field experiments have to be undertaken. In these, the natural levels of insect herbivores, associated with a range of sites of different successional age, were reduced by the judicious use of insecticides. Characteristics of the vegetation were then compared in control and insecticide-treated plots, enabling the impact of herbivory to be assessed.

Four sites, representing different stages in the succession from bare ground were used in the first, third, seventh and 15th years of succession at the start of the experiment in 1985. The first three sites were established by the natural colonization of bare ground (see Southwood et al. 1979). The oldest site was an area of permanent pasture, dominated by the grass, *Holcus lanatus*, and grazed by rabbits prior to the start of the experiment. Within each site, there were ten replicates (3 m × 3 m plots) of insecticide treatment and ten controls (sprayed with an equal volume of water). The non-persistent insecticide used (Malathion-60) was found to have no direct effect on the vegetation (Brown et al. 1987c) or on the activities of pollinating insects (Gange et al. 1989). Details of the rate, method of application and other precautions employed in the application of the insecticide were also given by Brown et al. 1987c).

### 10.3.1 Community Effects

Vegetation of the four sites was monitored by placing at random in each plot five linear 35 cm frames containing 10 equally spaced vertical 3.00 mm diameter point quadrat pins. All touches of living plant material were recorded to species. Five samples were taken for each season over a 3-year period (May–October) and provide information on the effects of insect herbivores on:

1. The number of plant species (species richness)
2. The total number of touches of vegetation, which provides a measure of cover abundance
3. The total number of touches of plant species belonging to different life-history groupings (e.g. annual forbs (=herbs) and grasses, perennial forbs and grasses (see Gibson et al. (1987) for details)

Results confirmed the prediction that insect herbivores, because of their special life-cycle traits, have a more rapid and greater impact in early succession. In the first year of treatment, there was only a significant effect of reducing natural levels of herbivory in the first-year site. Both plant species richness and cover abundance were enhanced in the insecticide-treated plots (Fig. 10.2). By contrast, it was only in the second year of treatment that effects of herbivory on cover abundance were seen in the older sites, and these were more evident in the third year (Fig. 10.3). Thus, despite the lower overall abundance of herbivores in very early succession, as colonization of the habitat occurs the impact is considerable. Herbivory affects plant species richness and cover abundance differently. From the second year of treatment in all sites, species richness became higher in control plots. The explanation of this, initially surprising, result is that different plant life-history groupings responded in different ways to herbivory. In the first year of succession, annual forbs were enhanced by a reduction in herbivory (Fig. 10.4a), which consisted mainly of the activities of weevils and aphids, while in early and mid succession perennial grasses showed the most significant response (Fig. 10.5). This effect on grasses was directly related to the build up of the sap-feeding Cicadellidae associated with grasses. The increased growth of the perennial grasses following insecticide treatment caused the forbs to decline both in cover abundance and species richness, since the grasses were more successful competitors.

This differential effect of herbivory on different plant life-history groupings leads to a considerable impact on not only the rate, but also the direction, of plant succession. In terms of the former, insect herbivores appear to have an effect similar to that of larger herbivores in slowing the rate of natural succession mainly by their impact on the perennial grasses.

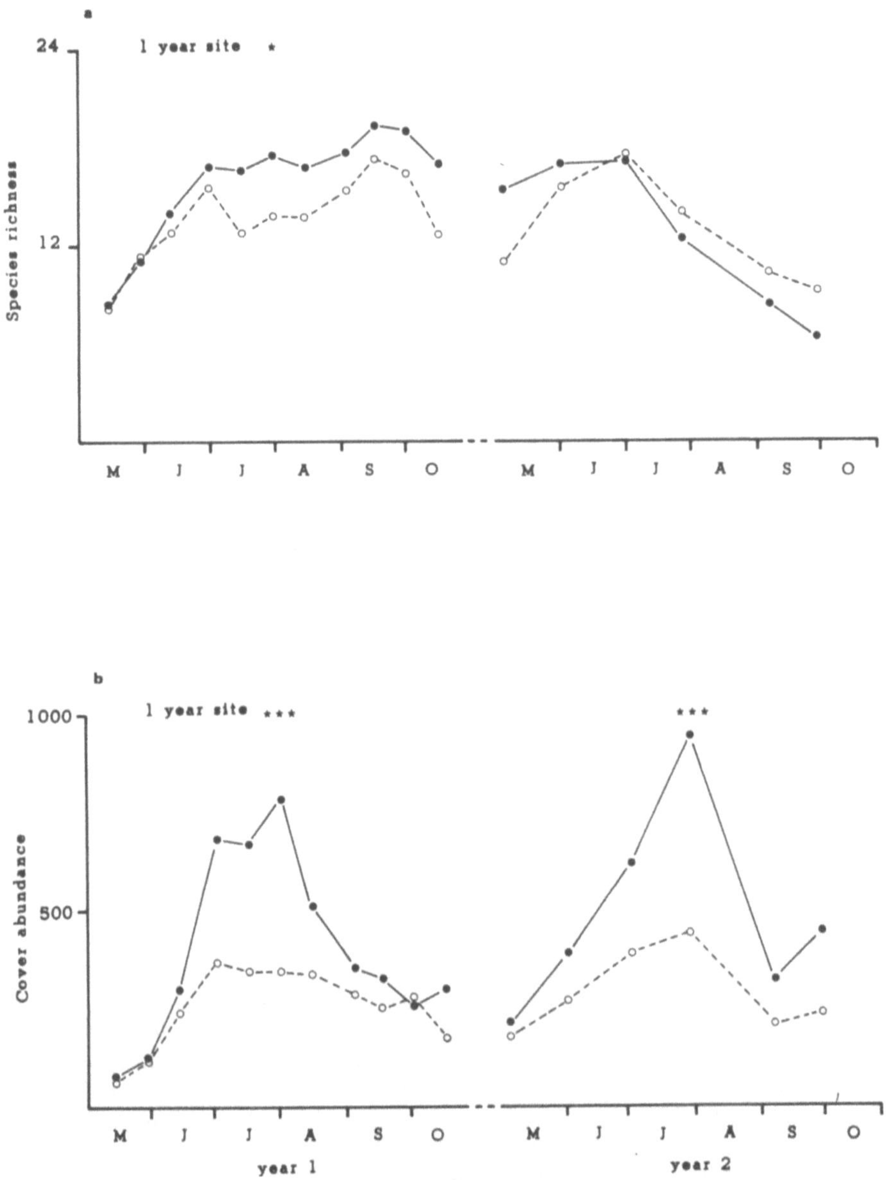


**Fig. 10.2.** The effect of insect herbivory on a ruderal plant community over 2 years: **a** species richness (mean/plot) and **b** cover abundance. (—●—) Insecticide-treated, (--○--) control. $^{*}P<0.05$, $^{***}P<0.001$.

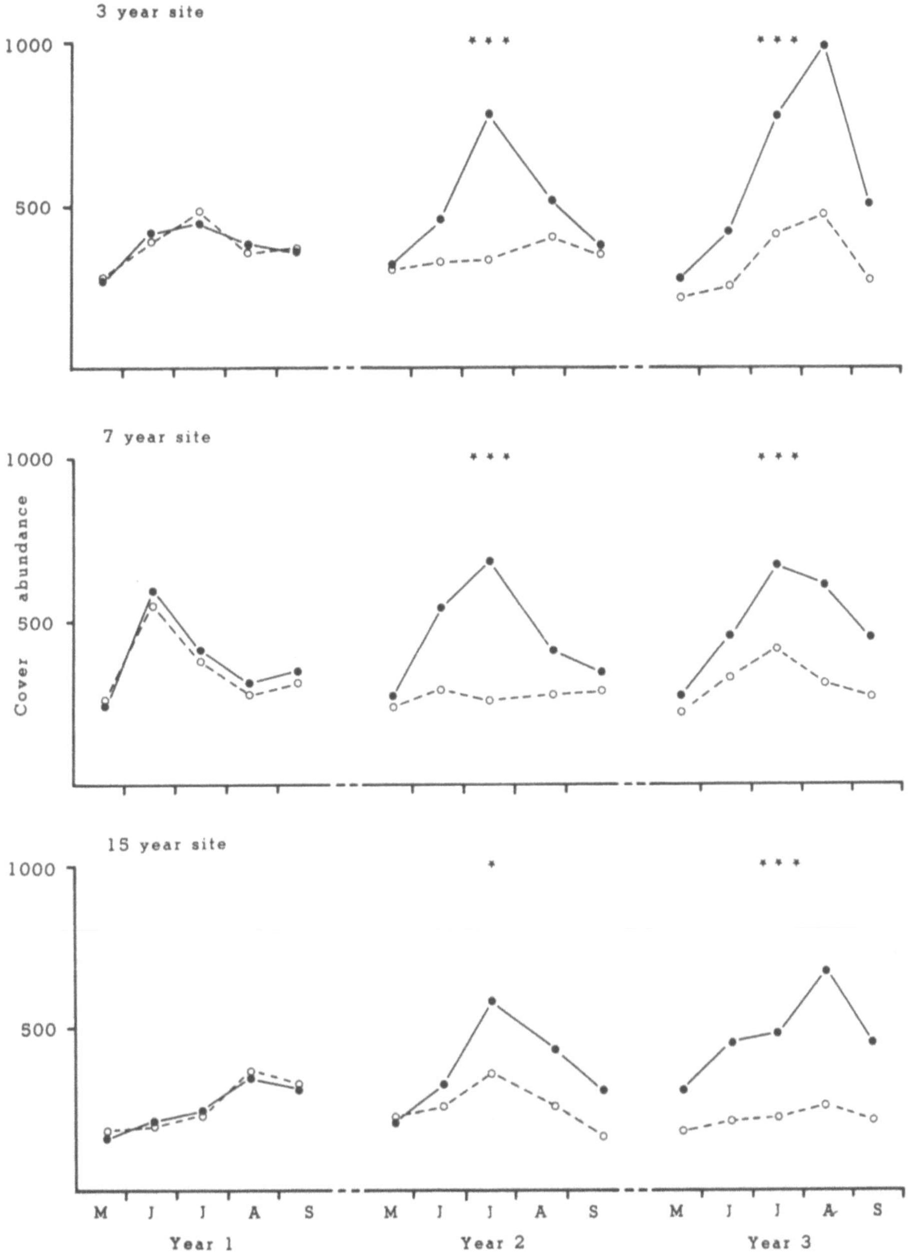


**Fig. 10.3.** The effect of insect herbivory on the cover abundance of vegetation in early and mid-successional plant communities over 3 years. Symbols as in Fig. 10.2.

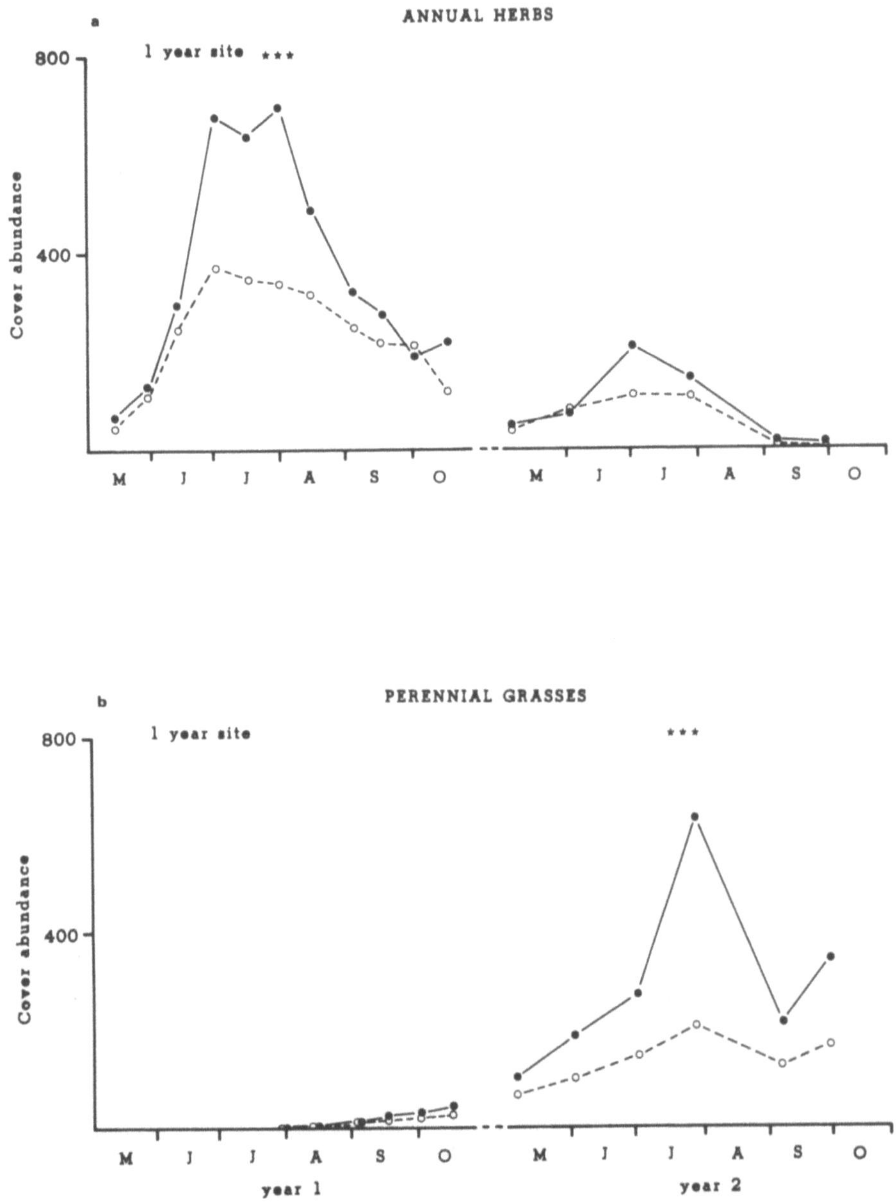


**Fig. 10.4.** The effect of insect herbivory on the cover abundance of vegetation in a ruderal plant community: **a** annual herbs and **b** perennial grasses. Symbols as in Fig. 10.2.

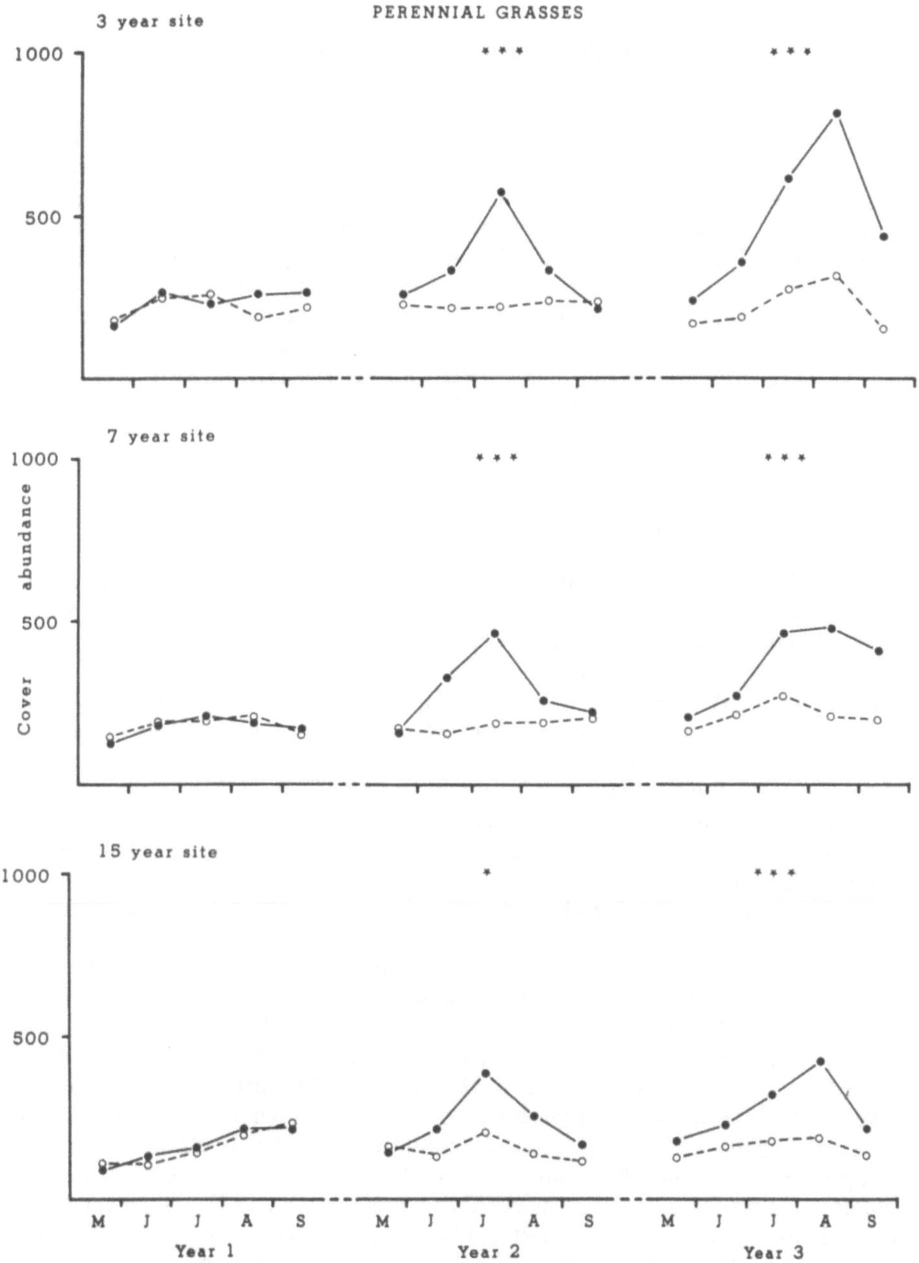


**Fig. 10.5.** Effect of insect herbivory on the cover abundance of perennial grasses in early and mid-successional sites. Symbols as in Fig. 10.2.

## 10.3.2 Comparative Studies

The use of comparative studies, which adopt the same experimental protocol, are vital in the search for generalizations as to the nature of the insect herbivores and their effects on a plant community. Such a study on calcareous soil at Wytham, Oxfordshire, where the insect herbivore fauna was similar in composition and abundance to that previously described (see Brown et al. 1988a), revealed consistent effects and trends (see Brown et al. 1988b). However, in another comparative study there was a dramatic difference in the guild structure and herbivore load between sites, with the latter being consistently higher in the Old World site, where native plant species predominated (Brown et al. 1987b). When manipulative field experiments, such as those just described, were undertaken, Hendrix et al. (1988) found that, whereas the effects of herbivores in the Old World were dramatic during early succession as we have already seen, those effects in the New World were negligible.

## 10.3.3 Effects of Insect Herbivores on Single Plant Species

To consider the impact of natural levels of herbivory on single species during succession, only plant species occurring at a range of different successional stages (including the very earliest stages of colonization) can be used. Such species are in fact rare, although two vetches, *Vicia hirsuta* (hairy tare) and *Vicia sativa* (common vetch), are good examples. These are annual forbs which rapidly colonize bare ground, but also persist into mid succession in certain situations. They are ideal subjects for discussion because, although the plants are similar in many respects, their insect fauna is very different. Thus, they provide a type of natural experiment in which the effects of different insect species, with different feeding strategies and other life-cycle traits, can be compared on two closely related plant species.

The insects associated with each species were sampled by Univac suction sampling of individual plants of the two vetch species at monthly intervals throughout the growing season. Although some species were found associated with both species, there was a trend for sap-feeding insects to predominate on *V. hirsuta*, while chewing-species were more common on *V. sativa* (Brown et al. 1987a). The ratios of sap-feeding to chewing insects on *V. hirsuta* and *V. sativa* were 4.3 : 1 ($n = 569$) and 1 : 9 ($n = 375$) respectively. Furthermore, the overall level of herbivores on *V. hirsuta* was substantially higher. This may well be explained by the ant-tending associated with the extrafloral nectaries in *V. sativa* (Koptur and Lawton 1988), since ants visiting extrafloral nectaries often give plants some protection against insect herbivores (Bentley 1977). However, hard-bodied insects, such as weevils and leaf beetles, are not deterred by ants (Koptur and Lawton 1988) and these were the dominant herbivores on *Vicia sativa*.

The effects of insect herbivores were assessed by the regular monitoring of 50 marked plants in both insecticide-treated and control plots of each species in three sites of different successional age (1 year, 2 years and 7 years). A number of vegetative and reproductive attributes were recorded. Several of these attributes displayed differences according to the age of the site and can be related directly to the increase in plant competition as succession proceeds. This can be clearly seen in the number of flowers produced on plants at different stages of succession. These successional trends are considered fully by Brown et al. (1987a). Here, only the results for the first year of succession are discussed, but are representative of the differences between insecticide-treated and control plants at later stages of succession.

Although there was no difference in plant height between treatments, the number of leaves showed a distinct trend (Table 10.1). In *V. hirsuta* there were consistently more leaves on insecticide-treated plots, while in *V. sativa* leaf production appeared to be stimulated by herbivory. This difference in response of the plants to herbivory is of interest, and suggests that, while leaf removal by chewing species (mainly weevils) may stimulate plants to produce more leaves, sap removal, such as occurs predominantly in *V. hirsuta*, results in a reduction in leaf number. Compensatory leaf growth is known to occur in other species and McNaughton (1983) considers that it may even enhance plant fitness, although this view is not widely held (e.g. Belsky 1986).

In both *Vicia* species, plants subjected to natural levels of herbivory lost their leaves and eventually died before those on which herbivory was reduced. By marking newly emerged leaves at weekly intervals with embroidery silk throughout the season, leaf life-expectancy or turnover was calculated (see Southwood 1978). Leaves of *V. sativa* were generally longer lived than those of *V. hirsuta*. In both species, leaves of insecticide-treated plants had a higher life expectancy than those of control plants (Table 10.1), irrespective of whether they are damaged or not. The percentage reduction in life expectancy can be considered as a "herbivory effect". In the 1 year site, in which herbivores had to colonize and populations build up, the effect was greater in mid season, while in the older sites it was highest in early season where herbivores may well have overwintered as immatures and were immediately able to exploit the new season's growth (see section 10.2.1). Visible damage to leaves of control plants was generally higher in *V. sativa* and was greatest on the 1 year site. This can be explained by the predominance of chewing-insects on this species, especially the weevil, *Sitona lineatus*. In both species, damage was more likely to occur later in the season on the shorter-lived leaves and supports current hypotheses that investment in anti-herbivore defenses reflects the degree of predictability of each plant as a resource

**Table 10.1.** The effects of reducing insect herbivory on vegetative and reproductive characters of plants of two species of *Vicia* in early succession (values are means ± SE)

| | *V. sativa* | | | *V. hirsuta* | | |
|---|---|---|---|---|---|---|
| | I | C | | I | C | |
| Maximum number of leaves | 25.1±2.65 | 45.9±3.87 | $P<0.001$ | 83.4± 4.06 | 52.1± 4.58 | $P<0.001$ |
| Life expectancy | 2.00 | 1.03 | | 1.48 | 0.85 | |
| Herbivory effect (%) | | | 48.5 | | | 42.5 |
| Pods/plant | 43.4±5.40 | 25.0±3.25 | $P<0.001$ | 117.5±14.83 | 96.1±21.66 | $P>0.05$ |
| Seeds/pod | 10.6±0.19 | 9.8±0.28 | $P<0.05$ | 1.9± 0.02 | 1.9± 0.03 | $P>0.05$ |
| Individual seed weight (mg) | 15.8±0.10 | 15.3±0.24 | $P>0.05$ | 9.3± 0.09 | 8.2± 0.13 | $P<0.001$ |
| Seedling recruitment | 0.5±0.06 | 0.2±0.001 | $P<0.001$ | 1.3± 0.50 | 0.6± 0.008 | $P<0.001$ |

I, insecticide-treated plants. C, control (natural levels of herbivory).

for herbivores (Coley 1980). In the early season, on longer-lived and therefore more apparent leaves ("sensu" Feeny 1976) damage was much lower.

In both *Vicia* species, more flowers were produced by insecticide-treated plants and gave rise to more pods per plant, although the difference was not significant in *V. hirsuta*, mainly because the number of pods was so variable (Table 10.1). In *V. sativa*, insecticide-treated plants produced longer pods containing more seeds than control plants, whereas in *V. hirsuta* the number of seeds, normally two per pod, did not vary (Table 10.1). However, individual seed weight was significantly higher in pods from insecticide-treated plants in *V. hirsuta* and did not differ in *V. sativa*. It would appear therefore that the two species respond to herbivory in different ways: *V. sativa* by producing fewer pods with fewer seeds and *V. hirsuta* by reducing individual seed weight.

Seed weight is known to have an influence on germination and plant establishment (Hendrix 1988; Parker and Salzman 1985; Louda 1982). Thus, the viability of seeds produced by insecticide-treated and control plants from the three sites of different successional age was tested under controlled conditions. In each case, a high proportion of seeds germinated, but there was no significant difference between treatments. However, when assessed in the field in the spring of the following year the number of plants establishing in an area of 500 $cm^2$ around marked plants did show a difference between treatments, although the numbers establishing were very small (Table 10.1). This result may be due to a combination of the direct effects of herbivory on seedlings and/or the indirect effects of herbivory on seed size. Larger seeds are known to have greater absolute growth rates of roots, stems and leaves at least initially (Weis 1982) and hence may be able to withstand greater environmental stress (Kinsman and Platt 1984), competition from neighbouring plants (Crawley and Nachapong 1985) and insect herbivory. A reduction in plant establishment under conditions of natural herbivory may well be one of the factors which modifies the direction and rate of plant succession.

## 10.4 Conclusions

The interaction between herbivorous insects with different life-cycle strategies and the host-plant community is complex (Fig. 10.6). On the one hand, the life-cycle strategies of insects change with the successional age of the plant community. In turn, these changes in certain traits, e.g. voltinism, dispersal ability, overwintering stage, reproductive potential and niche specialization modify the impact the herbivores have on the plant community. Thus, the life-cycles of herbivores have an indirect effect on the plant community. This chapter has shown that insect herbivores can alter both the direction and rate of plant succession and must therefore be considered a key force in the structuring of plant communities.

*Acknowledgements*. I am grateful to many colleagues for assistance with field work and particularly to Alan Gange for his participation in the work on insect herbivory. The work was financed by the Natural Environment Research Council and NATO.

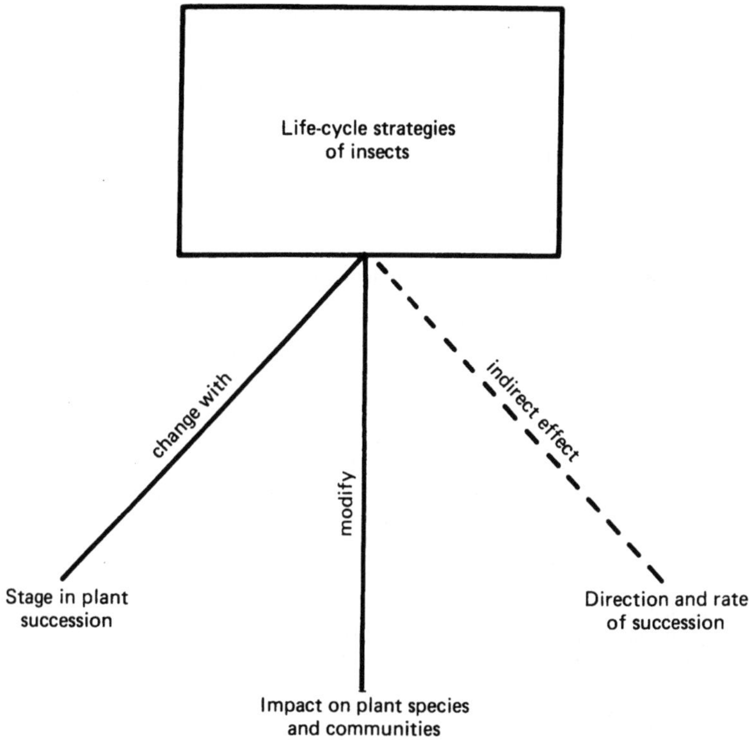


**Fig. 10.6.** Interactions between the life-cycle strategies of insects and plant community dynamics.

# References


Bazzaz FA (1968) Succession on abandoned fields in the Shawnee Hills, southern Illinois. Ecology 49: 924–936

Beckwith SL (1954) Ecological succession on abandoned farm lands and its relation to wildlife management. Ecol Monogr 24: 349–376

Belsky AJ (1986) Does herbivory benefit plants? A review of the evidence. Am Nat 127: 870–892

Bentley BL (1977) Extrafloral nectaries and protection by pugnacious body guards. Annu Rev Ecol Syst 8: 407–427

Brown VK (1985) Insect herbivores and plant succession. Oikos 44: 17–22

Brown VK (1986) Life cycle strategies and plant succession. In: Taylor F, Karban R (eds) The evolution of insect life cycles. Springer, Berlin Heidelberg New York, pp 105–124

Brown VK (1990) The effects of changes in habitat structure during succession. In: Bell SS, McCoy ED, Mushinsky H (eds) Habitat structure: the physical arrangement of objects in space. Chapman & Hall, London (in press)

Brown VK, Llewellyn M (1985) Variation in aphid weight and reproductive potential in relation to plant growth form. J Anim Ecol 54: 651–661

Brown VK, Southwood TRE (1983) Trophic diversity, niche breadth and generation times of exopterygote insects in a secondary succession. Oecologia 56: 220–225

Brown VK, Southwood TRE (1987) Secondary succession: patterns and strategies. In: Gray AJ, Crawley MJ, Edwards PJ (eds) Colonisation, succession and stability. Blackwell Scientific Publications, Oxford, pp 315–337 (Symposia of the British Ecological Society 26)

Brown VK, Gange AC, Evans IM, Storr AL (1987a) The effect of insect herbivory on the growth and reproduction of two annual *Vicia* spp., at different stages of plant succession. J Ecol 75: 1173–1189

Brown VK, Hendrix SD, Dingle H (1987b) Plants and insects in early old-field succession: comparison of an English site and an American site. Biol J Linn Soc 31: 59–74

Brown VK, Leijn M, Stinson CSW (1987c) The experimental manipulation of insect herbivore load by the use of an insecticide (Malathion): the effect of application on plant growth. Oecologia 72: 377–381

Brown VK, Gange AC, Gibson CWD (1988a) Insect herbivory and vegetational structure. In: Werger MJA, van der Aart PJM, During HJ, Verhoeven JTA (eds), Plant form and vegetation structure. S.P.B. Academic Publishing, The Hague, pp 263–279

Brown VK, Jepsen M, Gibson CWD (1988b) Insect herbivory: effects on early old field succession. Oikos 52: 293–302

Coley PD (1980) Effects of leaf age and plant life history patterns on herbivory. Nature 284: 545–546

Crawley MJ, Nachapong M (1985) The establishment of seedlings from primary and regrowth seeds of ragwort (*Senecio jacobaea*). J Ecol 73: 255–261

Feeny PP (1970) Seasonal changes in oak leaf tannins and nutrients as a cause of spring feeding by winter moth caterpillars. Ecology 51: 565–581

Feeny PP (1976) Plant apparency and chemical defense. In: Wallace J, Mansell R (eds) Biochemical interactions between plants and insects, pp 1–40 (Rec Adv Phytochem vol. 10)

Fox BJ, Fox M (1986) Resilience of animal and plant communities to human disturbance. In: Dell B, Hopkins KJM, Lamarte B (eds) Resilience of Mediterranean-type ecosystems. Dr W. Junk, The Hague, pp 39–64

Gange AC, Brown VK, Evans IM, Storr AL (1989) Variation in the impact of insect herbivory on *Trifolium pratense* through early plant succession. J Ecol 77: 537–551

Gibson CWD, Dawkins HC, Brown VK, Jepsen M (1987) Spring grazing by sheep: effects on seasonal changes during early old field succession. *Vegetatio* 70: 33–43

Hendrix SD (1988) Herbivory and its impact on plant reproduction. In: Lovett Doust J, Lovett Doust L (eds) Plant reproductive ecology: patterns and strategies. Oxford University Press, New York, pp 246–263

Hendrix SD, Brown VK, Gange AC (1988) Effects of insect herbivory on early plant succession: comparison of an English site and an American site. Biol J Linn Soc 35: 205–216

Kinsman S, Platt WJ (1984) The impact of a herbivore upon *Mirabilis hirsuta*, a fugitive prairie plant. Oecologia 65: 2–6

Koptur S, Lawton JH (1988) Interactions among vetches bearing extrafloral nectaries, their biotic protective agents, and herbivores. Ecology 69: 278–283

Louda SM (1982) Limitation of the recruitment of the shrub *Haplopappus squarrosus* (Asteraceae) by flower- and seed-feeding insects. J Ecol 70: 43–53

McNaughton SJ (1983) Compensatory plant growth as a response to herbivory. Oikos 40: 329–336

Moran VC, Southwood TRE (1982) The guild composition of arthropod communities in trees. J Anim Ecol 51: 289–306

Olson JS (1958) Rates of succession and soil changes on southern Lake Michigan sand dunes. Bot Gaz (Chicago) 119: 125–170

Oosting HJ (1942) An ecological analysis of the plant communities of Piedmont, North Carolina. Am Midl Nat 28: 1–126

Parker MA, Salzman AG (1985) Herbivore exclosure and competitor removal effects on juvenile survivorship and growth in the shrub *Gutierrezia microcephala*. J Ecol 73: 903–913

Root RB (1973) Organisation of a plant arthropod association in simple and diverse habitats: the fauna of collards (*Brassica oleracea*). Ecol Monogr 43: 95–124

Southwood TRE (1962) Migration of terrestrial arthropods in relation to habitat. Biol Rev 37: 171–214

Southwood TRE (1978) Ecological methods with particular reference to the study of insect populations, 2nd edn. Chapman & Hall, London

Southwood TRE, Brown VK, Reader PM (1979) The relationships of plant and insect diversities in succession. Biol J Linn Soc 12: 327–348

Tramer EJ (1975) The regulation of plant species diversity on an early successional old field. Ecology 56: 905–914

Vepsäläinen K (1978) Wing dimorphism and diapause in *Gerris*: determination and adaptive significance. In: Dingle E (ed) Evolution of insect migration and diapause. Springer, Berlin Heidelberg New York, pp 218–253

Waloff N (1983) Absence of wing polymorphism in the arboreal, phytophagous species of some taxa of temperate Hemiptera: an hypothesis. Ecol Entomol 8: 229–232

Watt AS (1960) Population changes in acidophilous grass-heath in Breckland, 1936–57. J Ecol 48: 605–629

Weis IM (1982) The effects of propagule size on germination and seedling growth in *Mirabilis hirsuta*. Can J Bot 60: 1868–1874

*Chapter 11*

# Migration and the Coordination of Life-Cycles as Exemplified by Lygaeinae Bugs

Christer Solbreck, David B. Anderson and Jonas Förare

## 11.1 Migration and Life-Histories

Flight ability has undoubtedly been an important contributing factor for the large success of insects as a group. Although some insects may travel considerable distances by means other than flight, flight is nevertheless by far the most common means for insects to move to new habitats. Migratory flights, by which insects move to new habitats for breeding or diapause, are performed during a limited period in adult life. During this period the insect is behaviourally and physiologically committed to flight, whereas other functions such as feeding and reproduction tend to be suppressed (Johnson 1969, 1976; Rankin 1985; Kennedy 1985).

In many insects, flight also serves the function of short everyday movements, for example between feeding stations. This kind of flight is often called trivial flight, and the flights tend to be shorter, less straightened out and more easily interrupted by feeding cues etc. Although the two types of flight may be viewed as end points of a continuum, it is often useful to make this distinction when discussing flight in a life-history context. This chapter is concerned mainly with insect flight in its migratory function.

In insects, migratory ability is manifested in a highly "compartmentalized" way. The locomotory apparatus, consisting of wings and flight muscles, is located in a separate compartment of the body, the upper part of the thorax. Furthermore, this apparatus used in migration is usually fully functional only during a short phase in adult life. The migratory phase is preceded by a teneral period and usually succeeded by a period of diapause or reproduction (Johnson 1969). Probably because of this distinct morphological and ontogenetic manifestation of migratory ability in insects, migration has long been included as an important component in the study of insect life-histories (Dingle 1972). It is now generally acknowledged that migratory capability is strongly coordinated with other life-history traits, with regard to both genetic architecture and phenotypic coordination. Migration is hence coordinated with other traits into a life-history syndrome (Dingle 1984, 1985a, b, 1986a, b).

This study further elaborates the theme of how migration is coordinated into insect life-history syndromes, using examples from our own studies of life-history variation in a group of seed-feeding bugs of the subfamily Lygaeinae (Heteroptera: Lygaeidae). Most species of Lygaeinae feed on poisonous plants, such as members of Asclepiadaceae, Apocynaceae and Senecioneae. The insects are usually aposematically coloured in red/orange and black, which makes them easy to observe in the field. Most species are easy to rear in the laboratory on a diet of dry seeds, which makes them good experimental objects. Data on several life-history traits (Table 11.1) are presently available on 13 species, most of which have been collected in the Mediterranean region (Table 11.2).

**Table 11.1.** Life-history traits studied in Lygaeinae bugs. All data refer to rearings of bugs on best-known seed diets at a temperature of 30 °C and a daylength of 16 or 17 h. This represents optimal temperature as well as diapause-inhibiting conditions. Larvae are raised at standardized low densities, and adults as pairs

| Trait | Measured as |
|---|---|
| Body weight | Live weight of newly emerged last-instar larvae |
| Egg volume | $\pi l w^2/6$, where $l$ = length and $w$ = width of egg measured with aid of microscope scale |
| Development time | Number of days from egg deposition to adult emergence |
| Growth rate | Indexed as (ln (last-instar weight) − ln (egg weight))/development time |
| Wing development | Front wings classified |
| Flight muscles | Occurrence of wing muscle histolysis determined by dissection |
| Adult survival | Survival from adult emergence, of reproducing females kept as female–male pairs. |
| Egg number | Mean number of eggs per female |
| Egg production rate | Mean number of eggs per female per day |
| Starvation tolerance | Mean survival (days) of first instar larvae isolated on day of emergence from egg and kept without food but with access to water |

**Table 11.2.** Lygaeinae species studied

| Species | Origin | Wing development[a] |
|---|---|---|
| *Tropidothorax leucopterus* (Goeze) | North Italy | M |
| *Spilostethus pandurus* (Scop.) | Sicily | M |
| *Lygaeus creticus* Lucas | Sicily | M |
| *Lygaeus equestris* (L.) | Sicily | M |
| *Graptostethus servus* (Fabr.) | Sicily | M |
| *Melanocoryphus albomaculatus* (Goeze) | Sicily | M |
| *Neacoryphus bicrucis* (Say) | Maryland, US | M |
| *Caenocoris nerii* (Germar) | Sicily | M |
| *Horvathiolus gibbicollis* (Costa) | Sicily | M/B |
| *Horvathiolus superbus* (Pollich)[b] | Sicily | M/B |
| *Lygaeosoma sardae* (Spinola) | Sicily | M/B |
| *Apterola kunckeli* Mulsant & Rey | Sicily | A |
| *Apterola rubicunda* (Stål) | Cyprus | A |

[a] M, macropterous; B, brachypterous; A, apterous.
[b] Brachypterous form rare.

We analyse the coordination of migration into life-histories at two levels. First we consider a case of *within-species variation*. A species may have several alternative developmental pathways coordinated with migratory ability. We explore this kind of variability and lability in the flight system of one species, *Horvathiolus gibbicollis*, and demonstrate the strong coupling to changes in several other life-history traits. This study also points out the multitude of life-history traits forming the overall life-history, and how these are coordinated.

Second, we consider broad patterns of *between-species variation* in migratory potential and associated life-history patterns. Lygaeinae bugs exhibit considerable variation in flight muscle and wing development, as well as in numerous other life-history traits. There is a distinct polarization of life-history patterns between species displaying wing reductions and those which are always fully winged. This association of whole suites of life-history traits with flight apparatus reductions illustrates the strong coordination of migratory ability with overall life-history syndromes.

Before presenting these case studies, however, we will give an outline of alternative developmental pathways connected to variation in the expression of migratory capability.

## 11.2 Three Steps in the Expression of Insect Migration by Flight

In comparison with other groups of flying animals, insects display an amazing variability with regard to development of migratory capability and of the flight apparatus. The relative roles of genetic and environmental factors in determining this variability varies from species to species (Harrison 1980; Dingle 1985a). Taking the winged insect with capacity for long flights as a starting point, various degrees of reductions in flight and migratory ability can be discerned.

These reductions can be viewed as taking place in three steps. In the first step, only variations in flight behaviour can be recognized. At this level migratory tendencies can be measured by various flight test methods such as chambers for free flight (Kennedy and Booth 1963) or by various tethered flight techniques (see e.g. Gatehouse and Hackett 1980). Migratory capability can thus be quantified, for example, by measuring flight duration, rate of climb or other orientation responses.

In the second step, internal changes, usually expressed as variation in the size of flight muscles, accompany changes in behaviour. Flight muscles are fairly large structures, which occupy a considerable part of thorax space. Muscles commonly appear as either fully developed or totally reduced, which makes their status easy to determine.

In the last step, changes also involve the external flight apparatus with varying degrees of wing reduction. This is the most easily observable step in the reduction of the flight apparatus. All degrees of wing reduction occur, from slightly shortened wings to almost total loss of wings (Fig. 11.1). In Heteroptera, where front wings also serve the function of protecting the abdomen, wing reductions are usually more pronounced in hind wings (Puchkova 1971).

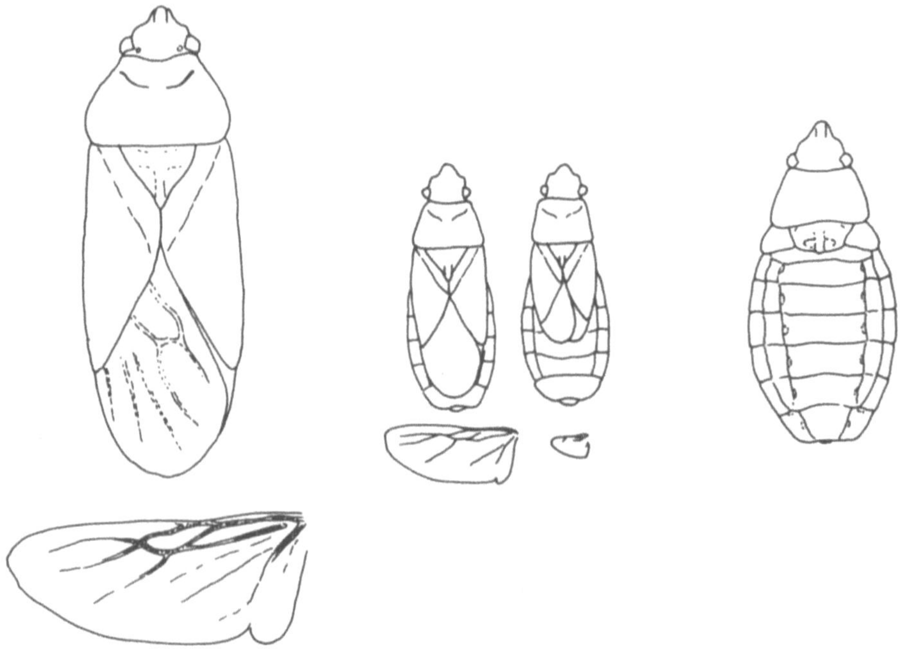

**Fig. 11.1.** Three species from Sicily, *Lygaeus equestris*, *Horvathiolus gibbicollis* and *Apterola kunckeli*, illustrating monomorphically long-winged, dimorphically long-winged/short-winged and monomorphically short-winged (apterous) Lygaeinae species. Hind wings shown×5.8.

The three steps outlined above also appear in similar sequence during the ontogeny of the individual insect. This is particularly evident when we look at how late in life developmental changes can be attained. Variations in flight behaviour are often determined by experiences late in the life of the adult insect. Changes in wing muscle development are normally determined earlier in adult life, during the teneral period. Wing development, finally, tends to be determined in earlier developmental stages. Final wing size is attained at the very start of adult life, and the course of wing development is usually apparent much earlier in life. In exopterygote insects, for example, adult wing development is usually evident from the size of wing pads already present in early larval stages.

Accordingly, for each successive step, the timing of "decisions" must be made earlier in life, and possibilities for reversal decrease. Thus, once formed in the adult stage, wings cannot be recycled into other bodily products. Flight muscle development, however, may retain its lability far into the adult stage (Johnson 1976), and, if flight muscles are histolysed, their energy can be funnelled into other functions such as reproduction. Flight behaviour, finally, is often amenable to change in response to current environmental conditions.

Fig. 11.2 illustrates, in simplified form, possible life-cycles of an insect, focusing on the three steps in flight apparatus reductions. An insect that migrates to a new habitat and breeds there – that is, reproduces "elsewhere" ($R_e$) – must develop wings, flight muscles and exhibit migratory behaviour. However, insects

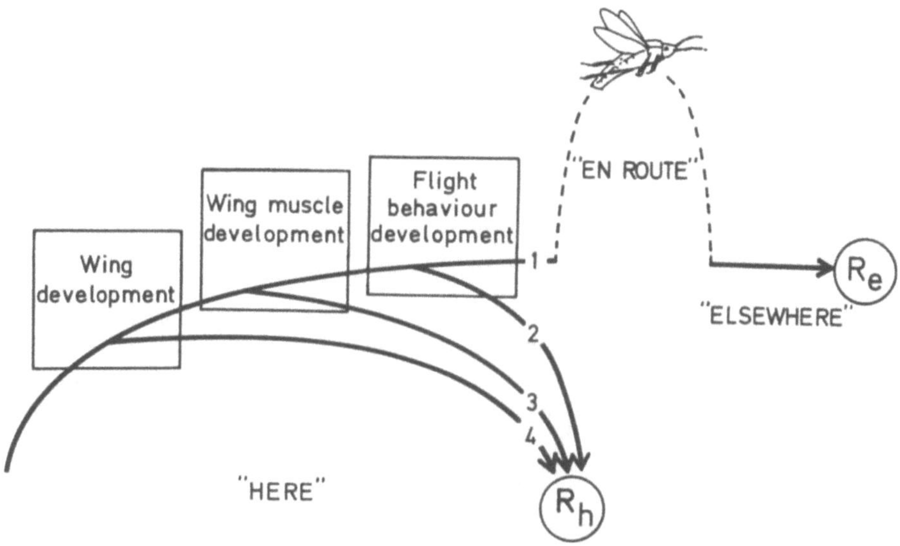


**Fig. 11.2.** Schematic representation of different developmental routes towards staying or migrating. For further explanations, see the text.

that stay and reproduce "here" ($R_h$), can do this by interrupting the chain of developmental events at different steps. Hence, there are at least three principally different life-cycles resulting in reproduction "here". One may ask why the two alternatives involving "unnecessary" wing development and flight muscle development occur as routes towards a non-migratory life. An important reason for this seems to be that these routes allow decisions to be made later in life, regarding whether or not to migrate.

The four life-cycles outlined will all have different life-history consequences. To understand the evolution of migration and life-histories, life-history patterns should be related to patterns of environmental conditions (the habitat templet) (cf. Southwood 1977; Solbreck 1978). This can be illustrated by indicating a spatial dimension in the life-cycle diagram (Fig. 11.2). The three life-cycles of non-migrants (2, 3 and 4) all lead to reproduction in one and the same habitat patch – "here". The life-cycle of the migrant (1), however, is divided spatially into three environments. First, pre-imaginal stages and early adult life occurs "here". Second, a short duration transport phase occurs – "en route". Finally, the insect will complete the life-cycle in a new habitat patch – "elsewhere". For sake of simplicity we are deliberately omitting more complex cases, for example with repeated migrations to and from hibernation sites.

According to this figure, for a full analysis of a case with a wing dimorphic species, one must consider conditions in three environments, as well as the life-history patterns of four life-cycle alternatives. To our knowledge no such study has yet been made, nor will we be able to present one here. However, we will address some problems concerning parts of this framework.

## 11.3 Within-Species Variation – *Horvathiolus gibbicollis*

*Horvathiolus gibbicollis* is a ground-living species of the eastern Mediterranean region. It has two wing morphs, one long-winged (macropterous) and one short-winged (brachypterous) (Fig. 11.1). We have studied this species in Sicily, where the short-winged morph is the most common form. All four life-cycles depicted in Fig. 11.2 occur in *H. gibbicollis*. However, our studies have thus far been limited to the three alternative life-cycles (2, 3, 4 in Fig. 11.2) occurring among bugs that stay.

Both genetic and environmental conditions affect the development of the different life-history alternatives in *H. gibbicollis*. Wing length seems to be totally genetically determined, with a dominant gene causing short wings. Flight muscle development, however, is environmentally controlled, with muscle histolysis taking place if females are both mated and have abundant food resources (Solbreck 1986). Factors affecting flight behaviour have not been investigated yet.

There are coordinated changes in several life-history traits accompanying the three alternative life-cycles. One difference is obvious if one merely looks at long- and short-winged bugs. The thorax is both shorter and narrower in short-winged bugs, allowing more resources for the abdomen. Under optimal conditions, when food is abundant, brachypterous bugs develop somewhat faster and start laying eggs about 1 day before macropters (Fig. 11.3a, B+ and M+). However, under conditions of moderate food shortage, onset of oviposition is delayed in macropters (M−) as compared with brachypters (B−). Whereas the rate of egg-laying has a somewhat higher peak in well-fed macropters (M+) than in well-fed brachypters (B+), brachypters (B−) are much more fecund than macropters (M−) when subjected to moderate food shortage. The reason for this is evident from Fig. 11.3c. Well-fed macropters (M+) do not have flight muscles, whereas macropters (M−) subjected to moderate starvation have fully developed flight muscles. When kept on the same low food level brachypters lay about ten times as many eggs as do macropters with flight muscles.

Another interesting change takes place in macropters (Fig. 11.3b). Whereas brachypters (except for the very first day of egg-laying) always lay eggs of the same average size, macropters start laying smaller eggs than do brachypters, which enables the former to reach a higher rate of egg production. As well-fed macropters histolyse their wing muscles, egg size increases. Egg size thus seems to be coordinated with wing muscle development. In individuals without wing muscles the ovaries occupy much of the thorax space, which suggests that there is competition for space between flight muscles and ovaries.

Egg size affects the survival of first-instar larvae emerging from the eggs. When subjected to total starvation, larvae from larger eggs survive longer than those from small eggs (Fig. 11.4). Even though variation around the regression line is large, there is a significant effect when a large enough sample is studied.

We can thus summarize the coordination of three of the life-cycles of *H. gibbicollis*. Wing development is coupled to the allocation of resources between thorax and abdomen, as shown by brachypters having a larger abdomen to thorax ratio than do macropters. Furthermore, brachypters have differently shaped egg-laying curves and initially lay larger eggs than do macropters. In macropters, resources are allocated between flight muscles and biomass of eggs produced

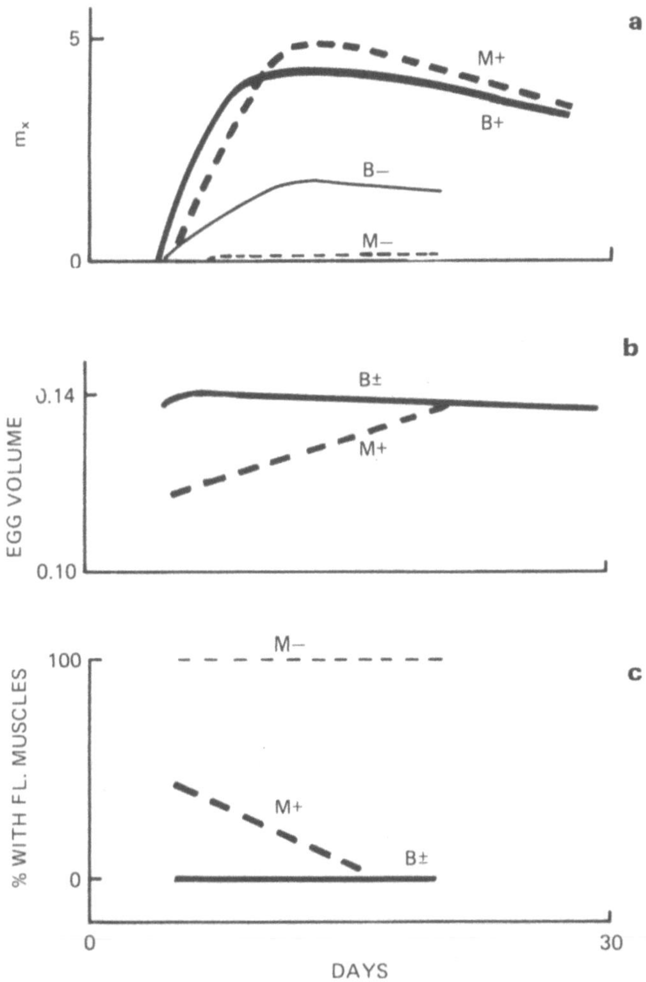


**Fig. 11.3.** Changes in $m_x$ curves, egg volume and flight (fl.) muscle development in long-winged (M) and short-winged (B) morphs of *H. gibbicollis* following adult emergence (day 0). Each morph was subjected to two food regimes, viz. abundant food (+) and moderate food shortage (−). (Somewhat simplified after Solbreck (1986) and H. Häggström and C. Solbreck (unpublished results).)

(Fig. 11.5). Thus, either muscles are large and eggs few or muscles are histolysed and eggs are numerous. Macropters furthermore show a temporal pattern of allocation between egg number and egg volume. Egg volume in turn will have consequences for the ability of young larvae to survive starvation.

The study of *H. gibbicollis* illustrates that a species may exhibit several alternative life-cycles with regard to migratory potential, and that many traits with obvious fitness consequences vary in a coordinated fashion for each alternative life-cycle. A similar richness of life-cycle alternatives, with associated differences in life-history traits, evidently occurs in many insect groups. Among the Hemiptera, aphids (Dixon 1985), planthoppers (Denno et al. 1989) and gerrids (Fairbairn and Desranleau 1987; Kaitala 1988; Vepsäläinen 1978; Vepsäläinen and Patama 1983) provide many examples. However, the specific

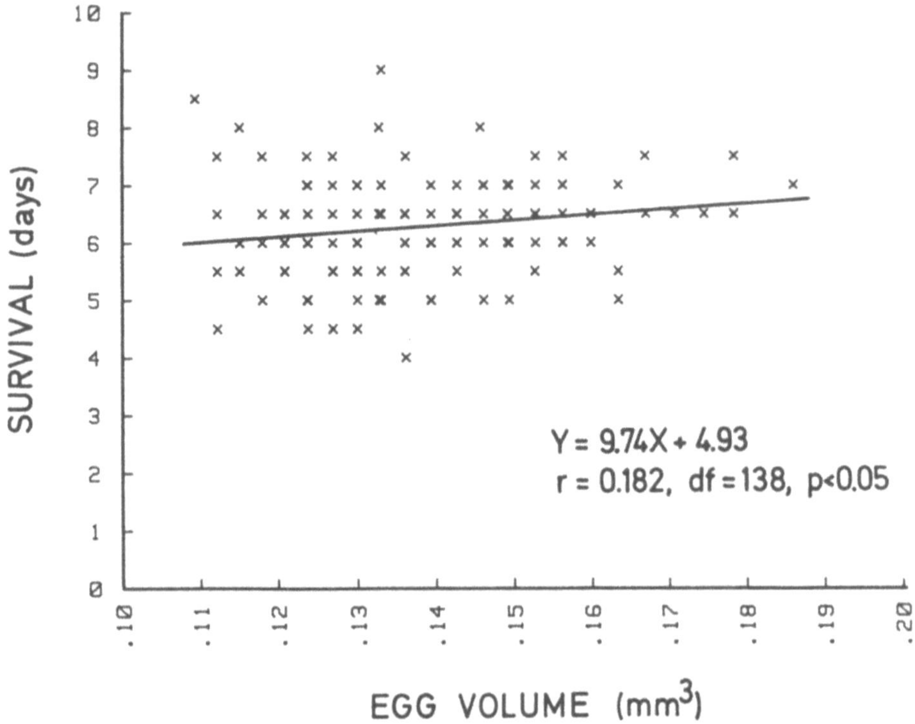


**Fig. 11.4.** Survival time of first-instar larvae of *H. gibbicollis* from eggs of different sizes. Overlapping points are not separated.

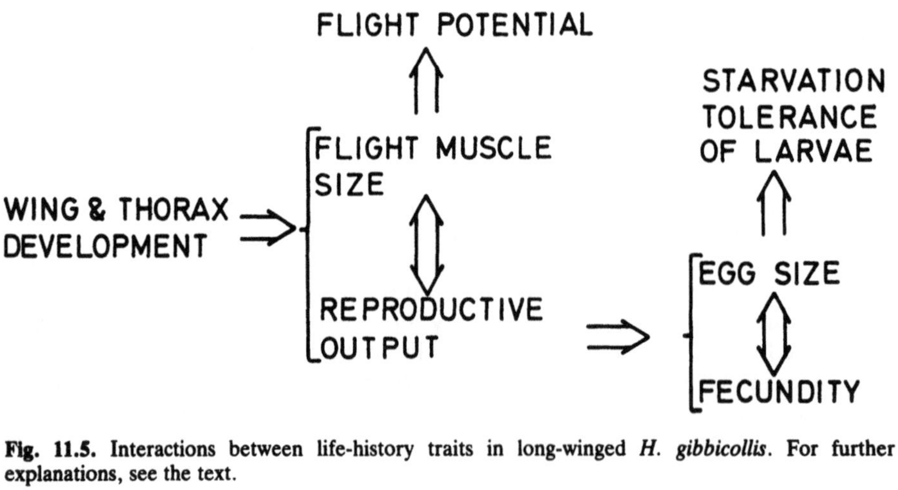


**Fig. 11.5.** Interactions between life-history traits in long-winged *H. gibbicollis*. For further explanations, see the text.

coordinations between life-history traits, as well as the role of genetic and environmental factors, in determining life-cycle alternatives vary considerably from species to species (Harrison 1980; Dingle 1985a; Roff 1986a). There still is a great need to add more detail in case studies on the coordination of migration into insect life-history syndromes. For example, the dynamics of flight muscle development and its life-history consequences seem to be important in many species (e.g. Kaitala 1988). It seems likely that flight muscles represent a larger energetic investment than do wings (cf. Roff 1986a), and in discussions of wing reduction the role of flight muscle development is usually not clear.

The life-cycle complexity with regard to genotypic variation and phenotypic plasticity is undoubtedly often important in insect population dynamics. When an insect population is subjected to environmental stress, potential genotypic and phenotypic variation may act as buffers at the population level (cf. Łomnicki 1988). Phenotypic plasticity in particular gives room for very rapid and drastic changes in the proportion of individuals with different migratory capabilities (cf. Johnson 1976; Dixon 1985). As indicated both by models (Roff 1986b) and selection experiments (Dingle 1985a, 1986b), genetic changes over a few generations may sometimes be rapid enough to deserve consideration in population studies.

## 11.4 Between-Species Comparisons

We have studied 13 Lygaeinae species, most of them from the Mediterranean region (Table 11.2). In this group there is considerable variation in several life-history traits (Table 11.1). Wing development, for example, shows extensive variation among species. There are monomorphically long-winged species, dimorphic species with both long- and short-winged forms, and even some monomorphically almost wingless species (Fig. 11.1, Table 11.2).

The life-cycles of species from these three groups can be described by referring to Fig. 11.2. Monomorphically long-winged species (denoted M) have two (routes 1 and 2 in Fig. 11.2) or three (routes 1, 2, 3) developmental alternatives, depending on whether or not wing muscle histolysis occurs. In wing-dimorphic species (M/B), all four alternatives occur, whereas in monomorphically short-winged species (A) only one alternative (route 4) occurs.

Most of the monomorphically long-winged species are large, whereas wing dimorphism and aptery occur mainly among small species (cf. Seidenstücker 1964). This association between size and tendency for wing reduction can also be seen among closely related species. Thus, in the genus *Horvathiolus*, with several species from the Mediterranean region, Asia minor and Africa, the frequency of short-winged forms is higher among the smaller species (Fig. 11.6).

Although there is large variation in weight between the species, development time varies little between most species (Fig. 11.7). Surprisingly the largest species has the same development time as the smallest one, which weighs less than a tenth of the large species. Among the monomorphically long-winged species (M) development time seems to be totally independent of body weight. This independence between body size and development time is attained by large

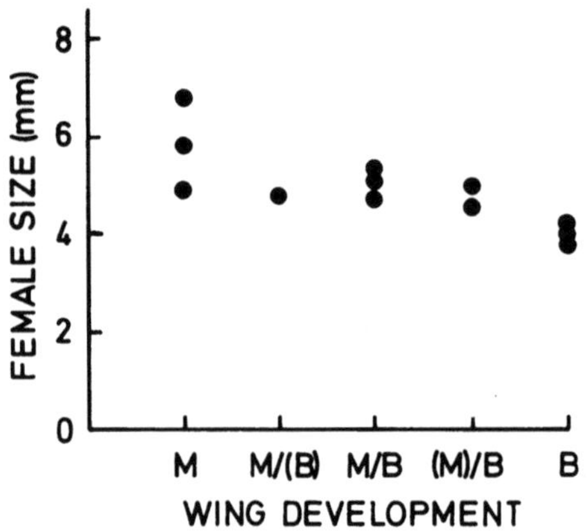


**Fig. 11.6.** Relationship between frequency of short-winged morph (from almost only macropters, M, to almost only brachypters, B) and adult size in 12 *Horvathiolus* species. (Data from Josifov 1965; Slater and Sperry 1973; C. Solbreck unpublished results.)

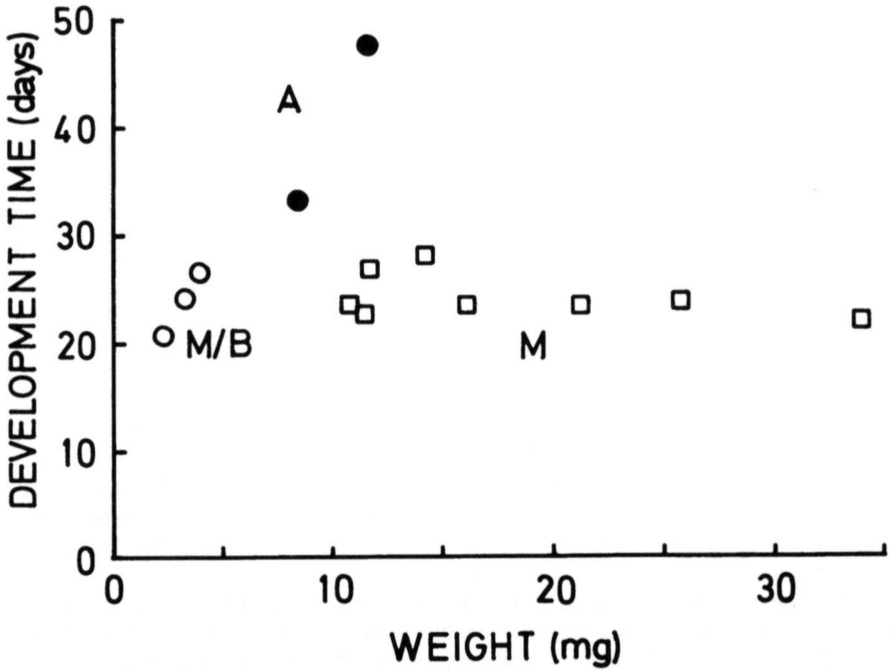


**Fig. 11.7.** Relationship between the weight of last-instar larvae and development time in long-winged (M), wing-dimorphic (M/B) and apterous (A) Lygaeinae species (species listed in Table 11.2). (●) A, (○) M/B, (□) M.

differences in growth rate. The large M species grow much faster than the small ones. However, among M/B and A bugs, generation time seems to increase with size (Fig. 11.7). The two apterous species at least have considerably longer development times than the other species. The data thus suggest that associations between size and development time are different in wing-reduced and monomorphically long-winged groups.

If larval growth rate and relative egg size are compared, large differences appear between the wing development groups (Fig. 11.8). The M species are characterized by high growth rates and relatively small eggs. Wing dimorphic and apterous species have relatively much larger eggs and slower growth rates.

Adult female survival (being a partial survival curve and hence denoted $s_x$) and fecundity ($m_x$) curves are also very different in the different groups, as illustrated by five examples (Fig. 11.9). *Spilostethus pandurus* and *Lygaeus equestris* are two monomorphically long-winged species, with comparatively low adult survival and high rates of egg production. Two species of the genus *Horvathiolus* (a genus with many wing-dimorphic species) have considerably higher adult survival and lower and more extended egg-laying curves. Finally, *Apterola kunckeli*, an apterous species, has extremely high adult survival and a very low and extended egg-laying curve. This species really has extraordinary $s_x$ and $m_x$ curves (obviously not due to diapause, which is induced by short days in this species). No adult mortality had occurred after 70 days of adult life at 30 °C, and egg-laying continued for 5 months in some individuals. Species with wing reductions thus seem to compensate for their lack of mobility by having an extended adult life as well as reproductive period.

Are there any costs for large size and high growth rate in M species? We subjected first-instar larvae of different species to total starvation. Larval survival

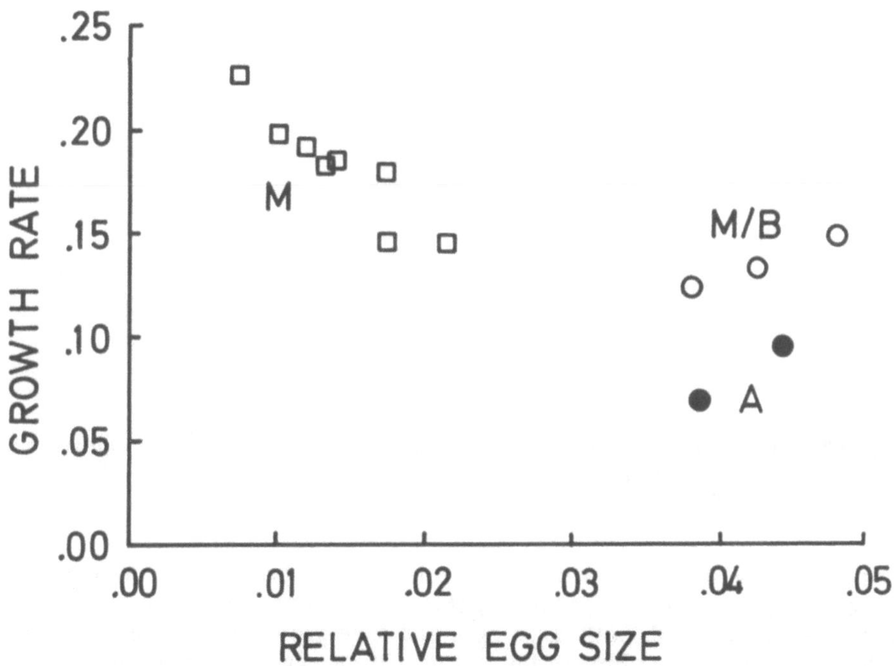


**Fig. 11.8.** Relationship between *relative* egg size (indexed as egg volume/last-instar weight) and growth rate in Lygaeinae species with different wing development. Symbols as in Fig. 11.7.

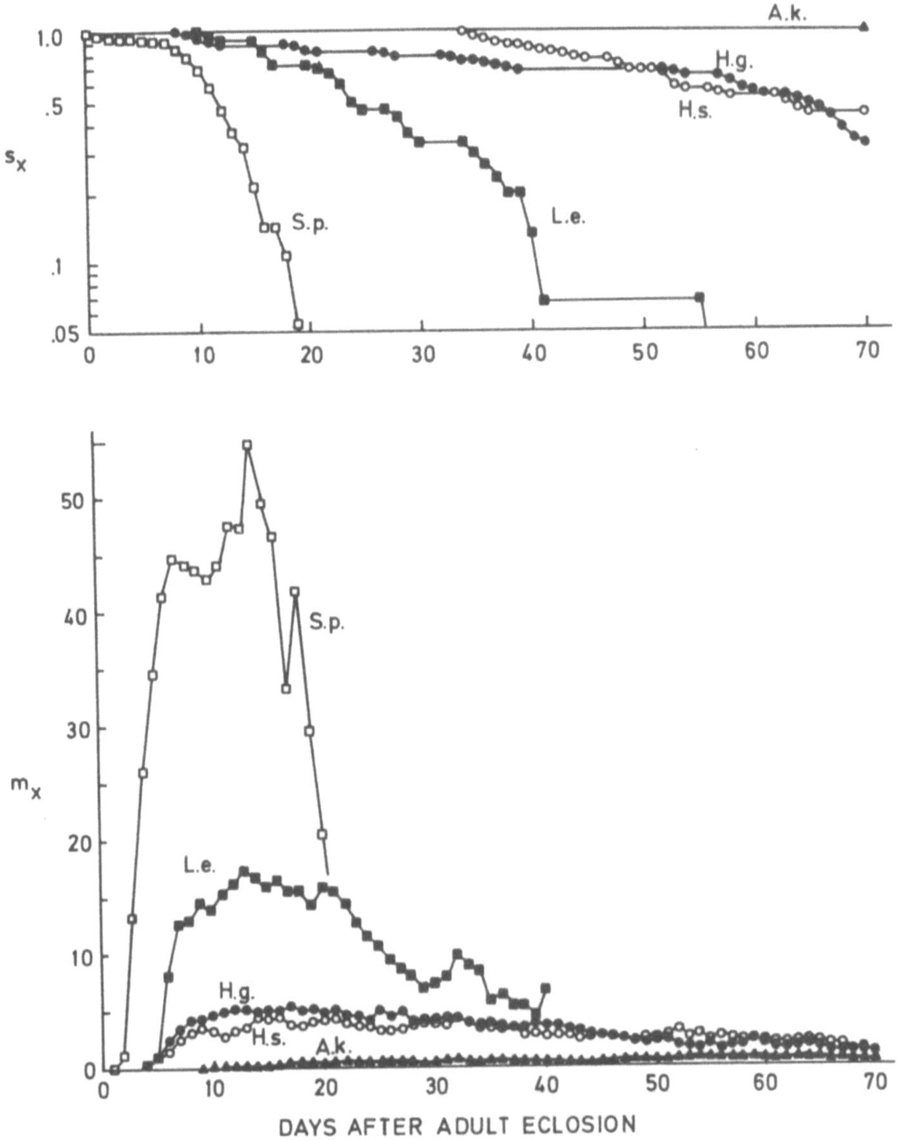


**Fig. 11.9.** Adult survival ($s_x$) and $m_x$ curves of *Spilostethus pandurus* (S. p.), *Lygaeus equestris* (L. e.), *Horvathiolus. gibbicollis* (H. g.), *H. superbus* (H. s.) and *Apterola kunckeli* (A. k.). For further explanations, see the text.

under food shortage is evidently correlated with both larval growth rate (negatively) and egg size (positively) (Fig. 11.10). For example, larvae of the apterous species, with both large eggs and slow growth rates, are extremely tolerant to starvation. First-instar larvae – from eggs less than half a cubic millimetre in size – can survive total starvation for almost 10 days at 30 °C.

Two sets of factors are of importance for the evolution of migratory (or non-

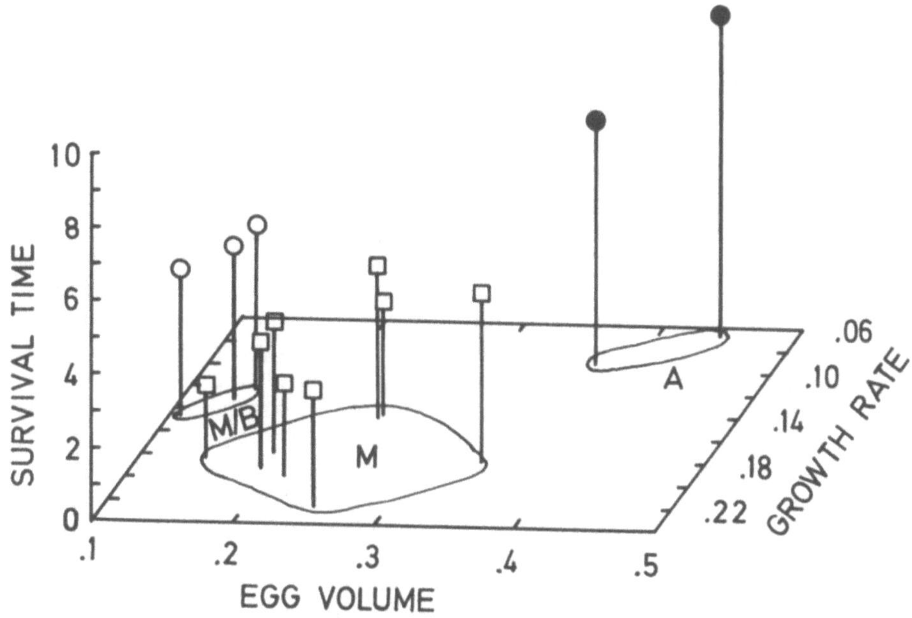


**Fig. 11.10.** Relationships between *absolute* egg size (mm$^3$), larval growth rate (per day) and survival time (days) of starved first-instar larvae of Lygaeinae bugs with different wing development patterns. Symbols as in Fig. 11.7.

migratory) ability. These are (a) environmental conditions and (b) coordination of life-history traits.

The environmental conditions affecting the evolution of wing development (an index of tendency to migrate) are evident from correlations between habitat quality and the frequency of wing morphs (Southwood 1962; Vepsäläinen 1978). Lygaeidae species with wing reductions are usually ground-living inhabitants of dry, stable habitats of low productivity (Slater 1975, 1977). There appears to be a strong historical component as well. Thus, the incidence of wing reductions seems to be related to the stability of an area also in terms of evolutionary time. This is evidenced by the high frequency of species with wing reductions among the evolutionarily old faunal elements of southwest Australia and South Africa (Slater 1975, 1977).

The Lygaeinae species with wing reductions are ground-living and feed on seeds on the ground, whereas species that largely feed up on the plants always are long-winged (cf. Slater 1977). A likely reason for the lack of wing reductions among arboreal species is that they need wings for everyday short flights in their structurally more complex habitats (Waloff 1983).

The role of life-history correlates in the evolution of migratory or non-migratory life-styles has received considerable attention. Several studies have been published concerning migration and life-histories in Lygaeinae bugs. Almost all of these, however, are concerned with species of the genera *Oncopeltus* and *Lygaeus*, where variation in migratory tendency is expressed mainly at the behavioural level. Few have been concerned with species showing wing muscle (Solbreck and Pehrson 1979) or wing dimorphism (Solbreck 1986).

This study widens the perspective of migration and life-histories in Lygaeinae bugs by considering several species where variation also includes flight muscle and wing development.

Wing reduction, which is one expression of migratory tendency, is thus strongly coordinated with several other life-history traits in Lygaeinae bugs. Although our data set is still limited, some tentative generalizations of life-history traits associated with wing reduction, and hence reduced migratory ability, will nevertheless be attempted (Fig. 11.11).

Species with wing-reduced forms tend to be small. A similar relationship has been observed in studies on monomorphically long-winged species, where flight duration has been used as an index of migratory tendency. Thus migration is associated with larger body size in *Oncopeltus* spp. (Lygaeinae) (Dingle et al. 1980) and in *Dysdercus* spp. (Pyrrhocoridae) (Derr et al. 1981). Furthermore, large size is associated with increased egg production both in species studied by us and in the earlier mentioned studies on *Oncopeltus* and *Dysdercus*. Furthermore, in selection experiments with *Oncopeltus fasciatus* there is a positive genetic association between body size, wing length and fecundity and a negative relationship with development time (Palmer and Dingle 1986; and see Chap. 1). An association between faster growth and migratory tendency was also noted by Chaplin and Chaplin (1981) and Zirkle et al. (1988).

Dingle et al. (1980) and Derr et al. (1981) suggest that larger size in species with higher migratory tendencies (measured as flight duration of long-winged species) is a buffer against vagaries of the environment, for example resulting in increased resistance to starvation. In addition to this positive effect of large size in the full-grown insect, our study shows a negative effect in young developmental stages. Large size is attained through higher growth rate. This high growth rate (in addition to small egg size) results in young larvae of monomorphically long-winged species being more sensitive to starvation. This lowered starvation resistance can obviously be tolerated only in species where larval food shortage can be prevented by parents responding to food shortage and flying to better places for egg-laying. Thus, trade-off patterns seem to be different in species with and without wing reduction. In wing-reduced species larger body size seems to be

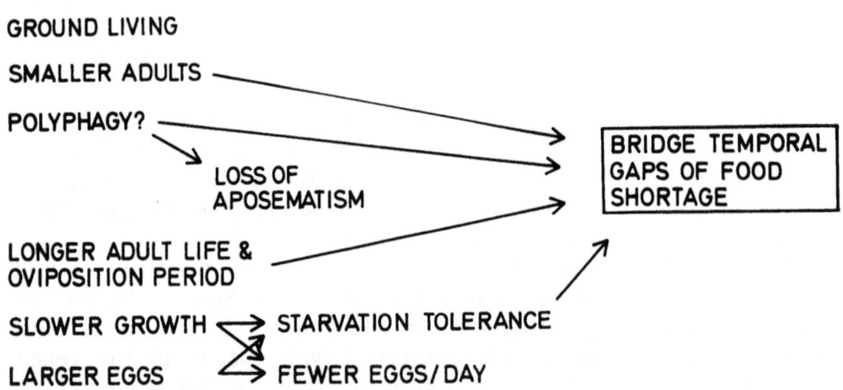


**Fig. 11.11.** Traits associated with wing reductions in Lygaeinae bugs. *Arrows* indicate functional relationships.

associated with longer development time, whereas in monomorphically long-winged species, increased body size is mainly correlated with increased growth rate.

Although data are still scanty, there do not seem to be any monophages among Lygaeinae species with wing reductions. Several of the monomorphically long-winged species are also highly polyphagous, but some species with highly restricted diets also occur in this group. This suggests that restricted diets are less frequent among species with wing reductions. There is also a tendency for loss of aposematism in genera showing wing reductions, and several cryptically coloured forms occur in this group of species. This is possibly related to their tendency to have a wider food-plant spectrum, possibly with fewer poisonous species in the diet.

Lygaeinae species showing wing reductions tend to have longer adult lives and longer oviposition periods. They have slower growth, and they lay relatively larger eggs, which results in fewer eggs being laid per day.

Young larvae in species with larger eggs and slower growth are more resistant to starvation. A longer oviposition period is presumably an adaptation to reduce the risk of total reproductive failure, and small adult size and lack of monophagy evidently reduce the need for total amount of food as well as for specific foods. The common "purpose" for all these life-history characteristics seems to be the need to survive periods of food shortage.

The coordination of migration with numerous other life-history traits is important to consider when discussing the evolution of wing apparatus reductions. Loss of flight ability has its dangers and hence is more likely to evolve where other life-history traits can compensate for these risks (Fig. 11.11). Thus, wing reductions are likely to originate only in certain constellations of life-history traits. In addition to the role of habitat stability both in an ecological (Southwood 1962, 1977) and an evolutionary time perspective (Slater 1977), life-history characteristics strongly modify the selective effects of the habitat template.

## 11.5 Conclusions

Three general conclusions emerge from this chapter.

First, our schematic figure of alternative life-cycles illustrates a common deficiency in most studies of migration and life-histories (including ours). The life-cycle of the migrant occurs in three different environments, "here", "en route", and "elsewhere". What happens in the latter two environments is rarely considered. The figure also clearly depicts the possibility of several alternative developmental routes among non-migrants.

Second, migration is indeed strongly integrated into the life-history of the insect. Life-histories are composed of many life-history traits that covary, forming adaptive syndromes. We are not the first to point this out; it has been stressed repeatedly, for example, by Dingle (1986a, b) and Parsons (1983).

Third, patterns of life-history coordination tend to be quite different in species with only migratory and those with non-migratory forms. Indeed migratory and non-migratory forms seem to originate in different constellations of life-histories, and to evolve along different dimensions in the "hyperspace of life-history variables". The reason for this seems to be that a non-migratory life puts strong demands on certain traits and combinations of traits. A common denominator seems to be "to bridge gaps in time".

## 11.6 Summary

Migration by flight forms an integral part of insect life-cycles by being coordinated with several other life-history traits. Migratory tendency may be expressed as variation in flight behaviour, in flight muscle development as well as in wing development. Migrating individuals need fully developed wings and flight muscles as well as behavioural preparedness for migration. For non-migrants, however, there are three partly overlapping developmental routes depending on whether reductions occur in the development of (a) migratory behaviour, (b) behaviour and flight muscles, and (c) behaviour, muscles and wings. These three developmental pathways, which may occur in the same species, are likely to have different life-history consequences.

The analysis of migration and the coordination of life-cycles may thus have to consider three life-history alternatives for non-migratory individuals, in addition to one alternative for migrants. Whereas the three non-migratory routes occur in one environment ("here"), the migratory life-cycle occurs in three different environments: "here", "en route" and "elsewhere". Thus far no study has considered all environmental and life-cycle components when migratory strategies are analysed.

The coordination of migration into life-history syndromes is illustrated by our studies on a group of seed-feeding bugs, the Lygaeinae. Within-species variability in migration and associated life-cycles is illustrated by the wing and flight muscle polymorphic species, *Horvathiolus gibbicollis*. Life-history consequences of the three life-cycle alternatives among non-migratory bugs are described.

Between-species comparisons are made on the basis of life-history studies of 13 Lygaeinae species with varying degrees of wing development. Species with wing reductions have life-histories quite different from those of monomorphically long-winged species. They have slower growth rates, longer lives, relatively larger but fewer eggs and have more starvation-tolerant larvae. Trade-off functions between life-history traits also seem to be different in taxa with and without wing reductions. Thus, in monomorphically long-winged species development time is uncorrelated with body size, in contrast to the situation among species displaying wing reductions.

*Acknowledgement*. We thank Christer Wiklund for comments on the manuscript.

### References

Chaplin SB, Chaplin SJ (1981) Comparative growth energetics of a migratory and nonmigratory insect: the milkweed bugs. J Anim Ecol 50: 407–420

Denno RF, Olmstead KL, McCloud ES (1989) Reproductive cost of flight capability: a comparison of life history traits in wing dimorphic planthoppers. Ecol Entomol 14: 31–44
Derr JA, Alden B, Dingle H (1981) Insect life histories in relation to migration, body size, and host plant array: a comparative study of *Dysdercus*. J Anim Ecol 50: 181–193
Dingle H (1972) Migration strategies of insects. Science 175: 1327–1335
Dingle H (1984) Behaviour, genes, and life histories: complex adaptations in uncertain environments. In: Price PW, Slobodchikoff CN, Gaud WS (eds) A new ecology. Novel approaches to interactive systems. John Wiley & Sons, New York, pp 169–194
Dingle H (1985a) Migration. In: Kerkut GA, Gilbert LI (eds) Comprehensive insect physiology, biochemistry, and pharmacology, vol 9. Pergamon Press, Oxford, pp 375–415
Dingle H (1985b) Migration and life histories. Contr Mar Sci [Suppl] 27: 27–42
Dingle H (1986a) The evolution of insect life cycle syndromes. In: Taylor F, Karban R (eds) The evolution of insect life cycles. Springer, Berlin Heidelberg New York, pp 187–203
Dingle H (1986b) Evolution and genetics of insect migration. In: Danthanarayana W (ed) Insect flight dispersal and migration. Springer, Berlin Heidelberg New York, pp 11–26
Dingle H, Blakley NR, Miller ER (1980) Variation in body size and flight performance in milkweed bugs (*Oncopeltus*). Evolution 34: 371–385
Dixon AFG (1985) Aphid ecology. Blackie, Glasgow
Fairbairn D, Desranleau L (1987) Flight threshold, wing muscle histolysis, and alary polymorphism: correlated traits for dispersal tendency in the Gerridae. Ecol Entomol 12: 13–24
Gatehouse AG, Hackett DS (1980) A technique for studying flight behaviour of tethered *Spodoptera exempta* moths. Physiol Entomol 5: 215–222
Harrison RG (1980) Dispersal polymorphisms in insects. Annu Rev Ecol Syst 11: 95–118
Johnson CG (1969) Migration and dispersal of insects by flight. Methuen, London.
Johnson CG (1976) Lability of the flight system: a context for functional adaptation. In: Rainey RC (ed) Insect flight. Blackwells, Oxford, pp 217–234 (Symposia of the Royal Entomological Society 7)
Josifov M (1965) Zur Systematik der Gattung *Melanocoryphus* Stål (Hem. Het. Lygaeidae). Acta Entomol Mus Natl Pragae 36: 311–334
Kaitala A (1988) Wing muscle dimorphism: two reproductive pathways of the waterstrider *Gerris thoracicus* in relation to habitat instability. Oikos 53: 222–228
Kennedy JS (1985) Migration, behavioural and ecological. Contr Mar Sci [Suppl] 27: 1–20
Kennedy JS, Booth CO (1963) Free flight of aphids in the laboratory. J Exp Biol 40: 67–85
Łomnicki A (1988) Population ecology of individuals. Princeton University Press, Princeton, NJ
Palmer JO, Dingle H (1986) Direct and correlated responses to selection among life-history traits in milkweed bugs (*Oncopeltus fasciatus*). Evolution 40: 767–777
Parsons PA (1983) The evolutionary biology of colonizing species. Cambridge Univesity Press, Cambridge
Puchkova LV (1971) The functions of the wings in the Hemiptera and trends in their specialization. Entomol Rev USSR 50: 303–309
Rankin MA (1985) Endocrine influence on flight behavior. Contr Mar Sci [Suppl] 27: 817–841
Roff DA (1986a) The evolution of wing dimorphism in insects. Evolution 40: 1009–1020
Roff DA (1986b) Evolution of wing polymorphisms and its impact on life cycle adaptation in insects. In: Taylor F, Karban R (eds) The evolution of insect life cycles. Springer, Berlin Heidelberg New York, pp 204–221
Seidenstücker G (1964) *Psileula mimica* n. gen. n. sp. und die mikropteren Lygaeinen. Reichenbachiana 2: 56: 199–207
Slater JA (1975) On the biology and zoogeography of Australian Lygaeidae (Hemiptera: Heteroptera) with special reference to the Southwest fauna. J Aust Entomol Soc 14: 47–64
Slater JA (1977) The incidence and evolutionary significance of wing polymorphism in lygaeid bugs with particular reference to those of South Africa. Biotropica 9: 217–229
Slater JA, Sperry B (1973) The biology and distribution of the South African Lygaeinae, with descriptions of new species (Hemiptera: Lygaeidae). Ann Transvaal Mus 28: 117–201
Solbreck C (1978) Migration, diapause and direct development as alternative life histories in a seed bug, *Neacoryphus bicrucis*. In: Dingle H (ed) Evolution of insect migration and diapause. Springer, Berlin Heidelberg New York, pp 195–217
Solbreck C (1986) Wing and flight muscle polymorphism in a lygaeid bug, *Horvathiolus gibbicollis*: determinants and life history consequences. Ecol Entomol 11: 435–444
Solbreck C, Pehrson I (1979). Relations between environment, migration and reproduction in a seed bug, *Neacoryphus bicrucis* (Say) (Heteroptera: Lygaeidae). Oecologia 43: 51–62
Southwood TRE (1962) Migration of terrestrial arthropods in relation to habitat. Biol Rev 37: 171–214

Southwood TRE (1977) Habitat, the templet for ecological strategies? J Anim Ecol 46: 337–365
Vepsäläinen K (1978) Wing dimorphism and diapause in *Gerris*: determination and adaptive significance. In: Dingle H (ed) Evolution of insect migration and diapause. Springer, Berlin Heidelberg New York, pp 218–253
Vepsäläinen K, Patama T (1983) Allocation of reproductive energy in relation to the pattern of environment of five *Gerris* species. In: Brown VK, Hodek I (eds) Diapause and life cycle strategies of insects. Dr W. Junk, The Hague, pp 189–207
Waloff N (1983) Absence of wing polymorphism in the arboreal, phytophagous species of some taxa of temperate Hemiptera: an hypothesis. Ecol Entomol 8: 229–232
Zirkle DF, Dawson PS, Lavie B (1988) An experimental analysis of the genetic relationships among life-history traits and emigration behavior in *Tribolium castaneum*. Oikos 53: 391–397

*Chapter 12*

# Competition and Selection in a Patchy and Ephemeral Habitat: The Implications for Insect Life-Cycles

Bryan Shorrocks

## 12.1 Introduction

Many insects exploit resources which are patchy, consisting of small, separate units, and are ephemeral in the sense that they support only one or two generations and are spatially unpredictable. Such sites, or "centres of action" (Elton 1949) include dung (Mohr 1943; Valiela 1974; Hanski and Koskela 1977, 1979; Koskela and Hanski 1977; Nealis 1977; Holter 1982; Peck and Forsyth 1982; Doube 1987), carrion (Kamal 1958; Denno and Cothran 1975; Hanski 1976; Beaver 1977; Hanski and Kuusela 1977; Kuusela and Hanski 1982; Kneidel 1983, 1984; Doube 1987), fruit (Atkinson and Shorrocks 1977, 1984; Atkinson 1981, 1985; Shorrocks 1982; Lachaise and Tsacas 1983), fungi (Burla and Bachli 1968; Dely-Draskovits 1973; Hackman and Meinander 1979; Shorrocks and Charlesworth 1980, 1982; Shorrocks 1982; Janssen et al. 1988), sap-flows (Shorrocks 1982; Janssen et al. 1988) decaying leaves (Heed 1968; Montgomery 1975; Begon and Shorrocks 1978; Beaver 1979a), flowers (Brncic 1983), dead wood (Montgomery 1975; Beaver 1979a, c) and small bodies of water held in parts of terrestrial plants (phytotelmata) (Service 1965; Beaver 1979b, 1980, 1983; Lounibos 1979, 1981; Lounibos and Machado-Allison 1983).

Such resources are often individually quite small yet "en masse" they can form habitats of considerable size. For example, Sugden and Robins (1979) reported an average density of 17.5 mature epiphytic tank bromeliads occurring per square metre of ground area in a cloud forest site at 800 m altitude in Columbia. A conservative estimate of 250 ml per individual bromeliad would suggest over 50 000 litres water/ha available for use by aquatic animals at this site. In their estimates of fungi present in a 12 ha English deciduous woodland, Shorrocks and Charlesworth (1982) recorded over 2500 fruiting bodies of just one species, *Phallus impudicus*, present on one sampling date.

Elton (1966) and Beaver (1977) have commented on the large number of species commonly found in these centres of action. Some examples are quite spectacular. For instance, Holter (1982) recorded 13 *Aphodius* species from cattle dung at one site in Denmark, Hanski and Koskela (1978) recorded 47 species of coprophagous beetle from cow dung in southern Finland, and Heed (1968) listed 46 species of drosophilid flies mining the decaying leaves of one tree species (*Cheirodendron gaudichaudii*) in Hawaii.

Even though there is this anecdotal evidence that sites of this kind support more diverse communities than other kinds of resource, there is evidence of competition between species. In Diptera, it is possible to detect the effects of larval competition because, above a certain threshold, increased crowding causes the production of smaller adults (Bakker 1961; Haupt and Busine 1968). Atkinson (1979, 1985) has shown that high densities of certain drosophilids breeding in fruit are associated with reduced body size of adult flies. However, it should be noted that in the Australian data (Atkinson 1985) this effect was significantly biased towards intraspecific interactions.

A range of explanations has been put forward to account for this species richness, including resource partitioning (Atkinson 1981; Brncic 1983), fugitive species (Beaver 1977; Kneidel 1983; Marino 1988), priority effects (Kneidel 1983), predator/parasitoid pressure (Rouault 1979; Carton et al. 1986; Janssen et al. 1988) and even lack of competition (Worthen and McGuire 1988). However, recently there has been increasing interest in an explanation involving "probability refuges" (Shorrocks and Rosewell 1986; 1987). This chapter reviews the latter explanation for species diversity and for the first time extends it to the maintenance of within-species genetic diversity. The chapter ends by examining the implications of such "probability refugees" for insect life-cycles.

## 12.2 Competition

The simulation model of competition on a divided and ephemeral resource (Shorrocks et al. 1979, 1984; Atkinson and Shorrocks 1981) allows a competitively inferior species to survive in probability refuges. These are sites or patches with no or few superior competitors that arise as a result of an aggregated distribution of individuals over breeding sites. An independent model proposed by Hanski (1981, 1983) also incorporates spatial variance and promotes coexistence. Probability refuges may occur even at equilibrium density, since aggregation increases crowding (Lloyd 1967), and global population density is limited by strong intraspecific competition in sites with high local density while low-density sites still exist.

The model developed from field studies of drosophilid flies (Atkinson and Shorrocks 1977; Shorrocks and Charlesworth 1980; Shorrocks 1982) for which the major, if not the only, period of resource limitation occurs in the larval stage (Carson 1971; Shorrocks 1977; Atkinson 1979, 1985). In the model, therefore, competition is confined to the larvae and described by the equations of Hassell and Comins (1976):

$$N_i\,(t+1) = \lambda_i N_i\,(t)[1+a_i(N_i\,(t)+\alpha_{ij}N_j\,(t)]^{-b_i}$$

$N_i\,(t)$ and $N_j\,(t)$ are the numbers of each species in a breeding site at time $t$, $\lambda_i$ is the net reproductive rate, $\alpha_{ij}$ is a competition coefficient and $a_i$ and $b_i$ are constants.

The parameter $a_i$ (equal to $(\lambda_i-1)/N_i^*$, where $N^*$ is site carrying capacity) is related to the population size per site at which density dependence starts to act and $b_i$ describes the type of competition. As $b_i$ increases, competition becomes less contest and more scramble (Nicholson 1954; Hassell 1975).

In the model, the eggs of both species are independently distributed over the available breeding sites according to a negative binomial distribution, which has an exponent, $k$, inversely related to the degree of aggregation. In the initial model this parameter was kept constant. The survivors of competition within each patch disperse, reproduce and die. However, there is no difference in dispersal ability between the two species; the inferior species is not a fugitive.

The detailed conclusions that emerge from these simulations, confirmed by the analytical model of Ives and May (1985) have been summarized by Shorrocks and Rosewell (1986). Fig. 12.1 illustrates the effect of the major parameters. Critical $\alpha$ is that value of the competition coefficient, for the superior species,

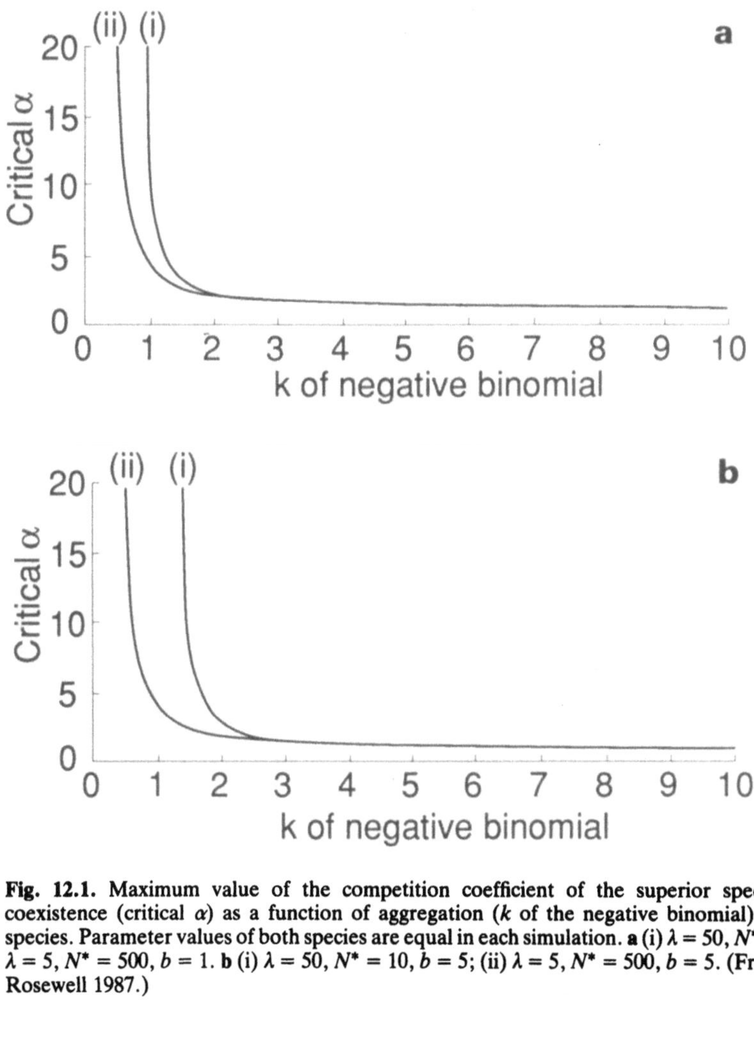


**Fig. 12.1.** Maximum value of the competition coefficient of the superior species that allows coexistence (critical $\alpha$) as a function of aggregation ($k$ of the negative binomial) of the superior species. Parameter values of both species are equal in each simulation. **a** (i) $\lambda = 50, N^* = 10, b = 1$; (ii) $\lambda = 5, N^* = 500, b = 1$. **b** (i) $\lambda = 50, N^* = 10, b = 5$; (ii) $\lambda = 5, N^* = 500, b = 5$. (From Shorrocks & Rosewell 1987.)

which just prevents coexistence. Coexistence is made easier with increasing aggregation (lower $k$) and lower values of $\alpha$ for the superior species. It is also made easier with increasing reproductive rate ($\lambda$) and increasing scramble competition ($b>1$) for the inferior species. Carrying capacity per breeding site ($N^*$) influences coexistence, since the inferior species will find it easier to invade when the superior species is at a lower relative density. Notice, however, that $k$ is the best predictor of coexistence. Given sufficiently small values ($k<1$), coexistence is virtually assured.

Fig. 12.2 shows the distribution of $k$ for 59 dipteran species (Shorrocks and Rosewell 1986; Rosewell et al. 1990) under the three different conditions of field observations, field experiments and laboratory experiments. There is no significant difference between the values of $k$ obtained from the three types of measurement ($\chi^2=4.13$) and the separate distributions are therefore combined in the figure. Fig. 12.3 shows the distribution of 52 values of $\alpha$ for seven drosophilid species (Shorrocks et al. 1984). Shorrocks and Rosewell (1986) also provide experimental data for the other parameters for drosophilid flies. These suggest typical values of $b$ between 1 and 2, of $N^*$ between 10 and 100, and of $\lambda$ between 5 and 10. For these flies there is now sufficient evidence to suggest that competitive exclusion will be a rare event. Similarly other workers (Hanski 1987; Ives 1988; Kneidel 1982) have provided evidence from both field and laboratory that

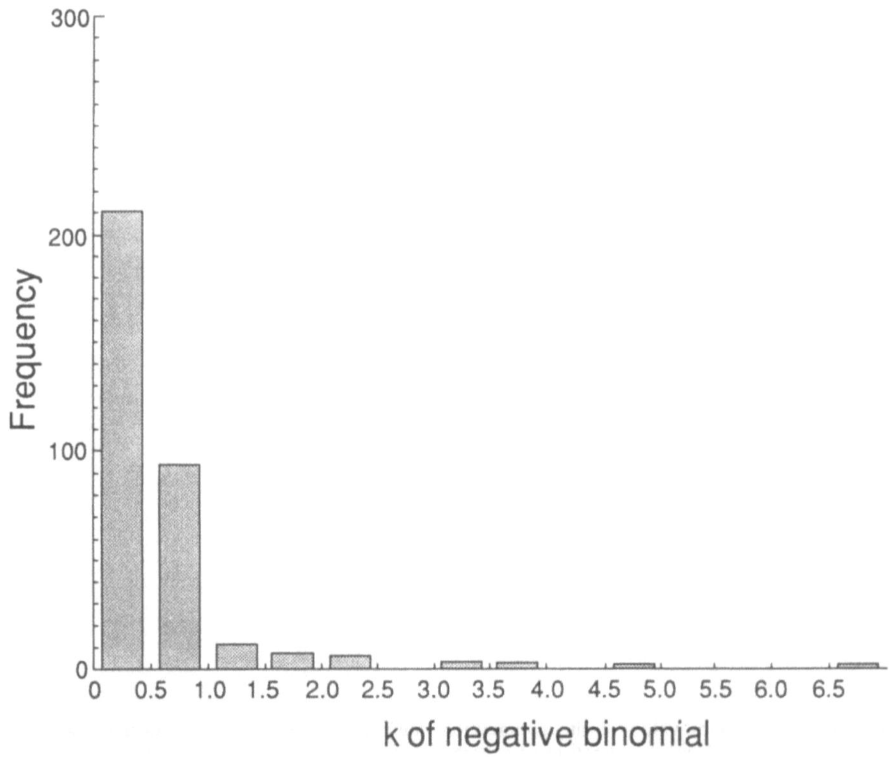


**Fig. 12.2.** Frequency histogram of values of $k$ of the negative binomial for 59 dipteran species. (From Rosewell et al. 1990.)

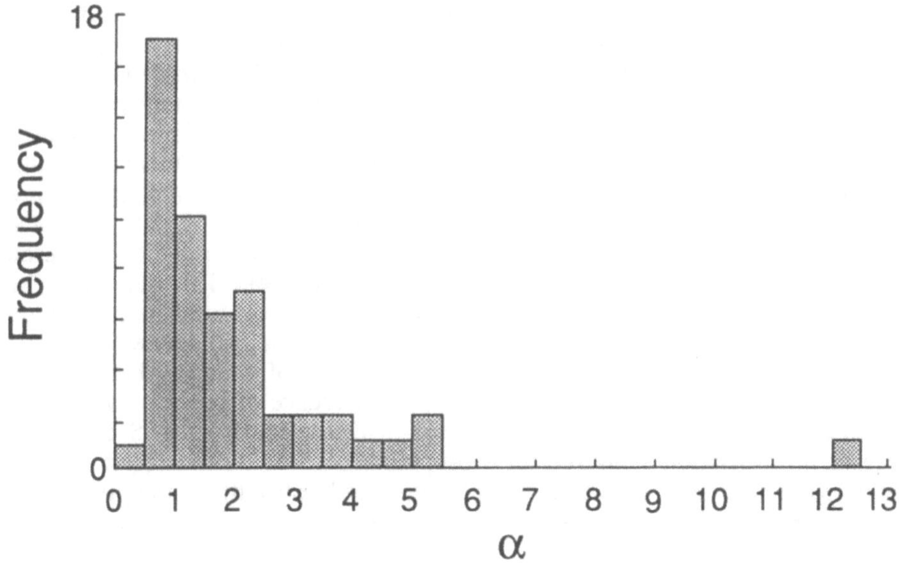


**Fig. 12.3.** Frequency histogram of values of $\alpha$ from laboratory experiments with *Drosophila* species. (From Shorrocks & Rosewell 1987.)

probability refuges could promote coexistence in other systems of insects exploiting patchy resources.

Shorrocks and Rosewell (1986) have examined some of the assumptions in the original simulation model: these include use of the negative binomial, constant $k$ and independent association between competitors. Although Atkinson and Shorrocks (1981) used the negative binomial as a convenient description of an aggregated distribution, Ives and May (1985) in their analytical model suggest that other aggregated distributions (variance >> mean) will have qualitatively similar results. The precise requirement of a negative binomial distribution is not a prerequisite for the probability refuge model and its prediction of enhanced coexistence. Nevertheless, Atkinson and Shorrocks (1984), Shorrocks and Rosewell (1986) and Rosewell et al. (1990) have fitted the negative binomial to drosophilid data and found it to be an excellent description.

There is plenty of field evidence that the degree of aggregation of animals changes with density (Taylor et al. 1978). Taylor (1961) has shown empirically that the variance ($v$) of the distribution of individuals over patches is related to the mean number of individuals per patch ($m$). From this relationship, Taylor et al. (1979) show that $k$ of the negative binomial will respond to density according to the equation $1/k = \alpha m^{\beta-2} - m^{-1}$. Shorrocks and Rosewell (1986) have fitted Taylor's Power Law to their drosophilid data and obtained values of $\alpha = 7.44$ and $\beta = 1.57$. Therefore as the mean numbers per patch increase, the value of $k$ will rise. Fig. 12.1 suggests that for values of $k<1$ coexistence would be very likely while for $k>1$ it would be more difficult, although not impossible. However, we must remember that we are talking about patchy and ephemeral sites and that $N^*$, at least for drosophilids, is probably less than 100. With $\alpha = 7.44$ and $\beta = 1.57$, a patch mean of 100 still only gives a $k = 0.98$ (Shorrocks and Rosewell 1988), suggesting that coexistence is still the most likely outcome for these flies. This result is not surprising, since natural values of *k* for drosophilids had already been measured (Shorrocks et al. 1984) and found to be mainly <1 (Fig. 12.2).

The original model assumed that the two species were distributed over patches at random with respect to each other; that is, they show neither positive association (attraction to the same site) nor negative association (resource partitioning). Positive association will make coexistence more difficult. Shorrocks and Rosewell (1986) and Shorrocks et al. (1990) were able to measure *k* and association between pairs of species. Only 8 out of 157 data sets gave significant association, all positive. As a convenient index of association they used the four-point correlation coefficient ($\Phi$) (Conover 1980). This index has the traditional range from +1 to −1. They ran the two-species simulation program using $\Phi$ to adjust the joint matrix of probabilities for the two species. Fig. 12.4 shows contours of critical $\alpha$ for various combinations of $\Phi$ and *k*. Three things may be concluded from this graph. First, the presence of positive association between the species does make it more difficult for coexistence. Second, the degree of association measured for drosophilid flies would still suggest coexistence as the most likely outcome. Third, aggregated distributions and probability refuges will also make coexistence easier when some degree of resource partitioning is present.

## 12.3 Selection

A number of models have been proposed (Levene 1953; Li 1955; Levins and MacArthur 1966; Maynard Smith 1966, 1970; Wallace 1968a; Bulmer 1972; Christianson 1974, 1975; Strobeck 1974; Yokoyama and Schaal 1985) in which environmental patchiness promotes genetic diversity. However, these models effectively describe the situation in which genotypes have different fitness values in different patches. They are therefore the genetic equivalent of the traditional resource-partitioning models in ecology. The probability refuge model described in the last section does not involve traditional resource refuges and this section will develop a parallel model for competition between genotypes.

Extrapolating from the interspecific competition model, detailed in the last section, it would appear that the type of selection most likely to be influenced by probability refuges is that which Wallace (1968b, 1975) and Christianson (1975) have called soft-selection. Let us imagine that in a certain drosophilid there are two larval genotypes, "slow" and "fast". These refer to the rate of feeding and therefore the rate of depletion of resources. Larvae that are slow can survive within a patch containing only slow genotypes. In a patch that contains fast genotypes, they may fail to obtain enough food before resources are depleted. In soft-selection, therefore, a genotype's fitness is dependent upon which other genotypes are present. Alternatively with hard-selection, a genotype's fitness is not dependent upon the presence of other genotypes. For example, suppose the above drosophilid had two larval genotypes, "warm" and "cold", which referred to the ability of the larvae to withstand low temperature. The fitness of the more susceptible, warm genotype would be a function of temperature, but not of how many cold genotypes were present in the patch. Notice that, in both types of selection, genotypes fail to leave offspring. Soft does not, in this instance, imply a

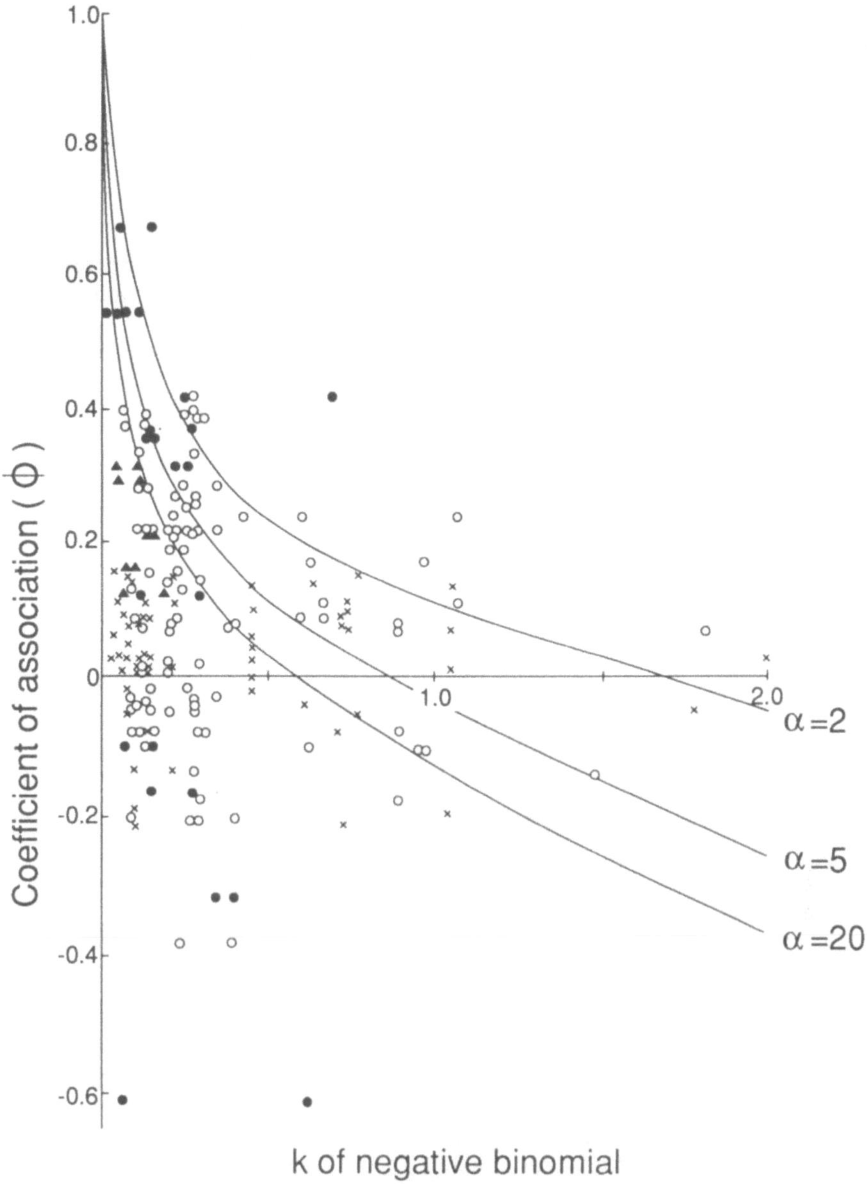


**Fig. 12.4.** Effect of aggregation and association between species on critical $\alpha$. The curves are contours of equal critical $\alpha$ ($\alpha = 2, 5, 20$) as a function of aggregation ($k$) and association ($\Phi$). Shown also are observed values of $k$ and $\Phi$ from various data sets. *Filled circles*, field observations, fungi; *open circles*, field observations, fruit; *crosses*, field experiments; *triangles*, laboratory experiments. (From Shorrocks et al. 1990.)

less severe type of selection. However, a consideration of a number of examples and some simple models leads to the conclusion (Wallace 1975, 1981) that soft-selection is both frequency- and density-dependent. Conversely, hard-selection is both frequency- and density-independent.

Despite the fact that many of the traditional calculations carried out by population geneticists do not involve "density" or "frequency" and are consequently involved with hard-selection, there are theoretical reasons for believing that soft-selection is the norm. Clarke (1973a, b) has presented a genetic model in which the competition between two phenotypes is represented by the equations:

$$dN_1/dt = N_1(w_1k_1/(k_1+w_1N_1+\alpha_1w_2N_2)-1)$$
$$dN_2/dt = N_2(w_2k_2/(k_2+w_2N_2+\alpha_2w_1N_1)-1)$$

where $w$ and $k$ are constants with effects similar to those of $r$ and $K$ in the familiar Lotka–Volterra model of two-species competition. The relative competitive

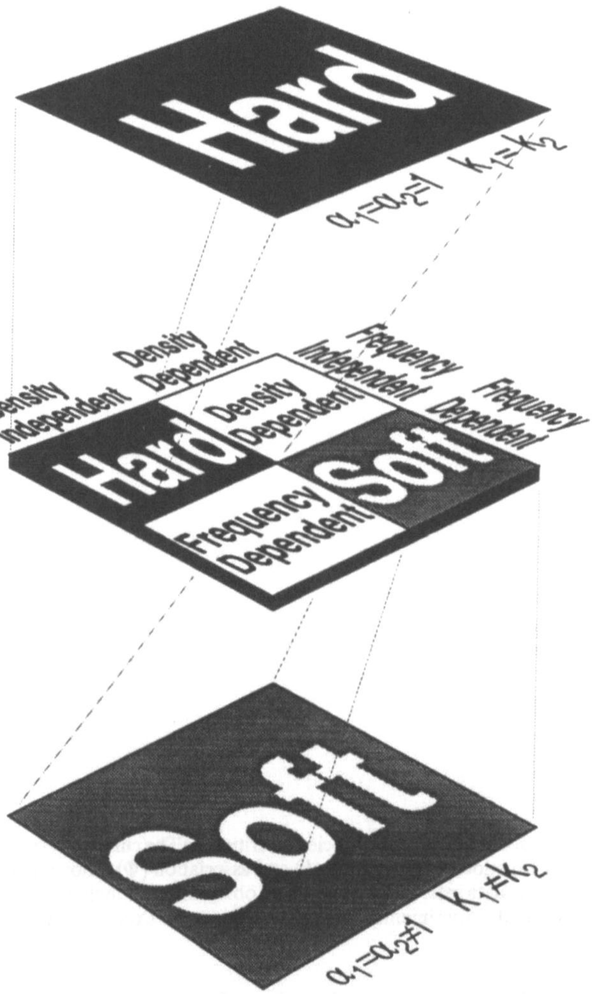


**Fig. 12.5.** A geometric representation of the nature of selection under various relationships of coefficients of competition ($\alpha_1$ and $\alpha_2$) and carrying capacities ($k_1$ and $k_2$) for individuals of phenotypes *M*- and *mm* under a model described by Clarke (1973a, b). (After Wallace 1981.)

ability of the two phenotypes (M- and mm) are represented by $\alpha_1$ and $\alpha_2$, $N$ is the number of fertile adults, fertility and fecundity of all genotypes are equal and young are produced in excess. Fig. 12.5 summarizes the consequences of Clarke's analysis. If $\alpha_1 = \alpha_2 = 1$ and $k_1 = k_2$, all selection is density- and frequency-independent and therefore hard. In other words the conditions for hard-selection to operate exclusively are rather restrictive. Most individuals in populations experience soft-selection of the type incorporated into the probability refuges below.

Essentially my genetic model is an extension of the two-species ecological model described in Sect. 12.2; however, there is an additional third species. These three "species" can be treated as genotypes, *AA* (genotype 1), *Aa* (genotype 2) and *aa* (genotype 3). Each genotype is independently distributed over the available patches according to a negative binomial distribution, survival in a patch is determined by the same equations of Hassell and Comins (1976) and emerging adults disperse, interbreed at random with respect to this locus, reproduce and die. All three genotypes have identical parameters, $\lambda = 5$, $N^* = 10$, $b = 1$ (therefore $a = 0.04$), which are within the range shown by Shorrocks and Rosewell (1986) to be typical of drosophilids under field conditions. The exponent $k$ of the negative binomial is also the same for each genotype but varied between simulations so that the importance of aggregation over patches can be assessed. With three "species" or genotypes there are six competition (fitness) values.

Two different situations have been analysed, both with complete dominance. Fig. 12.6 shows the results for selection acting only against genotype 3 (the

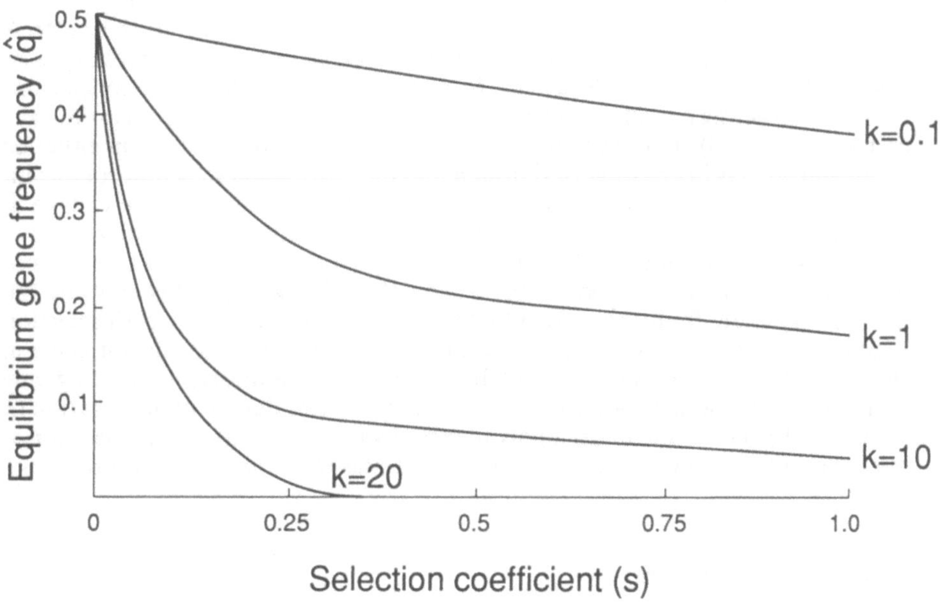


**Fig. 12.6.** The effect of aggregation upon selection. Each curve (different degrees of aggregation) represents the relationship between the intensity of selection against *aa* and the equilibrium frequency ($\hat{q}$) of allele *a*.

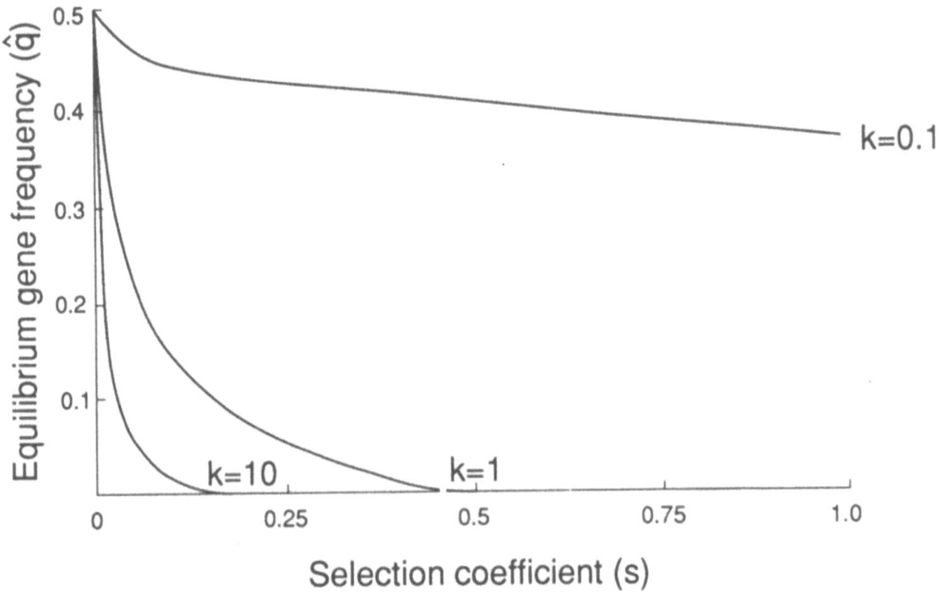


**Fig. 12.7.** The effect of aggregation upon selection. Each curve (different degrees of aggregation) represents the relationship between the intensity of selection against *Aa* and *aa* and the equilibrium frequency ($\hat{q}$) of allele *a*.

homozygous recessive). The fitness values are $\alpha_{21} = \alpha_{12} = 1$, $\alpha_{13} = \alpha_{23} = 0$ and $\alpha_{31} = \alpha_{32} =$ variable. In other words, *AA* and *Aa* are identical in their effects upon themselves and each other ($\alpha = 1$), and upon *aa* ($\alpha$ variable). Genotype *aa* has no effect upon *AA* or *Aa* ($\alpha = 0$). In contrast, Fig. 12.7 shows the conclusions for selection acting equally against *Aa* and *aa* ($\alpha_{21} = \alpha_{31} =$ variable, $\alpha_{32} = \alpha_{23} = 1$ and $\alpha_{12} = \alpha_{13} = 0$). In both figures the degree of selection imposed, on either *aa* or *Aa* and *aa*, has been converted from $\alpha$ into the more usual selection coefficient ($s$) used by population geneticists. This was achieved by equating the change in the frequency of allele *a* ($\Delta q$) that could be brought about by a particular value of both $\alpha$ and $s$ in an undivided environment simulation. For different degrees of aggregation over patches, both figures show the relationship between $s$ and the equilibrium allele frequency ($\hat{q}$). What is quite clear is that under both selection regimes, a degree of aggregation ($k<1$) is sufficient to keep the inferior allele in the population, despite very strong selection against it. With selection acting only on the recessive homozygotes (*aa*), even a very modest degree of aggregation ($k = 10$) is sufficient to maintain a polymorphic locus. In the latter case the inferior allele has a refuge in both heterozygotes and sites that are probability refuges.

## 12.4 Implications

Although originally discussed by den Boer (1968) some 20 years ago, there is a new and increasing awareness among population biologists that models and explanations framed in terms of averages are misleading (Łomnicki 1980; Begon 1982; Hassell and May 1985; Shorrocks and Rosewell 1988). Such explanations encourage one to commit the fallacy of the averages; that is, to assume that the mean of a function of a random variable necessarily equals the function of the mean of the random variable. The intensity of competition experienced by an individual or selection experienced by an allele is determined not so much by global averages but by local parameter values. This is particularly true of insects, such as drosophilids, whose larval stages inhabit resource patches that are ephemeral and discrete. The above two sections suggest that even in the absence of more traditional mechanisms which involve resource partitioning we should not be surprised to find high levels of both species and genetic diversity.

Traditionally, population biologists have placed great emphasis upon the measurement of competition coefficients and selection coefficients. However, an important ingredient in the patchy environment of drosophilid flies is the presence of aggregated clutches of eggs. Surprisingly, we still lack direct evidence for clutch size in drosophilids under field conditions (but see Shorrocks and Rosewell, 1988) and we have little information about the behavioural mechanisms involved in oviposition. Clutch size and the oviposition behaviour of female insects have received considerable attention from entomologists and ecologists interested in life-history evolution. In this context they are seen as the end point of a biological mechanism; that is, natural selection. However, we now see that they are also the starting point of a series of ecological processes which ultimately may result in considerable genetic and species diversity. Since these life-history parameters have such far-reaching ecological and genetic consequences, their causal mechanisms require urgent description and further study. In patchy and ephemeral habitats at least they may provide the key to much more than just life-history strategies.

## References

Atkinson WD (1979) A field investigation of larval competition in domestic *Drosophila*. J Anim Ecol 48: 91–102

Atkinson WD (1981) An ecological interaction between citrus fruit, *Penicillium* moulds and *Drosophila immigrans* Surtevant (Diptera: Drosophilidae). Ecol Entomol 6: 339–344

Atkinson WD (1985) Coexistence of Australian rainforest Diptera breeding in fallen fruit. J Anim Ecol 54: 507–518

Atkinson WD, Shorrocks B (1977) Breeding site specificity in the domestic species of *Drosophila*. Oecologia 29: 223–232

Atkinson WD, Shorrocks B (1981) Competition on a divided and ephemeral resource: a simulation model. J Anim Ecol 50: 461–471

Atkinson WD, Shorrocks B (1984) Aggregation of larval Diptera over discrete and ephemeral breeding sites: the implications for coexistence. Am Nat 124: 336–351

Bakker K (1961) An analysis of factors which determine success in competition for food among larvae of *Drosophila melanogaster*. Arch Neerl Zool 14: 200–281

Beaver RA (1977) Non-equilibrium "island" communities: Diptera breeding in dead snails. J Anim Ecol 46: 783–798

Beaver RA (1979a) Leafstalks as a habitat for bark beetles (Coleoptera: Scolytidae). Z Angew Entomol 88: 296–306

Beaver RA (1979b) Biological studies of the fauna of pitcher plants (*Nepenthes*) in West Malaysia. Ann Soc Entomol France 15: 3–17

Beaver RA (1979c) Non-equilibrium "island" communities: a guild of tropical bark beetles. J Anim Ecol 48: 987–1002

Beaver RA (1980) Fauna and food webs of pitcher plants in West Malaysia. Malay Nat J 33: 1–10

Beaver RA (1983) The communities living in *Nepenthes* pitcher plants: fauna and food webs. In: Lounibos LP, Frank JH (eds) Phytotelmata: terrestrial plants as hosts of aquatic insect communities. Plexus Publishing, Princeton, NJ, pp 129–159
Begon M (1982) Density and individual fitness: asymmetric competition. In: Shorrocks B (ed) Evolutionary ecology. Blackwells, Oxford, pp 175–194 (British Entomological Society Symposium)
Begon M, Shorrocks B (1978) The feeding and breeding sites of *Drosophila obscura* Fallen and *D. subobscura* Collin. J Nat Hist 12: 137–151
Brncic D (1983) Ecology of flower-breeding *Drosophila*. In: Ashburner M, Carson HL, Thompson JN Jr (eds) The genetics and biology of *Drosophila*, vol 3d. Academic Press, New York, pp 333–382
Bulmer MG (1972) Multiple niche polymorphism. Am Nat 106: 254–257
Burla H, Bachli G (1968) Beitrag zur Kenntnis der schweizerischen Dipteren, insbesondere *Drosophila*-Arten, die sich in Fruchtkorpen von Hutpilzen entwickeln. Vierteljahresschr Naturforsch Ges Zürich 113: 311–336
Carson HL (1971) The ecology of *Drosophila* breeding sites. In: Lyon HL (ed) Arboretum lecture no. 2. University of Hawaii, Honolulu, pp 1–27
Carton Y, Bouletreau M, van Alphen JJM, van Lenteren JC (1986) The *Drosophila* parasitic wasp. In: Ashburner M, Carson HL, Thompson JN Jr (eds) The genetics and biology of *Drosophila*, vol 3b. Academic Press, New York, pp 347–394
Christianson FB (1974) Sufficient conditions for protected polymorphism in a subdivided population. Am Nat 108: 157–166
Christianson FB (1975) Hard and soft selection in a subdivided population. Am Nat 109: 11–16
Clarke B (1973a). The effect of mutation on population size. Nature 242: 196–197
Clarke B (1973b) Mutation and population size. Heredity 31: 367–379
Conover WJ (1980) Practical non-parametric statistics. Wiley, New York
Dely-Draskovits A (1973) Systematical and ecological investigations of fly pests of mushrooms in Hungary. V. Drosophilidae (Diptera). Folia Entomol Hung 26: 21–29
den Boer PJ (1968) Spreading of risk and the stabilisation of animal numbers. Acta Biotheor 18: 165–194
Denno RF, Cothran WR (1975) Niche relationships of a guild of necrophagous flies. Ann Entomol Soc Am 68: 741–754
Denno RF, Cothran WR (1976) Competitive interactions and ecological strategies of sarcophagid and calliphorid flies inhabiting rabbit carrion. Ann Entomol Soc Am 69: 109–113
Doube BM (1987) Spatial and temporal organization in communities associated with dung pads and carcasses. In: Gee JHR, Giller PS (eds) The organisation of communities, past and present. Blackwells, Oxford, pp 255–280 (British Entomological Society Symposium)
Elton CS (1949) Population interspersion: an essay on animal community patterns. J Anim Ecol 37: 1–23
Elton CS (1966) The pattern of animal communities. Methuen, London
Hackman W, Meinander M (1979) Diptera feeding as larvae on macrofungi in Finland. Ann Zool Fenn 16: 50–83
Hanski I (1976) Breeding experiments with carrion flies (Dipt.) in natural conditions. Ann Entomol Fenn 42: 113–121
Hanski I (1981) Coexistence of competitors in patchy environments with and without predation. Oikos 37: 306–312
Hanski I (1983) Coexistence of competitors in patchy environment. Ecology 64: 493–500
Hanski I (1987) Colonization of ephemeral habitats. In: Gray AJ, Crawley MJ, Edwards PJ (eds) Colonization, succession and stability. Blackwells, Oxford, pp 155–185 (British Entomological Society Symposium)
Hanski I, Koskela H (1977) Niche relations among dung-inhabiting beetles. Oecologia 28: 203–231
Hanski I, Koskela H (1978) Stability, abundance, and niche width in the beetle community inhabiting cow dung. Oikos 31: 290–298
Hanski I, Koskela H (1979) Resource partitioning in six guilds of dung-inhabiting beetles (Coleoptera). Ann Entomol Fenn 45: 1–12
Hanski I, Kuusela S (1977) An experiment on competition and diversity in carrion fly community. Ann Entomol Fenn 43: 108–115
Hassell MP (1975) Density dependence in single species populations. J Anim Ecol 42: 693–726
Hassell MP, Comins HN (1976) Discrete time models for two-species competition. Theor Popul Biol 9: 202–221
Hassell MP, May RM (1985) From individual behaviour to population dynamics. In: Sibly RM, Smith RH (eds) Behavioural ecology. Blackwells, Oxford, pp 3–32 (British Entomological Society Symposium)

Haupt A, Busine JR (1968) The effect of overcrowding on the size of houseflies (*Musca domestica* L.). Trans R Entomol Soc London 120: 297–311
Heed WB (1968) Ecology of the Hawaiian Drosophilidae. Univ Tex Publ 6818: 387–419
Holter P (1982) Resource utilization and local coexistence in a guild of scarabaeid dung beetles (*Aphodius* spp.). Oikos 39: 213–227
Ives AR (1988) Aggregation and the coexistence of competitors. Ann Zool Fenn 25: 75–88
Ives AR, May RM (1985) Competition within and between species in a patchy environment: relations between microscopic and macroscopic models. J Theor Biol 115: 65–92
Janssen A, Driessen G, DeHaan M, Roodbol N (1988) The impact of parasitoids on natural populations of temperate woodland *Drosophila*. Neth J Zool 38: 61–73
Kamal AS (1958) Comparative study of thirteen species of sarcosaprophagous Calliphoridae and Sarcophagidae (Diptera). I. Bionomics. Ann Entomol Soc Am 51: 261–271
Kneidel KA (1982) Disturbance and diversity in communities of carrion-breeding Diptera. PhD thesis, University of North Carolina, Chapel Hill, NC
Kneidel KA (1983) Fugitive species and priority during colonization in carrion-breeding Diptera communities. Ecol Entomol 8: 163–169
Kneidel KA (1984) Competition and disturbance in communities of carrion-breeding Diptera. J Anim Ecol 53: 849–865
Koskela H, Hanski I (1977) Structure and succession in a cow dung-inhabiting beetle community. Ann Zool Fenn 14: 204–223
Kuusela S, Hanski I (1982). The structure of carrion fly communities: the size and the type of carrion. Hol Ecol 5: 337–348
Lachaise D, Tsacas L (1983) Breeding sites in tropical African drosophilids. In: Ashburner M, Carson HL, Thompson JN Jr (eds) The genetics and biology of *Drosophila*, vol 3d. Academic Press, New York, pp 221–332
Levene H (1953) Genetic equilibrium when more than one ecological niche is available. Am Nat 87: 131–133
Levins R, MacArthur RH (1966) The maintenance of genetic polymorphism in a spatially heterogeneous environment: variations on a theme by Howard Levene. Am Nat 100: 585–589
Li CC (1955) The stability of an equilibrium and the average fitness of a population. Am Nat 89: 281–296
Lloyd M (1967) Mean crowding. J Anim Ecol 36: 1–30
Łomnicki A (1980) Regulation of population density due to individual differences and patchy environment. Oikos 35: 185–193
Lounibos LP (1979) Mosquitoes occurring in the axils of *Pandanus rabaiensis* Rendle on the Kenya coast. Cahiers ORSTROM serie Entomol Med Parasit 17: 25–29
Lounibos LP (1981) Habitat segregation among African treehole mosquitoes. Ecol Entomol 6: 129–154
Lounibos LP, Machado-Allison CE (1983) Oviposition and egg brooding by the mosquito *Trichoprosopon digitatum* in cacao husks. Ecol Entomol 8: 475–478
Marino PC (1988) Coexistence on divided habitats: mosses in the family Splachnaceae. Ann Zool Fenn 25: 89–98
Maynard Smith J (1966) Sympatric speciation. Am Nat 100: 637–650
Maynard Smith J (1970) Genetic polymorphism in a varied environment. Am Nat 104: 487–490
Mohr CO (1943) Cattle droppings as ecological units. Ecol Monogr 13: 275–298
Montgomery SL (1975) Comparative breeding site ecology and the adaptive radiation of picture-winged *Drosophila* (Diptera: Drosophilidae) in Hawaii. Proc Hawaiian Entomol Soc 22: 65–103
Nealis VG (1977) Habitat associations and community analysis of south Texas dung beetles (Coleoptera: Scarabaeinae). Can J Zool 55: 138–147
Nicholson AJ (1954) An outline of the population dynamics of natural populations. Austr J Zool 2: 9–65
Peck SB, Forsyth A (1982) Composition, structure, and competitive behaviour in a guild of Ecuadorian rain forest dung beetles (Coleoptera: Scarabaeidae). Can J Zool 60: 1624–1634
Rosewell J, Shorrocks B, Edwards K (1990) Competition on a divided and ephemeral resource: testing the assumptions. I. Aggregation. J Anim Ecol 59: (in press)
Rouault J (1979) Rôle des parasites entomophages dans la compétition entre espèces jumelles de Drosophiles: approche expérimentale. CR Acad Sci 289: 643–646
Service MW (1965) The ecology of the tree-hole breeding mosquitoes in the northern Guinea savannah of Nigera. J Appl Ecol 2: 1–16
Shorrocks B (1977) An ecological classification of European *Drosophila* species. Oecologia 26: 335–345

Shorrocks B (1982) The breeding sites of temperate woodland *Drosophila*. In: Ashburner M, Carson HL, Thompson JN Jr (eds) The genetics and biology of *Drosophila*, vol 3b. Academic Press, New York, pp 385–428
Shorrocks B, Charlesworth P (1980) The distribution and abundance of the British fungal-breeding *Drosophila*. Ecol Entomol 5: 61–78
Shorrocks B, Charlesworth P (1982) A field study of the association between the stinkhorn *Phallus impudicus* Pers. and the British fungal-breeding *Drosophila*. Biol J Linn Soc 17: 307–318
Shorrocks B, Rosewell J (1986) Guild size in drosophilids: a simulation model. J Anim Ecol 55: 527–541
Shorrocks B, Rosewell J (1987) Spatial patchiness and community structure: coexistence and guild size of *Drosophila* on ephemeral resources. In: Gee JHR, Giller PS (eds) The organisation of communities, past and present. Blackwells, Oxford, pp 29–51 (British Entomological Society Symposium)
Shorrocks B, Rosewell J (1988) Aggregation does prevent competitive exclusion: a response to Green. Am Nat 131: 765–771
Shorrocks B, Atkinson WD, Charlesworth P (1979) Competition on a divided and ephemeral resource. J Anim Ecol 48: 899–908
Shorrocks B, Rosewell J, Edwards K, Atkinson WD (1984) Interspecific competition is not a major organizing force in many insect communities. Nature 310: 310–312
Shorrocks B, Rosewell J, Edwards K (1990) Competition on a divided and ephemeral resource: testing the assumptions. II. Association. J Anim Ecol 59: (in press)
Strobeck C (1974) Selection in a fine-grained environment. Am Nat 109: 419–425
Sugden AM, Robins RJ (1979) Aspects of the ecology of vascular epiphytes in Colombian cloud forests. I. The distribution of the epiphytic flora. Biotropica 11: 173–188
Taylor LR (1961) Aggregation, variance and the mean. Nature 189: 732–735
Taylor LR, Woiwod IP, Perry JN (1978). The density dependence of spatial behaviour and the rarity of randomness. J Anim Ecol 47: 383–406
Taylor LR, Woiwod IP, Perry JN (1979) The negative binomial as a dynamic ecological model for aggregation, and the density dependence of *k*. J Anim Ecol 48: 289–304
Wallace B (1968a) Polymorphism, population size, and genetic load. In: Lewontin RC (ed) Population biology and evolution. Syracuse University Press, Syracuse, NY, pp 87–108
Wallace B (1968b) Topics in population genetics. Norton, New York.
Wallace B (1975) Hard and soft selection revisited. Evolution 29: 465–473
Wallace B (1981) Basic population genetics. Columbia University Press, New York.
Worthen WB, McGuire TR (1988) A criticism of the aggregation model of coexistence: dipteran species are not distributed independently over ephemeral resources. Am Nat 131: 453–458
Valiela I (1974) Composition, food webs and population limitation on dung arthropod communities during invasion and succession. Am Midl Nat 92: 370–385
Yokoyama S, Schaal BA (1985) A note on multiple-niche polymorphisms in plant populations. Am Nat 125: 158–163

*Chapter 13*

# Facultative and Obligatory Diapause Responses in Three Species of Burnet Moth: A Characterization of Life-Cycle Phenologies by Field Observations and Laboratory Experiments

Wolfgang Wipking

## 13.1 Introduction

In temperate climates, phases of growth and dormancy of insect species must cope with the annually fluctuating weather conditions. Instead of quiescence, a prospective and hormonally controlled diapause prevents insect development outside of the season, even when environmental conditions are still advantageous. The induction, maintenance and termination of diapause as well as the resulting adaptive significance of various life-cycle strategies have been studied in detailed ecophysiological analyses (for recent reviews, see Danks 1987; Tauber et al. 1986; Zaslavski 1988). Many lepidopterous species respond to environmental factors during their larval instars, sometimes far in advance of the beginning of their diapause. Other species seem to develop independently of external cues (Friedrich 1983). Both extrinsic and intrinsic factors can regulate the life-cycles of these animals regardless of whether they are polycyclic species, with at least two generations within 1 year, or monocyclic species, with a single reproduction period every year (see e.g. Tauber and Tauber 1973).

Our knowledge is still limited, however, with regard to insect species with extended developmental periods of more than 1 year. These include, for example, species with an enlarged body size, so that individuals do not complete larval ontogenesis within 1 year (e.g. the cockchafer (Sorauer 1956) and several weevil-beetles (Kaestner 1973)), or small species with prolonged life-cycles of more than 1 year inhabiting biotopes with limited time for growth and reproduction (Barnes 1952; Matthes 1953; Harvey 1967; Lounibos and Bradshaw 1975; Falkovich 1979; Sotavalta et al. 1980; Friedrich 1983; Schweizerischer Bund für Naturschutz 1987). Because it is a small species and easy to breed in the laboratory, the burnet moth was chosen for investigation of the underlying physiological processes involved in such prolonged life-cycles.

## 13.2 Geographical Distribution of the Burnet Moth

All Palaearctic burnet moths belong to the genus *Zygaena* (97 species) (Lepidoptera, Zygaenidae). The geographical distribution of these species has been amply described (Naumann et al. 1984). The Mediterranean species inhabit grasslands and temperate mixed forests in the southern parts of their distribution range, but, near the northern border, they are restricted to isolated warm climatic outposts such as limestone habitats with southern exposure and steep hillsides (Wipking 1985).

## 13.3 Partivoltinism in the Burnet Moth

In all 13 tested species of the genus *Zygaena* (Wipking and Neumann 1986; Wipking 1987, 1988b), the prolonged larval stage could be linked either with maintenance of a prolonged diapause or with repetitive larval diapause at various larval stages in the second, third or even fourth year. This explains the overlapping and variable generation times recorded in *Zygaena* populations near the alpine or northern fringes of their geographical distribution ranges. By means of a life-cycle strategy that frequently inserts a diapause in a progressing larval ontogenesis, the populations are buffered against severe, unstable temperature

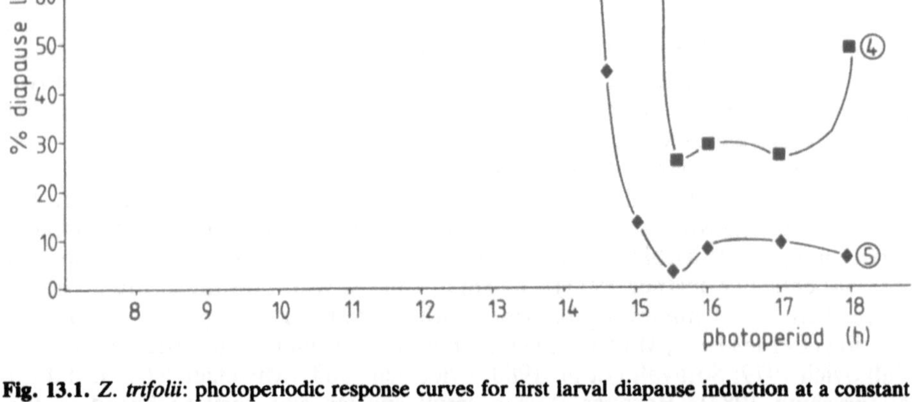


**Fig. 13.1.** *Z. trifolii*: photoperiodic response curves for first larval diapause induction at a constant temperature of 20 °C. *Curves 1–3*, obligatory diapause in the stocks from Köln (*1*), Angers (*2*) and Avignon (*3*); *curves 4 and 5*, facultative first diapause in the stocks from Marseille (*4*), and Valencia (*5*). (From Wipking 1988a.)

conditions, which might not allow either imago formation or successful reproduction in subsequent years. For instance, northern strains of *Zygaena trifolii* (Germany) can be principally univoltine, with an obligatory first diapause in autumn and final development in the following summer (Fig. 13.1, curve 1). However, only a few specimens are really monocyclic, because most larvae undergo one to three additional facultative diapauses resulting in life-cycles of more than 1 year.

## 13.4 A Case of a Bivoltine Population?

An additional and different mode of prolonged larval development in *Zygaena* species has been recently found in *Zygaena transalpina hippocrepidis* (= *Z. hippocrepidis*). As demonstrated in the following experimental results, a seasonal pattern with two distinct flight periods per year, characteristic for a bivoltine species, is achieved by different dormancy behaviour of individuals within the same generation.

The phenology of this species was studied at the French Atlantic coast near La Rochelle (46° N, 20 m NN; Fig. 13.2). Further information from the same sites include data on fully grown larvae and adult moths (Alberti 1958, 1959; Dutreix 1985, and personal communication). The adult *Z. hippocrepidis* appears from late May until the middle of June and then, in a strictly separate period, from the end of August until the middle of September. The main flight period is always that in June (Fig. 13.3). Because the longevity of the day-active moths is about 2 weeks, it is obvious that other individuals must reproduce in the second flight period 2 months later. However, from such field observations, one would expect that *Z. hippocrepidis* would be bivoltine, or at least that the later population would be the offspring of those from the June flight period. But in a provisional breeding experiment under conditions corresponding to La Rochelle field conditions (C. Dutreix, personal communication), all of the offspring of females collected in June entered into a larval diapause in July.

However, field observations show that the moths are abundant regularly in early summer as well as in early autumn. But how is this possible if their progenies always enter first into an obligatory development arrest?

Laboratory stocks, based on the progenies of wild females captured in June and in September 1987 in the vicinity of La Rochelle, were kept separately. To induce and maintain diapause, the offspring of the field-collected specimens were reared on fresh leaves of *Coronilla varia* (Fabaceae) in small plastic cages and under different laboratory conditions (see Figs. 13.1, 13.3–13.6 and Tables 13.1–13.3 for details). Every 2 days the moulting stage was checked. The moult to the specialized, decolorized, smaller, and non-feeding diapause morph (Pörschmann 1986) was used as a criterion for winter diapause.

### *September Adults*

The photoperiodic response curve of $F_1$ offspring from eight females collected in September 1987 shows that diapause is obligatory, irrespective of the experimen-

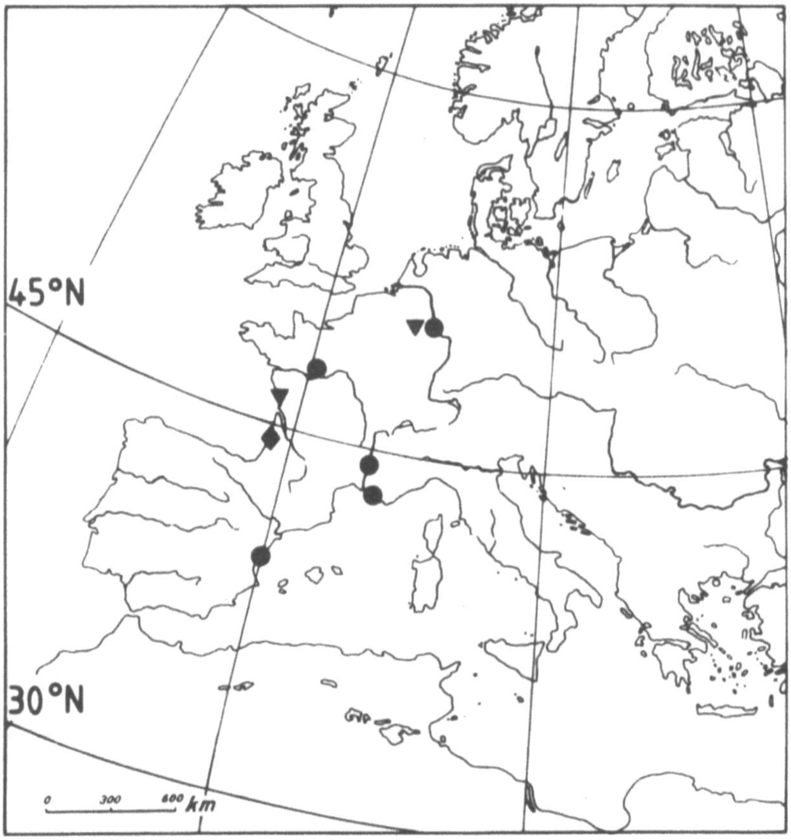


**Fig. 13.2.** The geographical origin of the species analysed. *Circles*, *Z. trifolii*, from north to south: Köln (Germany, 51° N), Angers (France, 47° N), Avignon (France, 44° N), Marseille (France, 43° N) and Valencia (Spain, 39° N). *Inverted triangles*, *Z. transalpina*, Bad Münstereifel (Germany, 51° N) and *Z. hippocrepidis*, La Rochelle (France, 46° N). *Diamond*, *Z. filipendulae*, Arcachon (France, 44° N).

**Table 13.1.** *Zygaena hippocrepidis*, La Rochelle stock

| | % of population | | | | | | | |
|---|---|---|---|---|---|---|---|---|
| Experimental procedure at LD 16 : 8 h | Mortality | Aestivation | $L_3D$ | $L_4D$ | $L_5D$ | $L_6D$ | Non-diapause development | *N* |
| 10 °C | 100 | 0 | 0 | 0 | 0 | 0 | 0 | 18 |
| 15 °C | 85 | 0 | 0 | 15 | 0 | 0 | 0 | 27 |
| 20 °C | 0 | 0 | 0 | 68 | 32 | 0 | 0 | 38 |
| 25 °C | 0 | 0 | 0 | 68 | 32 | 0 | 0 | 34 |
| 30 °C | 0 | 0 | 0 | 100 | 0 | 0 | 0 | 22 |

Relative frequencies of diapausing larvae and of non-diapausing, but dying, larvae related to different temperatures under long-day conditions (LD 16 : 8 h). Data from larvae of six females collected in June. *N*, number of larvae.

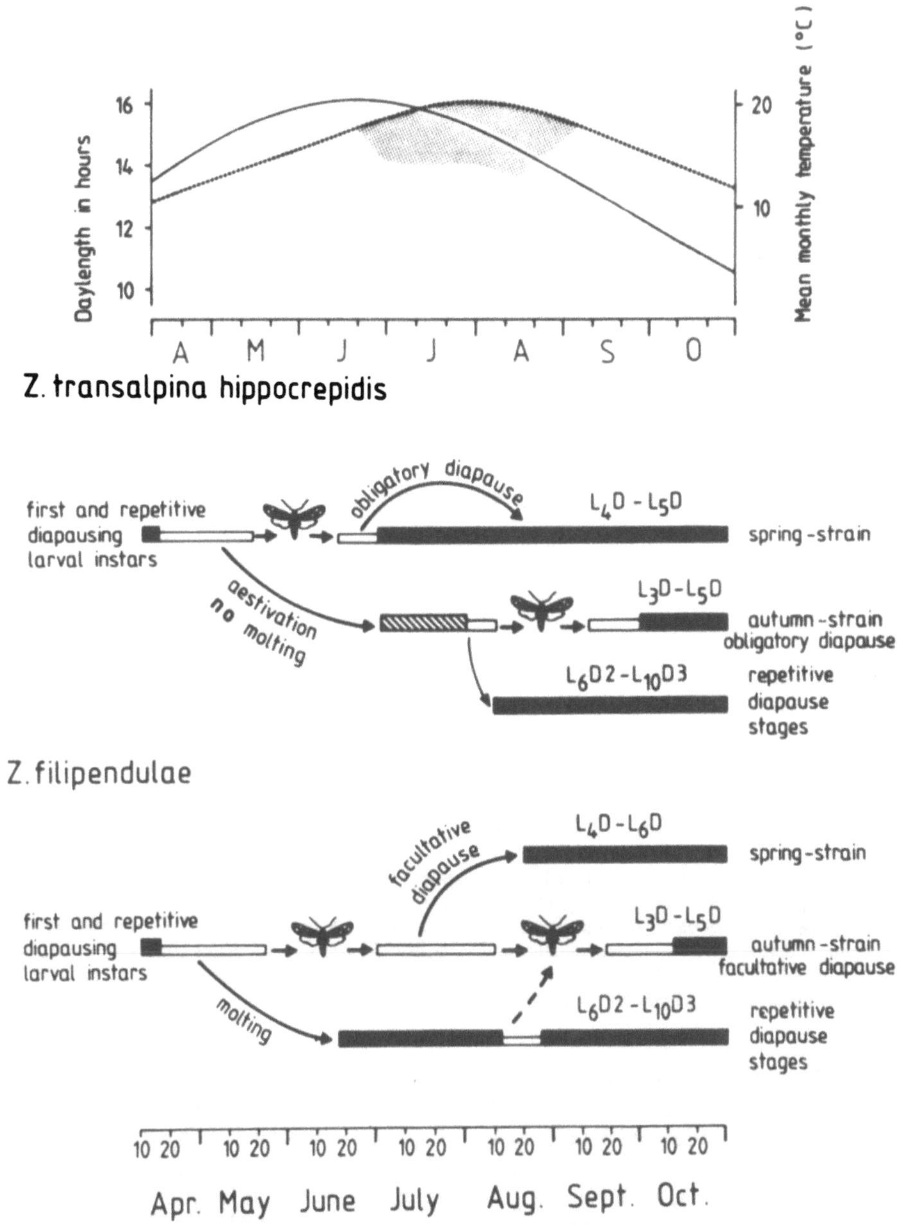


**Fig. 13.3.** Seasonal stages of burnet moth species *Z. hippocrepidis* and *Z. filipendulae* in the vicinity of La Rochelle (West France, 46° N).

*Upper panel*: local climatic conditions at La Rochelle. *Continuous line*, the variation of daylength including civil twilight periods between April and October; *dotted line*, mean monthly temperature; *shaded area*, period of summer drought. Data from Beck (1980) and Walter and Lieth (1960).

*Middle and lower panels*: comparative seasonal phenologies of *Z. hippocrepidis* and *Z. filipendulae*. *Wide black bars* indicate the periods of diapause, *the hatched bar* indicates the aestivation (waiting) period and the *open bars* indicate growth. *Moth emblems* characterize the moths' flight periods. Only the main development pathways are depicted.

The data for the early summer generation of *Z. filipendulae* were calculated from collection records and the laboratory experiments only.

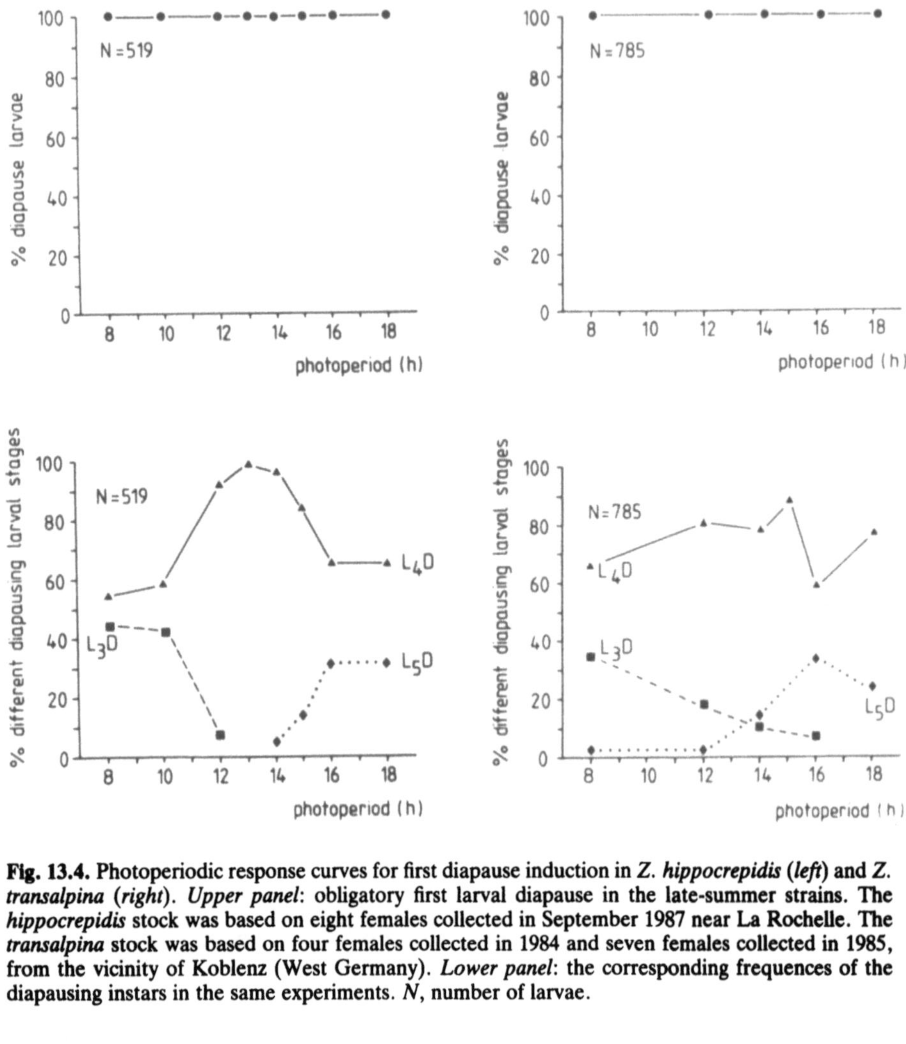


**Fig. 13.4.** Photoperiodic response curves for first diapause induction in *Z. hippocrepidis* (*left*) and *Z. transalpina* (*right*). *Upper panel*: obligatory first larval diapause in the late-summer strains. The *hippocrepidis* stock was based on eight females collected in September 1987 near La Rochelle. The *transalpina* stock was based on four females collected in 1984 and seven females collected in 1985, from the vicinity of Koblenz (West Germany). *Lower panel*: the corresponding frequences of the diapausing instars in the same experiments. *N*, number of larvae.

tal conditions (continuous photoperiods in the range between light–dark cycle (LD) 8 : 16 to 18 : 6 h; 20 °C) (Fig. 13.4 *left*). But diapause induction was not completely independent of the photoperiod. With increasing daylength, diapause was shifted to the older larval stages, so that in long-day conditions (e.g. LD 16 : 8 to 18 : 6 h) the $L_4D$- and $L_5D$-stages were the most frequent instars in diapause (where $L_4D$ and $L_5D$ are diapausing fourth and fifth instars, respectively; Fig. 13.4 bottom left).

### *June Adults*

It might be possible that the individuals of the June and September strains are equipped with different diapause characteristics (e.g. facultative first diapause versus obligatory diapause). Because an obligatory winter diapause might be advantageous only in the September strain, a second sample of offspring from six

June females was tested. To estimate temperature thresholds for diapause induction in the June strain, the larvae were raised at a 16 h daylength and several different temperatures (10, 15, 20, 25, 30 °C). As depicted in Table 13.1, the photoperiodic conditions were very similar to natural daylength during the larval period in late June (compare Fig. 13.3 above). At higher temperatures (20–30 °C), all individuals entered into a diapause of the winter type. At lower temperatures (10–15 °C), most of the larvae died in the fifth instar without any dormancy reaction.

It was pointed out previously that the duration of larval development might be extended by a repetitive larval diapause or by a prolonged diapause duration. It is practically impossible to establish the time when the diapause or the aestivation (waiting) period (see below) ends in the field. It is not possible to collect enough dormant larvae out of doors at various dates in order to test the variability in length of dormancy. Therefore the duration of dormancy can only be estimated from laboratory results combined with observations on the frequency of penultimate-instar larvae in the field (compare Fig. 13.3). The effect of photoperiod and temperature on repeated diapause induction after hibernation has therefore been studied in laboratory experiments with $L_4D$-stages which were the offspring of the September adults.

### *Different Individual Responses*

The photoperiodic response curve shows a facultative repetitive diapause of the long-day type ("sensu" Beck 1980). Most larvae (70%) developed without any further dormancy at long daylength (LD 18 : 6 h) (Fig. 13.5 *left*). The emerging moths could easily be correlated with the individuals swarming in the field during June. However, 30% of the larvae of the overwintering generation moulted into the specialized diapause morph in spite of long-day conditions. In the range LD 12 : 12 h (20 °C) and LD 16 : 8 h (20 °C) a substantial number of larvae (9%–35%) did not enter diapause but only stopped feeding for a couple of weeks (compare Tables 13.2 and 13.3) even when ample fresh leaves were offered. The latter dormancy response of *Z. hippocrepidis* is a mode for extended larval development which has not previously been detected in other *Zygaena* species. This mode occurs in addition to the normal dormancy responses and may be referred to as an aestivation (waiting) period, in contrast to winter and aestivation diapause.

**Table 13.2.** *Zygaena hippocrepidis*, La Rochelle stock

| | % population | | | | | | | | |
|---|---|---|---|---|---|---|---|---|---|
| Experimental procedure at LD 16 : 8 h | Mortality | Aestivation ($L_5$–$L_6$) | $L_4D$ | $L_5D$ | $L_6D$ | $L_7D$ | $L_8D$ | Non-diapause development | *N* |
| 10 °C | 60 | 0 | 0 | 0 | 30 | 0 | 0 | 10 | 10 |
| 15 °C | 33 | 0 | 0 | 0 | 0 | 20 | 0 | 47 | 15 |
| 20 °C | 0 | 35 | 0 | 0 | 8 | 22 | 0 | 35 | 26 |
| 25 °C | 0 | 37 | 21 | 0 | 0 | 6 | 10 | 26 | 19 |

The frequencies of repeated diapausing larvae at different temperatures in long daylength (LD 16 : 8 h). For further explanations see Table 13.1.

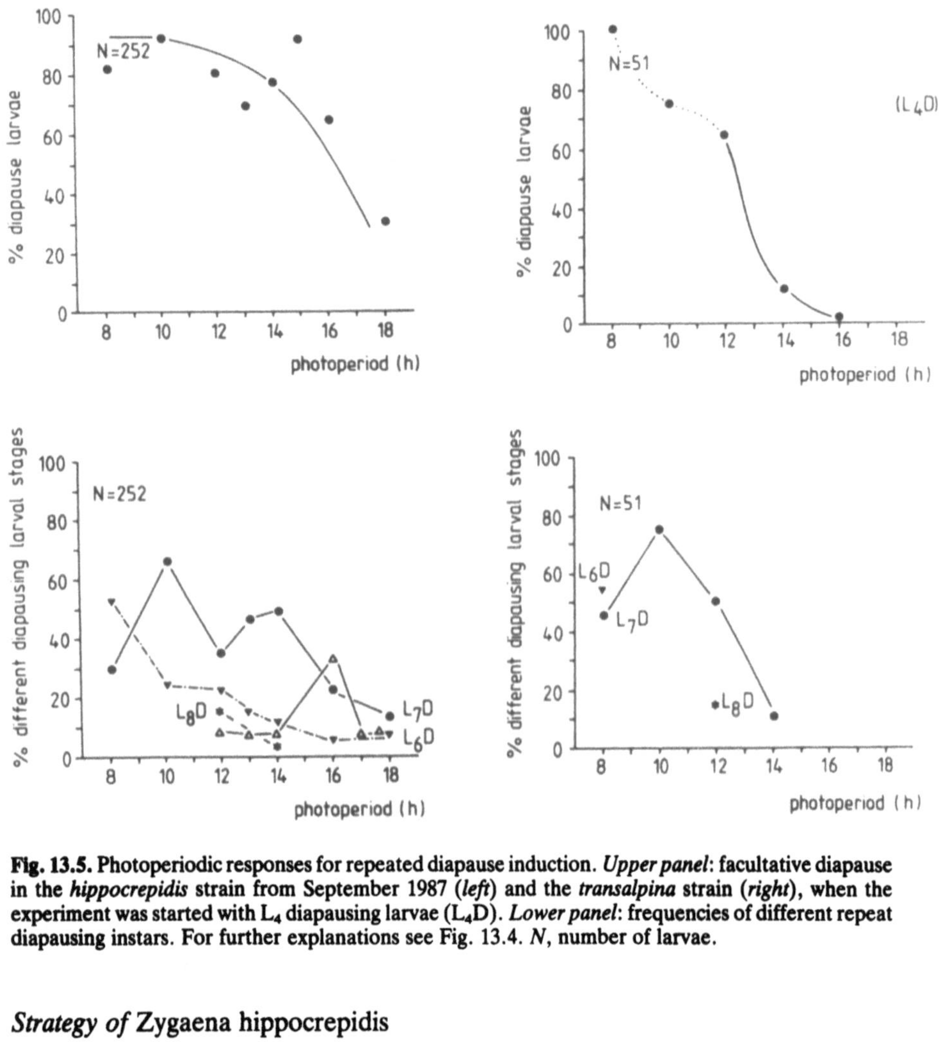


**Fig. 13.5.** Photoperiodic responses for repeated diapause induction. *Upper panel*: facultative diapause in the *hippocrepidis* strain from September 1987 (*left*) and the *transalpina* strain (*right*), when the experiment was started with $L_4$ diapausing larvae ($L_4D$). *Lower panel*: frequencies of different repeat diapausing instars. For further explanations see Fig. 13.4. *N*, number of larvae.

## *Strategy of* Zygaena hippocrepidis

In the following, the different dormancy responses of *Z. hippocrepidis* will be described in more detail. Diapause induction is always linked with the specific diapause moult to a decolorized smaller morph (Pörschmann 1986). In contrast, the aestivation (waiting) period is not associated with the moult to the characteristic diapause morph. In this phase, the feeding period of the larval instar is only interrupted and progress in ontogenesis suppressed. However, the larvae lose their greenish colour when entering this dormant phase, which might be attributed to an emptying of the gut. For about 4–10 weeks, feeding behaviour is irregular and infrequent (Table 13.3). In conclusion, the larvae exhibit a kind of premoulting character. After this aestivation (waiting) period, larvae either moult sporadically into a repetitive winter diapause or finish their larval development within 3 months. The laboratory experiments suggest that moths which swarm in September may be derived from individuals with such an aestivation (waiting) period during the $L_5$- and $L_6$-stages (non-diapausing fifth

**Table 13.3.** *Zygaena hippocrepidis*, La Rochelle stock

| Experimental procedure | Incidence of aestivation (%) | Aestivating larval stage | Minimum duration of aestivation | *N* |
|---|---|---|---|---|
| A. *At 20 °C* | | | | |
| LD 18 : 6 | 9 | $L_5$–$L_6$ | 36 | 21 |
| LD 16 : 8 | 35 | $L_5$–$L_6$ | 74 | 26 |
| LD 15.5 : 8.5 | 15 | $L_5$–$L_6$ | 69 | 27 |
| LD 14.5 : 9.5 | 34 | $L_5$–$L_6$ | 68 | 29 |
| LD 14 : 10 | 8 | $L_5$–$L_6$ | 77 | 24 |
| LD 13 : 11 | 8 | $L_5$–$L_6$ | 92 | 51 |
| LD 12 : 12 | 9 | $L_5$–$L_6$ | 127 | 22 |
| LD 10 : 14 | 0 | — | — | 15 |
| LD 8 : 16 | 0 | — | — | 12 |
| B. *At 25 °C* | | | | |
| LD 16 : 8 | 37 | $L_5$–$L_6$ | 77 | 7 |

The relative frequency and duration of the aestivating stage in relation to different environmental conditions.

and sixth larval instars). The few individuals which moult to a normal diapause under long-day conditions remain dormant for at least 12 weeks in the laboratory (LD 16 : 8 h, 20 °C). It may be supposed that all these individuals with a second diapause would not emerge under field conditions in the same year.

# 13.5 Conclusions

## 13.5.1 Adaptive Significance of Repeated Diapauses in Northern Populations

Our investigation shows that each *Zygaena* species has evolved its own characteristic life-cycle with a range of diverse optional responses. Within one species, individuals with a 1 year larval development might have increased reproductive success at optimal environmental conditions in subsequent years. Specimens with repetitive diapause reactions and an extended developmental period ensure the population's survival in years with climatic conditions that might not be sufficient for imago formation or successful reproduction. In addition to this, size ratios might be enlarged by mono- and semicyclic (2–4 years) cohorts with a wide range of co-occurring feeding and diapausing instars. This might be advantageous for northern populations, which are restricted to small and isolated locations with limited food resources. In this way, the polymorphic diapause strategy might be maintained in the population.

Nevertheless, in *Zygaena* species such as *Z. trifolii* a genetic effect could also be observed. When inbreeding experiments were conducted with moths emerging from the scarce $L_4D$-stages at extreme short-day conditions (LD 8 : 16 h, 20 °C), the frequency of the originally predominant $L_3D$ decreased rapidly. Within eight successive generations, the $L_4D$-stage became the most frequent diapause stage

(8% ($F_1$ generation) → 67% ($F_8$ generation)). However, the critical photoperiod for diapause induction remained the same as before (Wipking 1987).

### 13.5.2 Phenology, Synchrony and Voltinism at Southern Study Sites

In some species which are widely distributed in the Mediterranean climatic zone, the situation becomes more complicated. The climatic conditions of southern locations may allow two generations per year, but a long period of heat and aridity during the summer months may split and limit the timespan for larval development (for detailed information, see climate diagrams of Walter and Lieth (1960)). The host plants of the larvae wither during the period of summer drought. In this case, lush host plants are not available long enough to allow complete larval development within the same year, and consequently active larvae would starve to death. On the other hand, moths of a second flight period or an additional generation would not be abundant before late September at moderate temperatures, so that the larvae of this generation might not enter the first diapausing larval stage ($L_3$D-stage) at the right time. For reproduction twice per year under favourable conditions, high summer temperatures might be overcome either by additional aestivation (waiting) periods in older larvae after hibernation or by a summer diapause in young non-hibernating larvae.

*Pseudo-Bivoltine Species with Obligatory First-Diapause Reactions*

To make use of the favourable environmental requirements for larval development in May and September, *Z. hippocrepidis* has evolved a kind of pseudo-bivoltinism. The larval aestivation (waiting) period provides for reproduction twice in the same year, whereas the induction of a first obligatory diapause avoids emerging out of season. Similar life-cycle strategies have been observed in "bivoltine" populations of *Zygaena fausta* var. *niceae* from South France and Spain (Burgeff 1910, 1921) and *Zygaena alluaudi lamprotes* from Morocco (Hofmann and Reiss 1982). Last-instar larvae of both species were collected in the field in May, but only a few moths emerged in June. The aestivation (waiting) period was restricted to fully grown larvae, which finished feeding in May. The immobile and decolorized larvae remained dormant for more than 90 days until they pupated without feeding any more. In very good agreement with conditions encountered in the field, most moths emerged in September and the larvae of the next generation entered into a winter diapause at decreasing daylength in late autumn.

*Bivoltine Species with Facultative Diapause Responses*

Summer diapause, detectable by the characteristic diapause morph, is present in southern populations, e.g. of *Z. filipendulae* or *Z. trifolii*. Normally, populations of these species have two generations per year in the southern part of their distribution area with a first generation in May and a second generation flying in September. As shown by the laboratory experiments with *Z. trifolii*, the diapause is facultative (for critical daylengths see Fig. 13.1, curves 4 and 5). Despite long-day conditions, summer diapause may occur for the first time in the young non-hibernating larvae at high temperatures (Wipking 1988a). Specimens with a short summer diapause will follow the bivoltine developmental mode, but those with a longer diapause enter into a second diapause at a later instar after a short period of growth. However, most of the summer larvae will finish larval ontogenesis in the same year, which might be attributed to extended favourable periods for growth in spring and autumn.

A further factor to be considered is the striking difference between the field observations on the seasonal phenology of burnet moths and the corresponding laboratory results. It is curious that *Z. hippocrepidis* and *Z. filipendulae* occur syntopically and synchronously in two flight periods at La Rochelle, even though the underlying life-cycle strategies are quite different (Fig. 13.3, middle and below). *Z. filipendulae* really is bivoltine in the La Rochelle area (C. Dutreix, personal communication). In a laboratory experiment with a stock from Arcachon (46° N), the critical photoperiod for first diapause was >14 h of light (Fig. 13.6). An aestivation (waiting) period could not be detected in this species. The

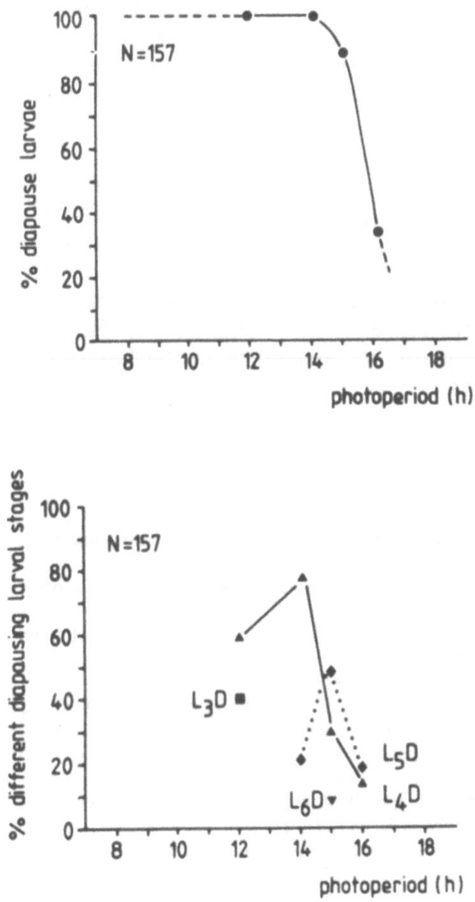


**Fig. 13.6** *Z. filipendulae*, Arcachon stock. *Upper panel*: photoperiodic response curve for first diapause induction. *Lower panel*: corresponding frequencies of different diapausing instars. For further explanations see Fig. 13.4. *N*, number of larvae.

additional facultative diapause reactions were always photoperiodically controlled (W. Wipking, unpublished results).

### *Sibling Species with Different Life-Cycle Phenologies*

Lastly, it should not be overlooked that, while the mode of an additional aestivation (waiting) period is realized in *Z. hippocrepidis*, it does not occur in the sibling-species *Zygaena transalpina astragali* (= *Z. transalpina*) from Germany. In laboratory experiments with *transalpina* larvae from Koblenz (Germany, 51° N) only the normal mode of diapause regulation with obligatory diapause first and facultative, repetitive diapause later on could be observed (Figs. 13.4 and 13.5, *right*). Consequently, the univoltine moths and the individuals with extended developmental periods of more than 1 year appear together in a single flight period in July (Maixner and Wipking 1985). An additional aestivation (waiting) period makes two flight periods possible in *Z. hippocrepidis*, which is the major difference between these species.

In conclusion, the burnet moth seems to provide excellent examples of life-cycle strategies that use quite different mechanisms of diapause regulation and the additional "aestivation" behaviour. To understand the life-cycle phenologies of Zygaenidae species, both field observations and laboratory results must be combined. The results obtained with burnet moths should inspire us to look carefully at life-cycle phenologies in other insect species that have not been subjected to corresponding laboratory experiments.

*Acknowledgements.* I am grateful to Prof. Dr D. Neumann for his support and valuable comments on an earlier draft of this paper. I thank Dr C. Dutreix (Nantes) for sending egg-batches of *Z. hippocrepidis* and for valuable comments on the seasonal phenology of burnet moths in the La Rochelle area. Thanks are also due to Mrs B. Röhler (Köln) for offering larvae of *Z. filipendulae*. I thank Dr F. Gilbert and Mrs S. Meyen-Southard for correcting the English text. Furthermore my thanks are due to Mrs R. Bieg for her assistance in the laboratory experiments and for preparing the figures.

## References

Alberti B (1958) Des problèmes dans le groupe de *Zygaena transalpina*. Bull Soc Entomol Mulhouse, pp 1–8

Alberti B (1959) Sur l'éxtension du groupe de *Zygaena transalpina* en France. Bull Soc Entomol Mulhouse, pp 35–36

Barnes HF (1952) Studies in fluctuations in insect populations. XII. Further evidence for prolonged larval life in the wheat blossom midges. Ann Appl Biol 39: 370–373

Beck SD (1980) Insect photoperiodism. Academic Press, New York

Burgeff H (1910) Beiträge zur Biologie der Gattung *Zygaena*. II. Z Wiss Insekt Biol 6: 39–44, 97–98

Burgeff H (1921) Beiträge zur Biologie der Gattung *Zygaena*. IV. Mitt Münch Entomol Ges 11: 50–64

Danks HV (1987) Insect dormancy: an ecological perspective. Biological Survey of Canada Monograph Series no 1. Biological Survey of Canada, Ottawa

Dutreix C (1985) Les collections entomologiques du museum d'histoire naturelle Nantes. Révision du genre *Zygaena*. Bull Soc Sci Nat Ouest de la France N.S. 7: 117–125

Falkovich MI (1979) Seasonal development of desert Lepidoptera of Soviet Central Asia and a historical analysis of the Lepidoptera fauna. Entomol Rev 58: 20–45

Friedrich E (1983) Handbuch der Schmetterlingszucht. Franckh (Keller & Co.), Stuttgart
Harvey GT (1967) On coniferophagus species of *Choristoneura* (Lep., Tortricidae) in North America. V. Second diapause as a species character. Can Entomol 99: 486–503
Hofmann A, Reiss G (1982) Beitrag zur Zygaenenfauna Nordafrikas. II. Atalanta 13: 136–153
Kaestner A (1973) Lehrbuch der Speziellen Zoologie 5, Insecta B (Spezieller Teil). Fischer, Jena, pp 534–574
Lounibos LP, Bradshaw WE (1975) A second diapause in *Wyeomyia smithii*: seasonal incidence and maintenance by photoperiod. Can J Zool 53: 215–221
Maixner B, Wipking W (1985) Zygaenidae Fabr. Mitt Arbeitsgem rhein-westf. Lepidopterol 4: 103–211
Matthes E (1953) Diapause, Bivoltinismus und zweimalige Überwinterung bei *Fumea crassiorella* (Lep., Psychidae). Mem Est Mus Zool Univ Coimbra 220: 1–16
Naumann CM, Feist R, Richter G, Weber U (1984) Verbreitungsatlas der Gattung *Zygaena* Fabricius 1775. Theses Zool 5. Cramer Verlag, Braunschweig
Pörschmann M (1986) Morphologische und habituelle Parameter im larvalen Entwicklungszyklus von *Zygaena trifolii* (Esper 1783). Diplom-Thesis, Universität Bielefeld
Schweizerischer Bund für Naturschutz (1987) Tagfalter und ihre Lebensräume. Fotorotar Verlag, Egg ZH
Sorauer P (1956) Handbuch der Pflanzenkrankheiten, vol 5. Parey, Berlin
Sotavalta O, Karvonen EI, Karvonen EE, Korpela S, Korpela J, (1980) The early stages and biology of *Acerbia alpina* (Lep., Arctiidae). Not Entomol 60: 89–95
Tauber MJ, Tauber CA (1973) Insect phenology: criteria for analyzing dormancy and the forecasting postdiapause development and reproduction in the field. Search Agric Cornell Univ. Agric Exp Stat Ithaca NY 3: 1–16
Tauber MJ, Tauber CA, Masaki S (1986) Seasonal adaptations of insects. Oxford University Press, New York
Walter H, Lieth H (1960) Klimadiagramm-Weltatlas. Fischer, Jena
Wipking W (1985) Ökologische Untersuchungen über die Habitatbindung der Zygaenidae. Mitt Münch Entomol Ges 74: 37–59
Wipking W (1987) Ökologische Untersuchungen über die Diapauseregulation bei westpaläarktischen Stämmen einer Schmetterlingsfamilie (Ins., Lepidoptera, Zygaenidae). PhD thesis, Universität zu Köln
Wipking W (1988a) Repeated larval diapause and diapause-free development in geographic strains of the burnet moth *Zygaena trifolii* Esp. I. Discontinuous clinal variation in photoperiodically controlled diapause induction. Oecologia 77: 557–564
Wipking W (1988b) Bivoltinismus und repetitive Larvaldiapause bei *Zygaena trifolii* Esp. Verh Dtsch Zool Ges 81: 334–335
Wipking W, Neumann D (1986) Polymorphism in the larval hibernation strategy of the burnet moth *Zygaena trifolii*. In: Taylor F, Karban R (eds) The evolution of insect life cycles. Springer, Berlin Heidelberg New York, pp 125–134
Zaslavski VA (1988) Insect development. Springer, Berlin Heidelberg New York

*Chapter 14*

# The Present and Future of Insect Life-Cycle Evolution

William E. Bradshaw

## 14.1 Introduction

The symposium in Vancouver and the chapters in this volume provide a diverse array of studies on insect life-cycle evolution. This array can be broken down into at least three levels of approach: (a) comparisons of genetic variation and covariation underlying life-cycle traits; (b) comparisons of variable phenotypes within and among populations of the same and/or related species; (c) comparisons of habitat-specific traits regardless of relatedness.

The first approach recognizes first that, for selection to affect a trait in a population, there must be heritable (genetic) variation for that trait, and, second, that the inheritance of life-cycle traits is generally quantitative rather than Mendelian in nature. Because life-history traits are directly associated with, and may be components of, fitness, it has long been held that selection, in maximizing fitness, should drive genetic variability for fitness traits to vanishingly low or zero levels above what could be accounted for by mutation alone (Fisher 1958; Falconer 1981). Much of the impetus of the genetic approach therefore revolves around explaining the persistence of phenotypic and genetic variation for life-history traits in laboratory and natural populations.

The second approach examines the phenotypes of conspecific or closely related insects across ecological gradients. By restricting analysis to closely related insects, one presumes that similar phylogenetic and/or ontogenetic constraints hold across populations. Phenotypic variability among populations therefore should represent a response to recent selection in those known, contrasting habitats rather than ancient selection in unknown environments for entirely different "bauplans". In some studies, the genetic underpinning of intra- and/or interpopulation variation are known. In other studies, the connection between genetic change and phenotypic evolution of life-cycle traits is not known but is at least potentially determinable.

The third approach makes comparisons among taxa independently of their relatedness and no pretence is made about the possibility of genetic connections. Rather, the focus is on the question of whether specific suites of life-history traits among broad taxa are generally, even if not universally, associated with specific habitats. Thus, along spatial and temporal ecological gradients, certain patterns of life-history syndromes begin to emerge.

Below, I shall discuss the contributions of the papers in this volume to these three levels of approach and the ways in which the levels are, and should continue to be, connected. Finally, I shall pose a series of questions whose pursuit might occupy our future endeavours until these questions are resolved or it is established that they are intractable and new questions supplant them.

## 14.2 Maintenance of Genetic Variability for Fitness Traits

Additive genetic variance (narrow-sense heritabilities) of life-history traits (high fitness components) persists over a wide variety of insects. Heritability for life-history traits may, generally, be less than heritability for morphological, physiological and behavioural traits, but it is consistently non-zero (Roff, Chap. 1; Istock 1982; Roff and Mousseau 1987; Mousseau and Roff 1987). Very generally, there is no longer a paradox between presumed lack of genetic variation and observed life-cycle variation. Sufficient genetic variation persists in present-day insect populations to explain the remarkable variety of life-cycle traits and syndromes presented in this volume.

The proposition that selection should drive genetic variation of high fitness traits to vanishingly low levels is now a straw man and no longer an appropriate focus for primary enquiry. The question does remain as to the processes that maintain such variation despite putative selection to the contrary. The answers to this question are still forthcoming and chapters in this volume have given us good direction.

Genetic variability for high fitness traits can be maintained above levels explained by mutation by frequency-dependent selection and/or environmental heterogeneity (Roff, Chap. 1). The effect of environmental heterogeneity in maintaining genetic variability may be enhanced through mechanisms discussed by Taylor (Chap. 9) and Shorrocks (Chap. 12). Taylor shows that the time for the mean and standard deviation of a phenotype to converge on the long-term optimum in response to stabilizing selection rapidly increases to hundreds of generations when environments fluctuate and are autocorrelated. When the mean of the optimal phenotype is also changing, gene frequencies may be perpetually chasing an elusive optimum. One can then ask whether populations in autocorrelated, seasonal environments whose mean varies over decades or centuries ever approach genetic equilibrium. Shorrocks (Chap. 12) extends his models of a "probability refuge" that serves to maintain potentially competing species in a community to include the interaction of potentially "competing" genes in a patchy environment. Shorrocks' models are especially enlightening because each genotype alone achieves the same fitness in all patches but fitness does vary according to the other genotypes in that patch. Consequently, even in the absence of multiple-niche polymorphism (Levene 1953; Haldane and Jayakar

1963; Levins and MacArthur 1966) aggregation of genotypes in transient, patchy environments can maintain high levels of genetic diversity or retard the elimination of inferior alleles. The combined effects of fluctuating, autocorrelated environments and aggregation of genotypes in patchy, ephemeral environments might then maintain high genetic diversity indefinitely.

Both Roff (Chap. 1) and Rose et al. (Chap. 2) propose that genetic variability of high fitness traits persists because of antagonistic pleiotropy. Rose et al. ask, in the most rigorously controlled laboratory populations of *Drosophila melanogaster*, whether antagonistic pleiotropy *can* account for trade-offs between adult longevity and reproductive effort early in adult life. Their answer demonstrates convincingly that antagonistic pleiotropy is real but at the same time, is not "exclusively important". Examination of a variety of life-histories in *D. melanogaster* and *Oncopeltus fasciatus* by Roff (Tables 1.3 and 1.4) would appear to confirm this conclusion.

The comparisons made by Roff (Chap. 1) illustrate several problems that must be confronted. First, genetic correlations with respect to fitness may be positive in recently inbred lines (Rose 1984) or when the determination of genetic correlations is made in environments that are different from the environment in which the organism underwent selection (Service and Rose 1985). Before we can entirely legitimize the comparative signs of genetic correlations, we need to answer two questions. First, what level of inbreeding and, second, how great a departure from the natural environment are likely to produce bias in the estimation of genetic correlations? Rose et al. (Chap. 2) used populations of *D. melanogaster* reared under the same laboratory conditions for over 250 generations. This technique permitted populations to approach genetic equilibrium and was essential for rigorously testing ways in which genotypes *can* be organized but says little about how genotypes *are* organized in natural populations. To avoid effects of novel environments, Bradshaw and Holzapfel (Chap. 3) looked for trade-offs (Fig. 3.11) in life-history traits among populations of pitcher-plant mosquitoes reared in the water-filled leaves of intact pitcher plants under conditions of near-natural light, temperature and feeding. Eventually, this technique will reveal a great deal about the phenotypic and genetic variation and covariation of traits within and between natural populations. The vast majority of insects do not, however, dwell in such highly circumscribed habitats that are transposable to controlled but near-natural conditions in the laboratory. How great an abstraction from nature are the laboratory cultures referenced in Roff's Tables 1.3 and 1.4? Would evidence for antagonistic pleiotropy have been more prevalent if rearing conditions were more natural? There are as yet no examples that I know of that compare genetic correlations over a series of environments from natural to artificial. Such studies are needed to provide a link between the work of Rose et al. (Chap. 2) and other studies, exemplified by Roff (Chap. 1), Scott and Dingle (Chap. 4), and Solbreck et al. (Chap. 11).

## 14.3 The Genetic Connection in Life-History Syndromes

The majority of chapters in this volume use a comparative approach to the evolution of life-cycle traits. The link between the existing patterns of insect life-cycles and the evolutionary sequence of genetic events is generally tenuous at best. At some point, this connection between genetic change and phenotypic evolution should be made wherever possible.

The gaps in the "genetic connection" occur at several levels. Rose et al. (Chap. 2) have shown clearly that antagonistic pleiotropy can and does exist in laboratory populations of *D. melanogaster*; Bradshaw and Holzapfel (Chap. 3) have shown a lack of detectable coevolution among demographic traits and between demographic and phenological traits in adjacent and widespread natural populations of *Wyeomyia smithii*. As yet there is little common ground. Bradshaw and Holzapfel do not show whether the underlying genetic correlations *within* populations are parallel to, or diverge from, the pattern *among* populations. Genetic correlations are usually interpreted as imposing a cost to the adaptive evolution of life-history traits (Hegmann and Dingle 1982) and the correlated response of a variety of traits to direct selection on body size in *O. fasciatus* (Palmer and Dingle 1986) is predictable from the underlying genetic variance–covariance matrix. Over longer-term selection, linkage groups may dissociate and this dissociation can lead to an accelerated response to selection (Falconer 1981, pp 195–196). Genetic correlations can differ in nearby (Groeters and Dingle 1987) or distant populations (Berven and Gill 1983). Just as selection could favour the evolution of co-adapted gene complexes that can form tightly linked "supergenes" (Ford 1971), so also could selection break up these linkage groups. Pleiotropic effects can be mediated by a single hormone. The titre of juvenile hormone is probably the common link between migration behaviour, wing muscle development, and wing development in gerrid (Roff, Chap. 1) and lygaeid (Solbreck et al., Chap. 11) bugs. Threshold titre could, however, vary independently, thus dissociating any one or two of these traits. The focus of natural selection can be a single trait, a group of genetically correlated traits, and/or the degree of genetic correlation itself (Bradshaw 1986). Clearly, genetic correlations do impose a "cost of correlation" (Dingle 1984; Palmer and Dingle 1986) in the immediate response to selection over a few generations, but we do not know whether genetic correlations impose significant impediments to the divergence of life-cycle traits over the range of a single species (Bradshaw and Holzapfel, Chap. 3; Scott and Dingle, Chap. 4; Masaki, Chap. 7; Solbreck et al., Chap. 11; Wipking, Chap. 13).

## 14.4 Emergent Life-Cycle Patterns

At some point in the broadening analyses of insect life-histories, making a genetic connection will become extremely difficult, if not impossible. Brown (Chap. 10), for example, looks for patterns of herbivore life-histories that change with plant successional stage. If one population of a single species persists through several successional stages *and* life-cycle traits of this population could be serially assessed without placing them in novel environments, then the genetic connection could, conceivably, be made. But, in any two given species the life-cycle traits during successional stages might change in opposite directions. By themselves, these two species would not tell us much about life-cycle strategies during succession but a survey of a broad number of taxa might. Thus, even without the genetic connection, the question of how life-cycle traits among diverse taxa covary along environmental gradients or over successional time remains legitimate. The emergent patterns then tell us something about life-cycles that facilitate or are necessary for existence in the particular spatial or temporal environments.

Without a genetic connection, we must be especially wary of ascribing cause and effect to emergent patterns. Do life-histories pre-ordain the suite of habitats in which an insect can sustain positive fitness or do life-histories of insects evolve continually to maintain or improve fitness in the habitats that they occupy? Brown's study (Chap. 10) shows an association of rapid dispersal ability among the insects prevalent at early rather than later stages of plant succession. This dispersal ability is then likely to be part of a broader syndrome of traits involving development time, growth rate and reproductive allocation (Roff, Chap. 1; Scott and Dingle, Chap. 4; Solbreck et al. Chap. 11). Insects capable of dispersing persist throughout succession but their frequency falls with advancing successional stages. Does this decline in ability to disperse mean that older successional communities select for non-dispersers or that, being capable of dispersal, dispersers simply leave and seek fresh habitats that they are more capable of exploiting? It is clearly exciting to think about a dynamic process such as coevolution or the habitat templet (Southwood 1977) driving the evolution of life-histories. It is much more mundane but sometimes more appropriate to think about organisms occupying habitats where they are able to make a living and then, when they cannot, moving on.

Focusing on mobile insects that tend to be generalists may provide a biased viewpoint of cause and consequence between the habitat and life-cycle syndromes, since host-plant specificity and reduced dispersal ability both increase with successional age (Brown, Chap. 10). Indeed, we can ask whether habitat-driven evolution is more intense among specialized, non-dispersers than among generalized, dispersal-capable insects. Even if generally true, counter-examples in this volume argue against it being universally true. Solbreck et al. (Chap. 11) found that wing-reduced, polyphagous lygaeids have larger body size and lay fewer, larger eggs than the monomorphic long-winged, monophagous lygaeids; by contrast, Gilbert (Chap. 6) found that among saprophagous syrphids, generalists are smaller and lay more, smaller eggs than specialists. There is therefore no universal tendency for dispersers to be more generalized as compared to non-dispersers. Finally, Berenbaum (Chap. 5) describes opportunistic host transfer to *Psoralea* (Leguminosae) by furanocoumarin-tolerant specialist papilionids and oecophorids that have otherwise probably coevolved with the Umbelliferae. Thus, even highly specialized, coevolved insects may opportunistically exploit novel hosts when they are available.

I do not despair at the variation in pattern of generalists versus specialists between lygaeids and syrphids. Rather, there is tremendous opportunity here to identify even more universal patterns by finding the common environmental causality between these two orders of insect. If the common ground remains elusive, I would make a plea that we preserve the inconsistency as a paradox rather than dismissing the differences to "phylogenetic constraints", a phrase that makes us think we know something but in reality may be simply a euphemism for "lygaeids and syrphids do it differently".

It is also important to know where answers are not going to be forthcoming. Berenbaum (Chap. 5) argues that even if host biochemical toxins and herbivore tolerances generate identical cladograms, there may still be no way to distinguish opportunistic colonization from mutual descent as an evolutionary process. Here, as Berenbaum's own work has shown, establishing the connection between genetic change and phenotypic evolution may be very necessary.

There are emergent patterns in insect life-history syndromes. These patterns are well illustrated in the chapters by Brown (Chap. 10), Gilbert (Chap. 6), and Solbreck et al. (Chap. 11). Emergent properties cannot and should not imply underlying genetic and/or physiological causality. Yet, emergent properties provide patterns of life-history variation across habitats and serve as analogues of models about selection in these habitats. Present emergent properties can therefore serve as guidelines for future testing of their predictions and their underlying assumptions by using reductionist approaches similar to those discussed above. Thus, lygaeid and gerrid life-cycle syndromes can be broken down from the diapause–migration flight (behavioural level) to the wing length and wing musculature versus ovarian or forelimb development trade-off (physiological level) or even to the mode of action of juvenile hormone on different developing tissues and lipid mobilization (cellular-biochemical level). Present emergent properties may just as well be refined, without implying genetic or physiological mechanisms, by continuing an expansionist approach in the future. Do emergent life-history patterns persist across taxa or across habitats? What is the common ground between wing-reduced lygaeids and saprophagous syrphids? Do similar life-history patterns during plant succession apply to carnivores, to detritivores, or in different climates, on different continents, or in different plant communities?

## 14.5 Where Do We Go From Here?

Life-history traits are the emergent phenotypes of underlying genotypes that may or may not respond to varying environments. It would seem to me to be a fruitful area to consider the physiological underpinnings of life-cycle syndromes, as shown in a general sense by Rose et al. (Chap. 2). Juvenile hormone has been implicated as a pivotal hormone in the life-cycles of Hemiptera (Roff, Chap. 1; Scott and Dingle, Chap. 4; Solbreck et al., Chap. 11). The Orthoptera considered by Scott and Dingle (Chap. 4) and by Masaki (Chap. 7) and the Lepidoptera considered by Wipking (Chap. 13) would appear to be good potential groups for similar studies. These authors all consider traits known to be under close hormonal control. What would be the various correlated responses to selection for low/high titre of hormones during development or for advanced or delayed production of the same titre? Answers to these questions would tell us the multitude of traits within a single life-cycle but both within and between stadia that could be affected by the timing and titre of single or a few hormonal events.

At the population level, there is good evidence that genetic correlations impose a "cost of correlation" in the immediate response to selection (Palmer and Dingle 1986). However, we do not know the extent to which genetic correlations have facilitated or impeded the adaptive divergence of even a single species. Observing the response to selection over the short term in nature is going to present several problems in interpretation. Natural environments will probably impose fluctuating selection so that detection of long-term optima will be elusive within two or tens of generations because of the "curse of autocorrelation" (Taylor, Chap. 9). Observed responses to short-term selection in nature are also likely to reconfirm the importance of genetic correlations as a constraint. Geographically disparate populations of the same species provide a special opportunity to examine the end points of longer-term selection. Do phenotypic correlations between disparate populations reflect static or changing genetic correlations for the same traits? When genetic correlations change among populations, do mean phenotypes change in the same way? Answers to these sorts of question may be forthcoming from a variety of ongoing studies including those of Roff (Chap. 1), Bradshaw and Holzapfel (Chap. 3), Scott and Dingle (Chap. 4) and Solbreck et al. (Chap. 11). Two others studies provide opportunities to examine life-history divergence across a sub- or sibling-species distance. Masaki (Chap. 7) outlines the responses to temperature and photoperiod that enable a subtropical cricket to enter an opportunistic diapause that is distinct from its temperate sibling-species. Since the two species are nearly identical morphologically, it is their diapause response to climate that distinguishes them. Similarly, Wipking (Chap. 13) describes contrasting life-cycle phenologies in two sibling-species of *Zygaena* on continental Europe where an additional aestivation (waiting) period is the major difference between the two species. Along with *W. smithii* (Bradshaw and Holzapfel, Chap. 3), there is a tremendous opportunity here to examine the genetic covariation within and between life-cycle syndromes involved in direct development (demographic traits) and syndromes involved in dormancy or migration (phenological traits) among three very different orders of insect.

In total agreement with Roff (Chap. 1), I would emphasize that "what is urgently required are empirical analyses of the genetic structure of ecologically important characters". I would like to add, however, that we should invest the enormous amount of required effort into species whose ecology is reasonably well known. "Ecological importance" is likely to vary enormously between species, even those with convergent life-styles. Host–pollinator mutalisms (Addicott et al., Chap. 8) provide a prime example. The constraints and opportunities operating on the fig wasp–fig interaction are far different from those operating on the yucca moth–yucca interaction. The constraints and opportunities affect the timing and variation of development and dispersal and also affect the degree of host–pollinator fidelity even though both systems could be classified as tight, obligate mutualisms.

The ecology of the species becomes even more important when studies are made of emergent life-cycle patterns. Species need to be classified as disperser/non-disperser, generalist/specialist, competitor-limited/free, uni-/multivoltine, egg/immature/adult overwintering stage, or predator/omnivore/detritivore/herbivore (Gilbert, Chap. 6; Brown, Chap. 10; Solbreck et al., Chap. 11). This classification is essential in determining how these life-cycle traits change across spatial and temporal habitats. At best, misclassification contributes to error and patterns that do, in fact, exist, may be missed. Worse, misclassification could result in the perception of a significant pattern when, in fact, one does not exist or is opposite from the one perceived.

To answer the initial question of where do we go from here, I would conclude that we must continue to integrate the three levels of approach described in this chapter. This integration will enable us to answer the following questions.

1. How does the underlying biochemistry contribute to the physiology that orchestrates the seasonal and developmental traits that we measure?
2. How does the genetic architecture of these traits vary between populations of the same and closely related species along ecological gradients?
3. Has the diversity of life-cycle traits that we see among closely related populations arisen because of, or despite, the underlying genetic architecture? What is the span in evolutionary time or in taxonomic level of genetic constraint and/or genetic facilitation?
4. How are life-cycle traits integrated into the ecology of insects? Are the fitness traits perceived by the investigator truly crucial to the insects in nature?
5. What are the major emergent patterns of insect life-cycles and how do we reconcile the exceptions?
6. Do environments direct the evolution of life-history or do established life-histories determine the ecology of a population, species or higher taxon?

*Acknowledgements*. First and foremost, I thank Dr C.M. Holzapfel for many years of discussion of insect physiology, ecology and evolution. I also thank Dr Valerie Brown for her efforts in organizing and running the symposium in Vancouver and Dr Francis Gilbert for taking on and doing such a fine job in editing this volume. I am grateful to all of the participants in the symposium for taking the time, effort and expense in coming to Vancouver and to the contributors to this provocative volume. Dr C. M. Holzapfel and Mr Jeffrey Hard read and provided helpful criticism of earlier versions of this chapter. Finally, I acknowledge my debt to the National Science Foundation for supporting Dr Holzapfel's and my work on insect life-cycles over the last 17 years.

## References

Berven KA, Gill DE (1983) Interpreting geographic variation in life history traits. Am Zool 23: 85–97

Bradshaw WE, (1986) Pervasive themes in insect life cycle strategies. In: Taylor F, Karban R (eds) The evolution of insect life cycles. Springer, Berlin Heidelberg New York, pp 261–275

Dingle H (1984) Behavior, genes, and life histories: complex adaptations in uncertain environments. In: Price PW, Slobodchikoff CH, Gaud WS (eds) A new ecology: novel approaches to interactive systems. Wiley, New York, pp 169–194

Falconer DS (1981) Introduction to quantitative genetics. Longman, London New York

Fisher RA (1958) The genetical theory of natural selection. Dover Publications, New York

Ford EB (1971) Ecological genetics. Chapman & Hall, London

Groeters FR, Dingle H (1987) Genetic and maternal influences on life history plasticity in response to photoperiod by milkweed bugs (*Oncopeltus fasciatus*) Am Nat 129: 332–346

Haldane JBS, Jayakar SD (1963) Polymorphism due to selection of varying direction. J Genet 58: 237–242

Hegmann J, Dingle H (1982) Phenotypic and genotypic covariance structure in milkweed bug life history traits. In: Dingle H, Hegmann J (eds) Evolution and genetics of life histories. Springer, Berlin Heidelberg New York, pp 177–188

Istock CA, (1982) The extent and consequences of heritable variation in fitness characters. In: King CR, Dawson PS (eds) Population biology: retrospect and prospect. Columbia University Press, New York, pp 61–96

Levene H (1953) Genetic equilibrium when more than one ecological niche is available. Am Nat 87: 331–333

Levins R, MacArthur R (1966) The maintenance of genetic polymorphism in a spatially heterogeneous environment: variations on a theme by Howard Levene. Am Nat 100: 585–589

Mousseau TA, Roff DA (1987) Natural selection and the heritability of fitness components. Heredity 59: 181–197
Palmer JO, Dingle H (1986) Direct and correlated responses to selection among life history traits in milkweed bugs (*Oncopeltus fasciatus*). Evolution 40: 767–777
Roff DA and Mousseau TA (1987) Quantitative genetics and fitness: lessons from *Drosophila*. Heredity 58: 103–118
Rose MR (1984) Genetic covariation in *Drosophila* life history: untangling the data. Am Nat 123: 565–569
Service PM, Rose MR (1985) Genetic covariation among life history components: the effect of novel environments. Evolution 39: 943–945
Southwood TRE (1977) Habitat, the templet for ecological strategies? J Anim Ecol 46: 337–365

# Species Index

# Subject Index

MIX
Papier aus verantwortungsvollen Quellen
Paper from responsible sources
FSC® C105338

If you have any concerns about our products,
you can contact us on
**ProductSafety@springernature.com**

In case Publisher is established outside the EU,
the EU authorized representative is:
**Springer Nature Customer Service Center GmbH**
**Europaplatz 3, 69115 Heidelberg, Germany**

Printed by Libri Plureos GmbH
in Hamburg, Germany